CW00791766

Vision and Action

This book deals with vision and action as an interactive, multisensory visual process. It is becoming apparent to perception scientists and visual psychophysicists that perceptual senses do not function in isolation. The visual processes involved in moving, reaching, grasping, and playing sports turn out to be a complex interaction. For example, the action of moving the head provides useful cues to help interpret visual information. Simultaneously, vision provides important information about those actions and their control. This becomes a reiterative process, and it is this process that is the focus of this volume.

This book is derived from an international conference held at York University where a handful of scientists, carefully selected from around the world as leaders in each of the several facets of vision and action, met to discuss the topic. Examples of the types of action considered vary from moving the eyes and head and body, as in looking around or walking, to complex actions such as driving a car, catching a ball, or playing table tennis.

Graduate students and researchers in vision science will be interested, as well as those involved in using visual processes in computer animations, display design, or the sensory systems of machines. Physiologists and neuroscientists interested in any aspect of sensory or motor processes will find this a very useful and broad-ranging text.

Laurence Harris is a neuroscientist with a background in sensory processes. He is presently a Professor of Biology and Psychology at York University, Ontario. Michael Jenkin is a computer scientist with interests in computational vision, mobile robotics, and artificial intelligence. He is currently chair of the Department of Computer Science at York University. They have edited two previous books: *Spatial Vision in Humans and Robots* and *Computational and Psychophysical Mechanisms of Visual Coding,* both published by Cambridge University Press.

Vision and Action

Edited by

Laurence R. Harris and Michael Jenkin

PUBLISHED BY THE PRESS SYNDICATE OF THE UNIVERSITY OF CAMBRIDGE
The Pitt Building, Trumpington Street, Cambridge CB2 1RP, United Kingdom

CAMBRIDGE UNIVERSITY PRESS
The Edinburgh Building, Cambridge CB2 2RU, United Kingdom
http: //www.cup.cam.ac.uk
40 West 20th Street, New York, NY 10011–4211, USA
http: //www.cup.org
10 Stamford Road, Oakleigh, Melbourne 3166, Australia

© Cambridge University Press 1998

This book is in copyright. Subject to statutory exception
and to the provisions of relevant collective licensing agreements,
no reproduction of any part may take place without
the written permission of Cambridge University Press.

First published 1998

Printed in the United States of America

Typeset in Computer Modern in TEX and Century Oldstyle in Quark XPress

A catalogue record for this book is available from
the British Library

Library of Congress Cataloguing-in-Publication Data

Vision and action / edited by Laurence R. Harris, Michael Jenkin

p. cm.

Includes biographical references and index
ISBN 0 521 63162-9
1. Vision. 2. Motion perception (Vision). 3. Visual perception. 4. Eye – Movements.
I. Harris, Laurence (Laurence Roy), 1953– .
II. Jenkin, Michael (Michael Richard MacLean), 1959– .
QP475.V568 1998
612.8'4 – dc21 97-50612
CIP

ISBN 0 521 63162 9 hardback

To our wives ... again

Contents

Contributors

Benjamin T. Backus
Department of Psychology University of California, Berkeley, CA

Dana H. Ballard
Computer Science Department, University of Rochester, Rochester, NY, USA.

Martin S. Banks
School of Optometry University of California, Berkeley, CA

K. I. Beverley
Natural Resources Canada, Geological Survey of Canada, 9860 West Saanich Road, P.O. Box 6000, Sidney, B.C., Canada V8L 4B2.

V. Cornilleau-Pérès
Laboratoire de Physiologie de la Perception et de l'Action, CNRS-Collège de France, Paris.

J. Douglas Crawford
Department of Psychology, York University, Toronto, Ontario, Canada, M3J 1P3.

J. Droulez
Laboratoire de Physiologie de la Perception et de l'Action, CNRS-Collège de France, Paris.

Martha Flanders
Department of Physiology, University of Minnesota, MN, USA.

Melvyn A. Goodale
Department of Psychology, University of Western Ontario, London, Ontario, Canada, N6A 5C2.

R. Gray
Department of Psychology, York University, Toronto, Ontario, Canada, M3J 1P3.

Antje Grigo
Dept. Zoology and Neurobiology Ruhr–University Bochum, D–44780 Bochum, Germany.

Angela Haffenden
Department of Psychology, University of Western Ontario, London, Ontario, Canada, N6A 5C2.

S. J. Hamstra
Dept. of Surgery, University of Toronto, 100 College St., Toronto, Ontario, Canada M1G 1L5.

Laurence R. Harris
Department of Psychology, York University, Toronto, Ontario, Canada, M3J 1P3.

X. H. Hong
Department of Psychology, York University, Toronto, Ontario, Canada, M3J 1P3.

Michael Jenkin
Department of Computer Science, York University, Toronto, Ontario, Canada, M3J 1P3.

Jan J. Koenderink
Helmholtz Instituut, Universiteit Utrecht, The Netherlands.

R. Kohly
Department of Psychology, York University, Toronto, Ontario, Canada, M3J 1P3.

Agnieszka Kopinska
Department of Psychology York University, Toronto, Ontario, Canada, M3J 1P3.

Michael F. Land
Sussex Centre for Neuroscience, School of Biological Sciences, University of Sussex, Brighton BN1 9QG, UK.

Markus Lappe
Dept. Zoology and Neurobiology, Ruhr–University Bochum, D–44780 Bochum, Germany.

Andrew McCallum
Justsystem Pittsburgh Research Center, Pittsburgh PA, USA, 15213

F. A. Miles
Laboratory of Sensorimotor Research National Eye Institute National Institutes of Health Bethesda, MD 20892, USA

S. Mitra
Center for the Ecological Study of Perception and Action, U-20, University of Connecticut, Storrs, CT, 06269-1020, USA.

Lance M. Optican
Laboratory of Sensorimotor Research, National Eye Institute, Bethesda, MD, USA.

A.L. Paradis
Service Hospitalier Frédéric Joliot, CEA, Orsay, France.

C. V. Portfors
Portfors, Neurobiology, Northeastern Ohio Universities College of Medicine, 4209 State Route, P.O. Box 95, Rootstown, OH 44272-0095, USA.

Christian Quaia
Laboratory of Sensorimotor Research, National Eye Institute, Bethesda, MD, USA.

Rajesh P. N. Rao
Salk Institute, La Jolla, CA, USA.

D. Regan
Department of Psychology, York University, Toronto, Ontario, Canada, M3J 1P3.

M. A. Riley
Center for the Ecological Study of Perception and Action, U-20, University of Connecticut, Storrs, CT, 06269-1020, USA.

Garbis Salgian
Computer Science Department, University of Rochester, Rochester, NY, USA.

R. C. Schmidt
Center for the Ecological Study of Perception and Action, U-20, University of Connecticut, Storrs, CT, 06269-1020, USA.

John F. Soechting
Department of Physiology, University of Minnesota, MN, USA.

Demetri Terzopoulos
Department of Computer Science, University of Toronto, Toronto, Ontario, Canada, M5S 3G4.

M. T. Turvey
Center for the Ecological Study of Perception and Action, U-20, University of Connecticut, Storrs, CT, 06269-1020, USA.

Andrea J. van Doorn
Helmholtz Instituut, Universiteit Utrecht, The Netherlands.

A. Vincent
Transport Canada, Transport Development Centre(CDC), 800 Rene Levesque Blvd. W, 6th. Floor, Montreal, Quebec, Canada, H3B 1X9.

Daniel C. Zikovitz
Department of Biology, York University, Toronto, Ontario, Canada, M3J 1P3.

Preface

This book is based on a conference on Vision and Action, the fourth conference of the York Centre for Vision Research organized by I. P. Howard, D. M. Regan and B. J. Rogers in June 1997 and sponsored by the Human Performance Laboratory of the Institute for Space and Terrestrial Science at York University.

The York Vision Conference, and this book, would not have been possible without the advise and support of Ian P. Howard, David Martin Regan and the Human Performance in Space Laboratory of the Institute of Space and Terrestrial Science (ISTS). Behind any successful endeavour is the person who really runs things, and none of this would have been possible without Teresa Manini.

Vision and Action

This book deals with vision and action as an interactive, multisensory visual process. It is becoming apparent to perception scientists and visual psychophysicists that perceptual senses do not function in isolation. The visual processes involved in moving, reaching, grasping, and playing sports turn out to be a complex interaction. For example, the action of moving the head provides useful cues to help interpret visual information. Simultaneously, vision provides important information about those actions and their control. This becomes a reiterative process, and it is this process that is the focus of this volume.

This book is derived from an international conference held at York University where a handful of scientists, carefully selected from around the world as leaders in each of the several facets of vision and action, met to discuss the topic. Examples of the types of action considered vary from moving the eyes and head and body, as in looking around or walking, to complex actions such as driving a car, catching a ball, or playing table tennis.

Graduate students and researchers in vision science will be interested, as well as those involved in using visual processes in computer animations, display design, or the sensory systems of machines. Physiologists and neuroscientists interested in any aspect of sensory or motor processes will find this a very useful and broad-ranging text.

Laurence Harris is a neuroscientist with a background in sensory processes. He is presently a Professor of Biology and Psychology at York University, Ontario. Michael Jenkin is a computer scientist with interests in computational vision, mobile robotics, and artificial intelligence. He is currently chair of the Department of Computer Science at York University. They have edited two previous books: *Spatial Vision in Humans and Robots* and *Computational and Psychophysical Mechanisms of Visual Coding,* both published by Cambridge University Press.

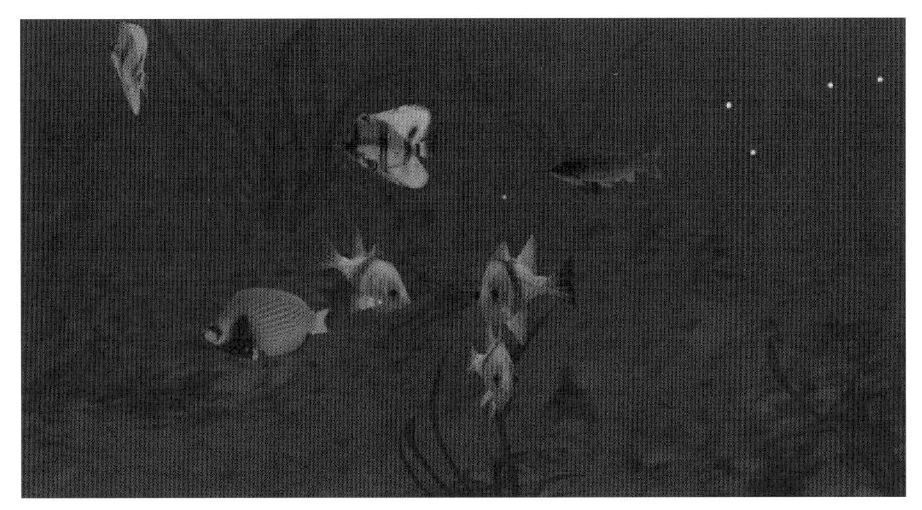

Figure 11.6 Artificial fishes in their physics-based virtual world.

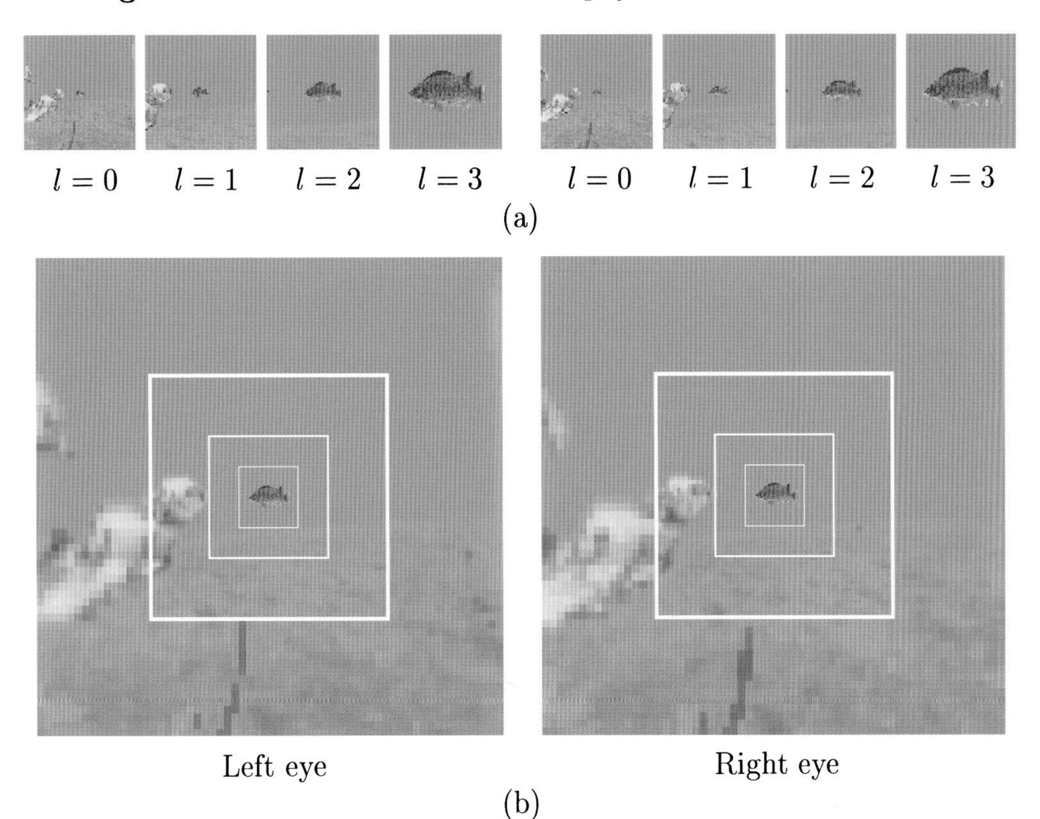

Figure 11.7 Binocular retinal imaging.

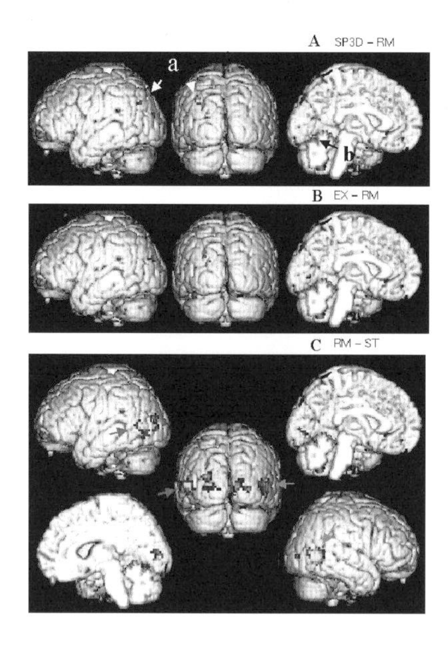

Figure 12.6 Results of the group analysis.

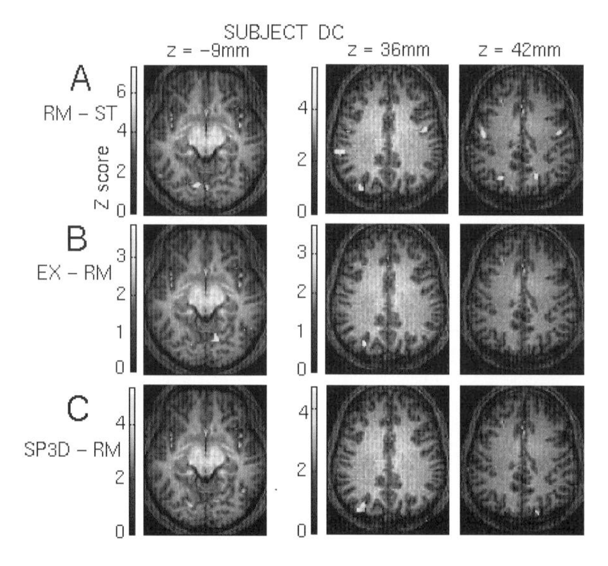

Figure 12.7 Localisation of the activity foci on individual data.

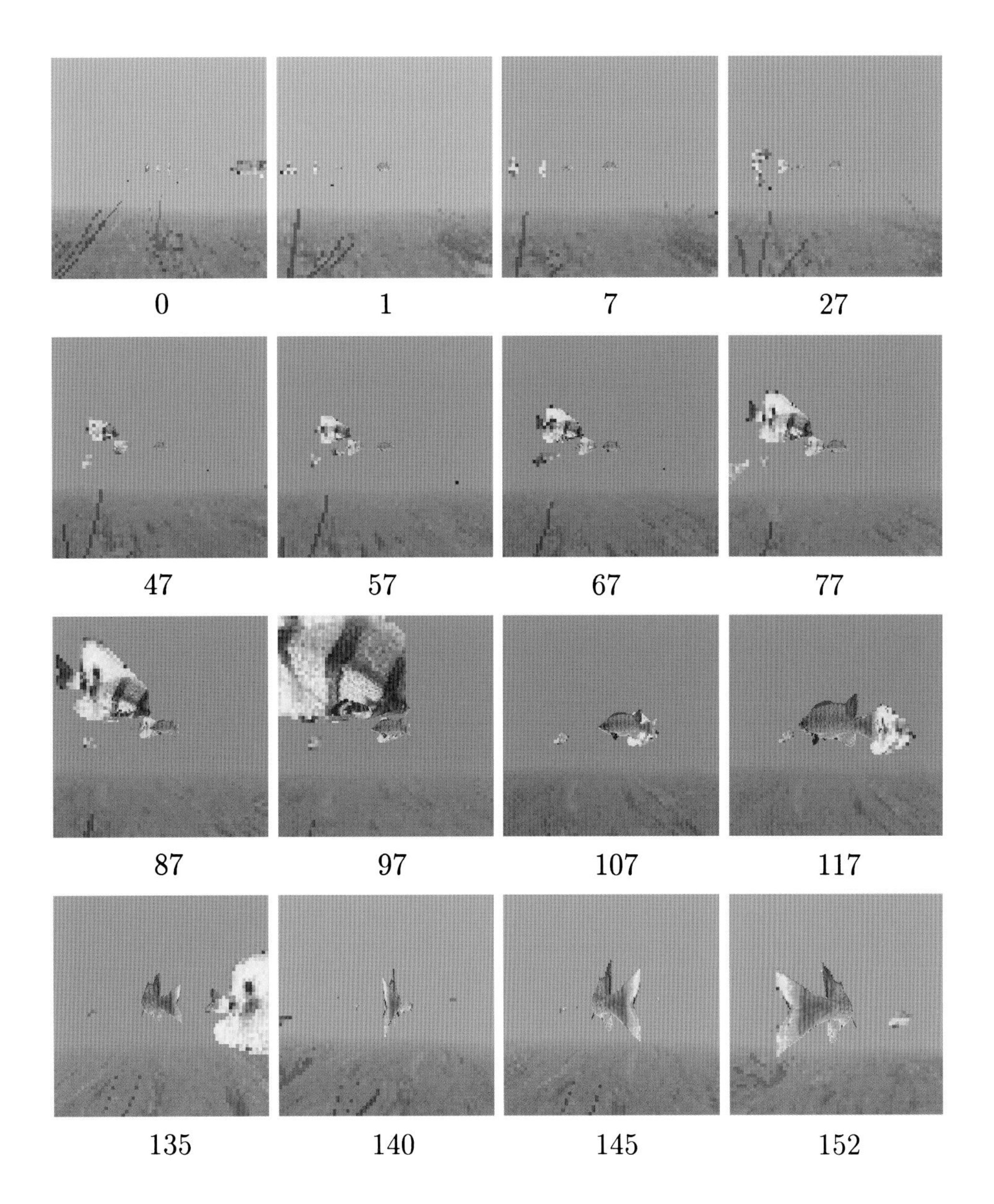

Figure 12.9 Retinal image sequence from the left eye of the active vision fish as it detects and foveates on a reddish fish target and swims in pursuit of the target (monochrome versions of original color images). The target appears in the periphery (middle right) in frame 0 and is foveated in frame 1. The target remains fixated in the center of the fovea as the fish uses the gaze direction to swim towards it (frames 7–117). The target fish turns and swims away with the observer fish in visually guided pursuit (frames 135–152).

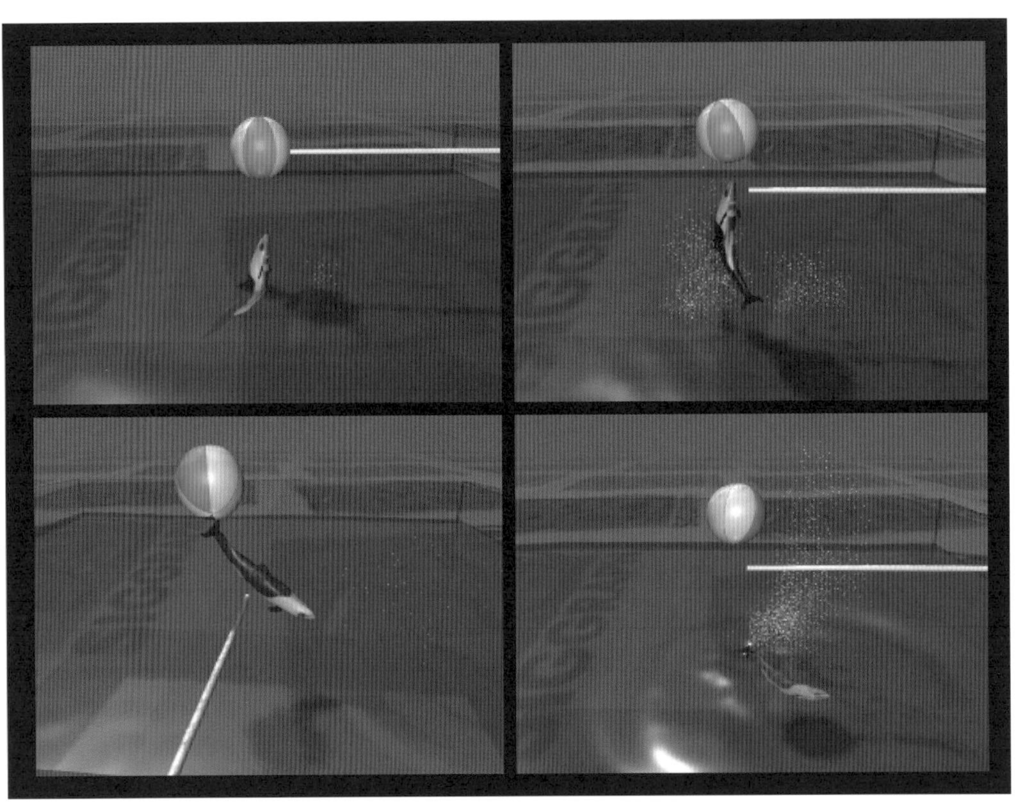

Figure 12.12 "SeaWorld" skills learned by an artificial dolphin.

1

Vision and action

Laurence R. Harris and Michael Jenkin

1.1 Introduction

The chapters in this book were written by leading vision researchers who were invited to a conference on "Vision and Action" at York University in 1997. Overall, these chapters convey to us, the editors, and hopefully to you, the reader, an interesting and consequential message. The message is that at least some of us are ready to take the study of vision back into the real world. As soon as the step out of the laboratory is taken, even in the imagination, the inadequacies and misleading features of the "psychophysical paradigm" in which subjects are clamped in front of a sparse stimulus presented on a two-dimensional screen in a darkened room, become clear. Vision in the real world is an interactive, multisensory process: the eyes actively explore the three-dimensional world by means of integrated eye, head and body movements, and information about this exploration is integral to the visual process. And many lines of evidence are converging to confirm that there is no single visual process in any case.

Introspection regarding what might be happening as we see suggests initially that visual information allows us to build up a unified internal representation of the external world. But on closer examination we realize how little of the visible world we can remember in any detail, making us doubt the reliability of this internal construct. The impression we have that we are aware in detail of our surroundings has been referred to as the "Grand Illusion". A more thorough analysis, in which we consider acting in the world, makes us realize that much of our use of vision is carried out without any corresponding perception. Batters hit balls without being able to describe the information they use; people navigate around complex rooms without bumping into things and without even noticing the objects and furniture they avoid. So the idea of a solid real world derived from a single sense, built up unerringly over time and available as a single resource to any system that needs to have access to it, does not actually match our experience.

In fact, as this book emphasizes, the "visual process" turns out to be a diverse bundle of processes with very disparate goals, bound together only by the coincidence that they all derive some of their information from the optic array.

1.1.1 Replacing the "psychophysical paradigm"

The psychophysical model owes its origin to predominantly-German, nineteenth century attempts to make psychology a respectable science. To this end people, the subjects of psychological research, were modeled as physical entities governed by Laws of the type that had been so successful in predicting the behaviour of the physical world. As examples we can list Emmert's Law (1881), Bloch's Law, Weber's Law (1834), Ekman's Law, Ricco's Law, Korte's Law (1915), Fechner's Law (1860), Hering's Law (1942), Donder's Law and Listing's Law (1866). Notice that, also following from the precedent set in the physical sciences, these 'Laws' are named not after the phenomena they purport to describe or govern, but after the protagonist, leading to many a frustrating and trivial series of undergraduate exam questions. We wonder if it is significant that some of the greatest figures of the time, e.g. Mach, Helmholtz and Exner, do not seem to have Laws named after them. Especially as it turns out that, except under very restricted circumstances, none of these so-called Laws is actually true. Some of them (e.g. Ekman's and Fechner's) are actually incompatible with each other.

Let us take Weber's Law as a statement taught as fundamental psychophysical fact. Weber's Law (1834) states that the smallest increment of something needed to be detected as an increment (e.g. the increase in luminance of a patch of light) is a constant ratio of the background from which the change is noted. This ratio is called the Weber fraction and the Law states that it is constant for a given task. Now we can quibble with Weber's Law on the basis that it only holds true for some values of the background. In particular the Law breaks down when the background is close to threshold (see Engen, 1971). But there is a more profound sense in which Weber's Law, and all the other Laws, fail. In order for Weber's Law to work at all, there is the implicit notion of controls: everything else has to be constant. Almost anything can affect the Weber fraction for luminance, for example. Factors that can be expected to alter the ratio include the dark adaptation state of the subject, the wavelength of the stimulus, the length of time and what happens in between the stimuli being compared, whether the stimuli are moving, whether the person is moving, etc... If these factors are not 'controlled' or kept constant, then the Law does not hold. Does this invalidate the Law in the "real world"?

Let us look at Listing's Law as a further example. To be fair, Listing's Law was actually called that by Helmholtz so we can perhaps think of it not so much as self aggrandizement as passing the buck. In any event, Listing's Law was the name given to a principle that can be used to describe the exact positioning of the eye (see von Helmholtz, 1911 and Crawford's and Optican's chapters in this volume). The Law was framed as a direct parallel to, almost a parody of, Laws of the physical universe such as Boyle's Law that had proved so successful in "taming inanimate nature". It states that the resting position of the eye can be predicted from a knowledge of a standard reference position and a plane. In order to reduce human behaviour to a set of such mathematically-tangible Laws, the fundamental modus operandum of experimenters has been to keep everything else constant and controlled. But as soon as normal behaviour is allowed and the eyes are permitted to converge at different distances and the head and body are released from constraint, the Law breaks down.

There are still important things to be discovered from experiments with subjects with immobile head and eyes staring fixedly at an oscilloscope. But the time has come also to go, sometimes literally, out into the field (see Koenderink's chapter this volume). Some of the most important aspects of processing turn out to be the interactions between the parameter under investigation and factors that were previously controlled (see Harris' chapter this volume and Harris, Jenkin and Zikovitz, 1998 for examples of unexpected interactions between auditory and visual localizations and head movement).

1.2 Multiple visual systems

The idea of a single visual system is no longer tenable. Of course it never was. There was always a realization that there were some visual processes going on 'below' the perceptual level. The primitive, brain-stem controlled visual processes involved in self-contained functions such as pupil control and breeding cycles were always regarded as separate from 'perceptual processes', for example. But other 'primitive' processes were less clear. What about orienting, for example? After all, we can voluntarily choose to orient or, to some extent, not to orient to a visual feature, whereas we cannot voluntarily override pupil size.

It seemed that as philosophers of vision, our choices were twofold: to regard vision either as an integrated whole embracing both 'higher'(such as object recognition) and 'lower'(such as object localization) aspects or to think of vision as comprising two parallel, independent systems. The idea of "two visual systems" was formally proposed by Schneider in 1969. He assigned the function of localizing objects in space to the superior colliculus of the brain stem and the task of identification to the cortex. The idea was immediately attacked, ostensibly for being too simplistic but really for daring to divide an 'integral' system. One of the authors' PhD thesis (Harris,

1978 see also Harris, 1980) concerned the superior colliculus and the Cambridge examiners were confounded when he supported this division, rather than rising to their invitation to attack the 'two visual system' notion. To have this idea made concrete, polarized scientific opinion. People were quick to point out that perceptual systems need to know the location of objects as an integral part of their recognition, so it would not be convenient to have that aspect processed entirely by the brain stem colliculus. At the very least the colliculus would have to report its findings to 'higher centres'.

Later Ungerleider and Mishkin (1982) raised the division of 'what' and 'where' into the cortex with their description of ventral and dorsal streams of visual processing extending beyond the primary visual projection area into the parietal and inferotemporal regions respectively. And again the idea was criticized vigorously for being too simplistic, and many studies have since emerged claiming to undermine this division by documenting an enormous overlap in the processing of each stream (see Regan, 1989 and Milner & Goodale, 1995 for useful reviews).

1.3 Two classes of visual processes

In writing this introduction and trying to provide a superstructure for the topic of Vision and Action, we have found it useful to follow the Gibsonian division of visual processes into two broad classes based on the information they extract: *proprioceptive vision* is that part of vision that provides information about the position and movement of parts or all of the self; *world vision* extracts information about the world (Gibson, 1977; see also Harris and Jenkin, 1997; Nakayam 1985). We have used the term 'world vision' instead of Gibson's 'exteroception' to emphasise that we are defining the system in terms of the use to which the information is put. Our division is based on the nature of information extracted and does not map exactly onto the egocentric and allocentric division that is sometimes used (e.g. Tamura *et al.*, 1990). The terms egocentric and allocentric refer to the frame of reference used for a given process (see also Harris' chapter this volume): egocentric means information coded with respect to the self and allocentric means with respect to external space. The divisions might overlap in that proprioceptive vision will sometimes use a frame of reference based on the self and thus be egocentric. But it not necessary or even usual that proprioceptive vision is coded egocentrically or that world vision be coded allocentrically (see below).

Neither does the division between proprioceptive and world vision map onto unconscious and conscious visual processes. All vision is essentially unconscious. Consciousness does not have access to any of the visual processes but only to the output of these processes. When using world vision to carry out the perceptual task of face recognition, for example, it is the output of the face recognition task that becomes conscious, not the visual processing that led to the task being achieved. And we are often (but not always) aware of our self motion, to which proprioceptive vision is an important contributor.

1.3.1 Proprioceptive vision: providing knowledge of self

Proprioception (from the Latin proprius: one's own) provides information about the body's own position and movement. Proprioceptive organs include muscle spindles, Golgi tendon organs and the vestibular system. But a very important source of one's own position and movement and the position and movement of one's own members comes from vision. Although the proprioceptive system is defined as providing information about the position and motion of the self, it is unusual for this information to be in an egocentric reference frame (see Harris, 1997). The position of a finger, for example, cannot be coded with respect to itself (relative to which it can of course have no movement), but only with respect to something else. If that something is another part of the self (e.g. the rest of the hand), then the coding is said to be egocentric, if that something is not part of the self (e.g., an earth-fixed target) then the coding is allocentric. But both codings represent examples of proprioceptive vision. There are many different examples of proprioceptive vision which share only the fact that they are describing the position of parts of the self. For example, the proprioceptive contribution of vision to the task of forming a precision grip (see Goodale's chapter this volume) requires a knowledge of how far the fingers are apart: an egocentric coding. This proprioceptive visual information shares very little with the extraction of self motion information from optic flow, measured allocentrically, to guide locomotion (see Grigo's and Miles' chapters this volume).

1.3.2 World vision: subserving perceptual tasks

World vision reports about the structure and features of the world. This includes the location of objects as well as higher-level perceptual aspects, such as shape and three dimensional structure. It can also include information about the self other than the proprioceptive information described above. For example, one can view the pattern of wrinkles on one's hand independently and in addition to any knowledge about the hand's movement and position.

World vision can be in an egocentric reference frame, remembering that there are many different egocentric frames from which to choose. World vision can also be in an allocentric reference frame, as when considering the relationships of the various features of a single object.

The category of world vision is too huge and unwieldly to be of much use on its own, except to distinguish it from proprioceptive vision. Probably both dorsal and ventral streams deal with world vision (although they may also use proprioception). Should proprioceptive and world vision be viewed as separate visual systems? With that we seem to have come back to the question of how many visual systems there might be.

1.4 Task-based visual systems

The question of whether there might be one, two or more visual systems might be to ask the wrong question. Let us take a step back. *Why* might there be different visual systems? Milner and Goodale (1995; and see Goodale *et al.*, 1991, 1992 and Goodale's chapter this volume) point out that there are many different tasks for which vision is used, many of which might require the extraction of often quite different pieces of information. It therefore makes sense to have many parallel systems to process the information for each class of task. The branch of the optic nerve into the pupil control centres clearly controls a discrete function. Could applying functional divisions, considering the multiple uses to which vision contributes, clarify some of the problems in describing the organization of the higher visual processes? A functional division does not require rigorous division of different aspects of vision. No longer is it relevant if multiple divisions all use stereopsis. Now the aim is not to carry out the calculations of stereopsis somewhere centrally, providing results to contribute to the Grand Illusion. Instead, each division extracts the information relevant to the task with which it is dealing. All the divisions share the use of the optic array by definition, so why not also share (or even duplicate) the use of higher-order features such as colour or motion extracted from that array. Processes such as stereopsis are critical for carrying out many varied tasks including three-dimensional object recognition, visually-guided reaching and grasping and the control of locomotion. So we perhaps shouldn't be surprised to find systems using stereoscopic information appearing in many parts of the brain. This dupicity does not necessarily represent redundancy. The information extracted from a comparison between the two eyes' images might be very different for these various tasks - just as the information useful for each task extracted from the luminance array might be different.

Indeed, considering a division in terms of tasks raises us well above the idea of two (or more) visual systems. Now they are not "visual" systems at all but *functional* systems that just happen to be using, amongst other things, information from the optic array. The question now arises: how many independent task-defined systems might there be? A consideration of this is beyond the scope of this introductory chapter. But it is important to recognize that we are capable of performing many simultaneous tasks, each using "vision" along with inputs from many other sources. Consider the tasks performed when putting on a watch while walking along. Some of these tasks are overt actions, others are perceptual. Tasks include directing the eyes from one part to another, aligning the pin and the hole of the buckle, recognizing the pin as distinct from surrounding objects, knowing the position of the hand that holds each component, as well as simultaneously processing the direction of locomotion well enough not to step off the side-walk and deciding when it is required to look up to obtain different proprioceptive information.

1.4.1 Dissociation between different task-based visual systems

Evidence that vision for action and vision for perceptual tasks are handled by individual independent systems is provided by the phenomenon of blindsight (Weiskrantz, 1986). When blindness results from cortical damage, sufferers are found to be able still to orient their eyes and head and even to point to visual targets in response to verbal instructions, although they are perceptually unaware of the objects. Mel Goodale (Goodale *et al.*, 1991, 1992 and Goodale's chapter this book) has shown independent processing of these processes when reaching for objects. Normal subjects can have their perceptual judgements altered independently of the performance of visual proprioceptive loops involved in positioning their fingers for the task.

1.5 Vision and action

Allowing of two classes of tasks (action and perception) and two classes of visual processes (proprioceptive and world vision) generates four combinations:

(i) Proprioceptive vision that contributes to action.
e.g. Visually-guided reaching (see Soechting's and Goodale's chapters this volume); visually-guided navigation

(ii) Proprioceptive vision that contributes to perception
e.g. knowledge of self motion and position (see Harris, 1997; Miles' chapter this volume)

(iii) World vision that contributes to action
e.g. directing orienting movement such as eye movement control (see Optican and Crawford's chapters this volume); Driving a car (see Land's and Harris's chapters this volume) ; detecting an approaching object and ducking (see Regan's chapter this volume); many sporting actions such as hitting a ball (see Regan's chapter this volume);

(iv) World vision that contributes to perception
e.g. Face recognition, object recognition, space perception (see Koenderink's chapter) 3D shape perception (see Cornilleau-Pérès' chapter)

1.6 Vision needed for action

Although, as a blind person will probably tell you, no motor acts *require* visual feedback before they can take place at all, there are many examples where vision can be used to make actions a lot more effective. For example vision is often used to guide motions of various kinds. Many complex actions, such as inserting a peg in a hole, require complex sensing. Fitting a peg into a hole requires a complex interaction between action, force feedback, and vision. It is not a one-shot process in which a view of the world can be processed so as to permit a plan to be executed without ongoing sensing. The action takes place within ongoing sensory input on which it may often rely.

1.6.1 Proprioceptive vision needed for action

Actions need information: they need information about the starting conditions, the desired finishing condition, and the difference between the two. Once they are underway, it is often useful to have ongoing information to adjust and fine-tune the action. Proprioceptive vision contributes to all these stages. Proprioceptive vision provides the same information that other proprioceptors do, it signals the difference between where you are and where you want to be. In doing so, proprioceptive visual information normally interacts with other proprioceptive, non-visual sources of information about the same things. These non-visual sources can, if necessary act on their own, although obviously with less effect. For example the muscle and joint proprioceptors of the hand and fingers can normally guide a finger to a target, even if the target is specified in a non-visual modality. In a very touching clinical case (Cole, 1996; see Mitra's chapter this volume) a quadriplegic is described who has lost the use of this non-visual input and is forced to use vision exclusively for the control of all his actions even to uncurl his fingers. Although his behaviour indicates this was possible under vision-only control, his problems highlight the inefficiency of vision alone, and the importance of multisensory convergence in the control of action.

Vision as a source of proprioceptive information about the body's performance contributes to all actions that are carried out with the eyes open. It is very important in the control of fine motor actions. But it is also important for actions such as walking, jumping, running, reaching, and also less obvious actions such as the eye and head movements involved in maintaining a stable retina. In the control of eye movement, proprioceptive vision that indicates eye drift plays an especially important role in the control loop because other proprioceptors indicating instantaneous eye motion and position seem relatively ineffective in this role. If deprived of the proprioceptive visual information concerning eye position, the eyes wander (see Carpenter, 1988 and Jeannerod, 1988 for reviews).

Following on from the important role of vision as a source of feedback in the control of actions, proprioceptive vision also has a longer term, teaching role in calibrating muscular movement to the demands of various tasks. After performing pointing movements with good visual feedback for a while, performance is improved and to some extent can be maintained in the absence of vision. If the feedback is experimentally manipulated, then the system can be retrained (see Jeannerod, 1988 and Howard, 1982 for reviews).

1.6.2 World vision contributes to action

The visual perceptions of the attributes of the world and objects in the world are integral to the initiation and control of actions. For actions to relate to the world they need information about the world that often originates from what we are calling world vision.

- **Sports** It would be difficult to imagine how games such as football can be played without the ability to perceive the game field and other players. Although vision is not the only possible sense for the perception of a sports field, it is certainly very effective.
- **Driving** Although it might be argued that many humans cannot master the skills required to drive, after a fashion both humans and machines can operate motor vehicles. Vision is a key component of driving, (see chapters by Land and Harris in this volume). They both indicate the importance of vision and the visual frame in driving.

1.7 Action needed for vision

A single snapshot view of the world provides a very under-constrained version of events. Even monocular viewers can disambiguate their view by moving their heads from side to side. This action provides depth information. But action is even more crucial to vision than this. Objects which do not move on the retina fade. Clearly much of vision requires action.

Vision does not exist or develop in isolation. It is part of action and part of exploration. We explore with a glance as much as with an outstretched arm. When we are asleep we close our eyes. The first thing we do on waking is to open them again: our first action. Matching the visual input with the expected or anticipated input is as much a part of vision as matching the actual and desired positions of a limb is of motor control.

1.7.1 Action needed for proprioceptive vision

Proprioception refers to knowledge concerning both the position and its time-varying derivations, velocity, acceleration etc... These higher-order derivatives are only useful if a person changes position over time, that is, in the presence of action. Self motion creates such a time-varying pattern and therefore provides proprioceptive information concerning that motion. We also need to have a representation of time (see Ballard's chapter, this volume). It is the interaction between an action and the vision it creates that is useful.

1.7.2 Action needed for world vision

The pioneering experiments of Held and Hein (1963) and others (see Lund, 1978 for a review) showed that the development of vision critically depends on being able to interact with the environment. In these classic experiments, described in many textbooks, cats were brought up with identical visual experience to their litter mates, but without being able actively to explore their world. These passively-exposed animals grew up functionally blind.

Also, using non-retinal information about self motion theoretically makes it much easier to decide which visual movements are due to self motion and which are due to actual object motion in the scene. Visual motion due to self motion can be *predicted* from the information provided by extra-retinal sources. Recordings in the parietal cortex have characterized cells whose properties are ideal to provide this service (Tanaka & Saito, 1989; Duffy & Wurtz, 1991a, Duhamel, Colby & Goldberg, 1992). The chapter by Cornilleau-Pérès boldly questions this theory, which was fast becoming dogma, since self motion in her experiments does not seem to help performance. Psychophysical experiments on large-scale integration of visual motion support a vision-only system being organized in such a way as to extract the visual consequences of self motion without extra-retinal help (Harris & Lott, 1995, 1996).

1.8 Conclusion

This book emphasizes the critically important, bi-directional interactions between vision and action. These interactions are so essential that vision cannot exist without them. A whole area of vision research devoted to this topic has blossomed in the 1990's (see Aloimonos, 1995; Blake & Yuille, 1992). In retrospect it seems obvious that vision is an active process. Our eyes never just passively take in visual information but actively search it out. And the information sought is minimal, highly relevant to the task in hand and held in memory in the most transient manner. This type of 'minimal vision' is far from the model used by computational visual scientists who, following from the pioneering work of Marr (1982), take a snap-shot image and regard working on it as a project similar to the processes of vision attempting to extract information from the retinal image. The minimal information, task-based visual process matches the needs of the particular task (Land & Forneaux, 1997). The study of vision will never be static again.

Acknowledgments

This chapter sponsored by an NSERC, Canada collaborative grant to Harris and Jenkin and to the generous contributions of the Centre for Research in Earth and Space Technology (CRESTech) of Ontario. We would like to thanks Carolee Orme for her helpful comments on this manuscript.

References

Aloimonos, Y. E. (1993). *Active Perception*, Lawrence Erlbaum Associates, Hillsdale, NJ.

Blake, A, Yuille, A. (1992). *Active Vision*, MIT Press, Cambridge, MA.

Carpenter, R. H. S. (1988). *Movements of the Eyes*, Pion, London.

Cole, J. (1995). *Pride and a Daily Marathon*, MIT Press, Cambridge, MA.

Duffy, C. J., & Wurtz, R. H. (1991). "Sensitivity of MST Neurons to Optic Flow Stimuli 1. A Continuum of Response Selectivity to Large-Field Stimuli", *J. Neurophysiol.*, 65: 1329-1345.

Duffy, C. J., & Wurtz, R. H. (1991). "Sensitivity of MST Neurons to Optic Flow Stimuli 2. Mechanisms of Response Selectivity Revealed by Small-Field Stimuli", *J. Neurophysiol.*, 65: 1346-1359.

Duhamel, J.-R., Colby, C. L., & Goldberg, M. E. (1992). "The updating of the representation of visual space in parietal cortex by intended eye movements", *Science*, 255: 90-92.

Engen, T. (1971). "Psychophysics: Discrimination and detection", in J. W. Kling & L. A. Riggs (Eds), *Woodworth and Schlosberg's Experimental Psychology (3rd Edition)*, Holt, New York, NY.

Gibson, J. J. (1977). "On the analysis of change in the optic array in contemporary research on visual space anhd motion perception", *Scand. J. Psychol.*, 18: 161-163.

Goodale, M. A. & Milner, A. D. (1992). "Seperate visual pathways for perception and action", *Trends in Neurosci.*, 15: 20-25.

Goodale, M. A, Milner, A. D., Jakobson, L. S. & Carey, D. P. (1991). "A neurological dissociation between perceiving objects and grasping them", *Nature*, 349: 154-156.

Harris, L. R. (1997). "The coding of self motion", in L. R. Harris & M. Jenkin (Eds) *Computational and Psychophysical Mechanisms of Vsual Coding*, 157-183, Cambridge University Press, Cambridge, MA.

Harris, L. R. (1978). "The superior colliculus and movements of the head and eyes in cats", Unpublished PhD Thesis, Cambridge University.

Harris, L. R. (1980). "The superior colliculus and movements of the head and eyes in cats", *J. Physiol.*, 300: 367-391.

Harris, L. R. & Jenkin, M. (1997). "Computational and psychophysical mechanisms of vsual coding", L. R. Harris and M. Jenkin (Eds), *Computational and*

Psychophysical Mechanisms of Vsual Coding, 1-19, Cambridge University Press, Cambridge, MA.

Harris, L. R., Jenkin, M. & Zikovitz, D. C. (1998). "Vestibular cues and virtual environments", Proc. IEEE Virtual Reality Annual International Symposium (VRAIS), 133-138.

Harris, L. R. & Lott, L. A. (1995). "Sensitivity to Full-Field Visual Movement Compatible with Head Rotation - Variations Among Axes of Rotation", *Visual Neurosci.*, 12: 743-754.

Harris, L. R. & Lott, L. A. (1996). "Sensitivity to Full-Field Visual Movement Compatible with Head Rotation - Variations with Eye-in-Head Position", *Visual Neurosci.*, 13: 277-282.

Held, R. & Hein, A. (1963). "Movement-produced stimulation in the development of visually-guided behaviour", *J. Camparitive and Physiol. Psych.*, 56: 872-876.

Howard, I. P. (1982). *Human Visual Orientation*, John Wiley, New York, NY.

Jeannerod, M. (1988). *The Neural and Behavioural Organization of Goal-Directed Movements*, Oxford University Press, Oxford, UK.

Land, M. F. & Furneaux, S. (1997). "The Knowledge-Base of the Oculomotor System", *Proc. Roy. Soc. Lond. B*, 352: 1231-1239.

Lund, R. D. (1978). *Development and Plasticity of the Brain*, Oxford University Press, New York, NY.

Marr, D. (1982). *Vision*, W. H. Freeman Press., San Francisco, CA.

Milner, A. D. & Goodale, M. A. (1995). *The Visual Brain in Action*, Oxford University Press, Oxford, UK.

Nakayama, K. (1985). "Biological Image motion processing: a review", *Vis. Res.*, 25: 625-660.

Regan, D. M. (1989). *Human Brain Electrophysiology: Evoked Potentials and Evoked Magnetic Fields in Science and Medicine*, Elsevier, New York, NY.

Tamura, R., Ono, T., Fukuda, M. & Nakamura, K. (1990). "Recognition of egocentric and allocentric visual and auditory space by neurones in the hippocampus of monkeys", *Neurosci. Letters*, 109: 293-298.

Tanaka, K. & Saito, H. (1989). "Analysis of motion of the visual field by direction, expansion/contraction, and rotation cells clustered in the dorsal part of the medial superior temporal area of the macaque monkey", *J. Neurophysiol.*, 62: 626-641.

Ungerleider, L. G. & Mishkin, M. (1972). "Two cortical visual systems", in D. J. Ingle, M. A. Goodale & R. J.W. Mansfield (Eds), *Analysis of Visual Behaviour*, 549-586, MIT Press, Cambridge, MA.

von Helmholtz, H. (1911). *Handbuch Der Physiologischen Optik. (3rd Edition)*, Voss, Leipzig.

Weiskrantz, L. (1986). *Blindsight: a Case Study and Implications*, Oxford University Press, Oxford, UK.

2

Eye movements as a probe for the early cortical processing of 3-D visual information

F. A. Miles

Abstract

Recent studies have revealed the existence of three visual tracking mechanisms that generate eye movements at ultra-short latencies (<60 ms in monkeys, <85 ms in humans) to compensate for translational motion of the observer. All operate in machine-like fashion and function as automatic servos, using pre-attentive parallel processing to initiate eye movements before the observer is even aware that there has been a disturbance. Two of these mechanisms sense the observer's motion by decoding the pattern of optic flow, one dealing with the problems of the observer who looks off to one side and the other with the problems of the observer who looks straight ahead. The third reflex can be regarded as a complement to the second but utilizes a different set of cues related to motion in depth and concentrates specifically on the problem of maintaining binocular single vision by sensing changes in the relative alignment of the images on the two retinas (binocular disparity). Despite their rapid, reflex nature, all three tracking systems rely on cortical processing - two utilizing binocular stereomechanisms - and evidence from monkeys supports the hypothesis that all are mediated by the medial superior temporal (MST) area of cortex. Remarkably, MST seems to represent the first stage in cortical motion processing at which the visual error signals driving each of the three reflexes are fully elaborated at the level of individual cells.

2.1 Introduction

As we move about the environment, visual and vestibular mechanisms help to stabilize our gaze on particular objects of interest by generating eye movements to offset our head movements. The traditional approach emphasized mechanisms that deal with rotational disturbances of the observer and only recently have translational disturbances been considered. These translational stimuli have uncovered new visual and vestibular reflexes with ultra-short latencies and the general picture that has emerged is of two vestibulo-ocular reflexes, the rotational vestibular ocular reflex (RVOR) and translational vestibular ocular reflex (TVOR), that compensate selectively for rotational and translational disturbances of the head respectively, each with its own independent visual backup mechanism. These concepts and the data supporting them have been extensively reviewed elsewhere (Miles

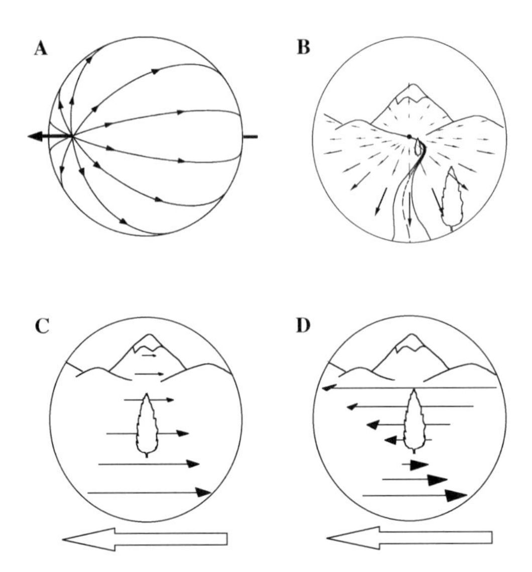

Fig. 2.1. Patterns of optic flow experienced by a translating observer. (A) The retinal optic flow can be considered to be distributed over the surface of a sphere and created by projection through a vantage point at the center. (After Miles *et al.*, 1991.) (B) A cartoon showing the observer's limited field of view and the centrifugal pattern of optic flow experienced by the observer who looks in the direction of heading - the black dot at the foot of the mountain. (After Busettini *et al.*, 1997.) (C) The optic flow experienced by the moving observer who looks off to the right but makes no compensatory eye movements so that the visual scene appears to pivot about the distant mountains (effective infinity). The speed of image motion is inversely proportional to the viewing distance. (D) Again, the observer looks off to one side but here attempts to stabilize the retinal image of a particular object in the middle ground (tree), necessitating that she track to compensate for her own motion, thereby reversing the apparent motion of the more distant objects and creating a swirling pattern of optic flow. The scene now appears to pivot about the tree. (C, D after Miles *et al.*, 1992b.)

et al., 1992a; b; Miles & Busettini, 1992; Miles, 1993; 1995; 1997; 1998). My main concern here is with the new visual mechanisms - three different ones have been discovered so far - which seem to constitute a family of machine-like, ultra-rapid reflexes that utilize relatively low-level, preattentive cortical processing of complex visual stimuli and function largely independently of perception. These three visual reflexes all have special features to help stabilize gaze during translational disturbances of the observer.

2.2 The optic flow associated with translational disturbances

A passive observer who undergoes pure translation experiences optic flow in which streams of images emerge from a focus of expansion straight ahead and disappear into a focus of contraction behind, the overall pattern resembling the lines of longitude on a globe (Figure 2.1A). The direction of flow at any given point is dictated entirely by the observer's motion but the speed of the flow at any given point also depends on the viewing distance at that location. Thus, nearby objects move across the field of view much more rapidly than more distant ones (motion parallax: Gibson, 1950; 1966). Given the observer's restricted field of view, the patterns of motion actually experienced, and the eye movements required to actively compensate for them, depend on where the observer chooses to look. If the observer looks straight ahead she/he experiences an expanding world (as in the cartoon in Figure 2.1B) and, insofar as the radial pattern of flow is associated with a change in viewing distance, the observer must converge her/his eyes if the object of interest in the scene ahead is to stay imaged on both foveas. Of course, the amount of convergence required to maintain binocular alignment is inversely related to the viewing distance, hence the greatest challenge comes with near viewing. However, if the passive observer looks off to one side the sensation is of the visual world pivoting around the far distance (as in the cartoon in Figure 2.1C): objects move across the field of view at a rate that is in inverse proportion to their viewing distance so that only the retinal images of the most distant objects (here, the mountains) are stable on the retina. In order to fixate objects in the middleground (like the tree in Figure 2.1C), the observer must produce eye movements that compensate for her/his motion. The magnitude of those compensatory eye movements needs to be inversely proportional to the viewing distance. If the observer succeeds in this then her/his visual world will now pivot about the object of regard (the tree in Figure 2.1D). The optic flow that results is due to a combination of translational motion (due to the observer's bodily motion) and rotational motion (due to the observer's compensatory eye movements). It will be clear from this that the translating observer can stabilize only the images in one particular depth plane and we shall see that the problem confronting the system here is to ensure that the "plane of stabilization" coincides with the plane of fixation where the two lines of sight intersect at the object of regard.

2.3 Ocular following

The first of the three ultra-rapid visual tracking mechanisms that I shall consider - usually referred to as ocular following - helps to stabilize the retinal images of objects in the plane of fixation and has special built-in features for dealing with the visual problems posed when the moving observer looks off to one side, as in C and D of Figure 2.1. The visual task confronting the visual stabilization mechanisms here is to single out the motion of particular elements in the scene - such as the mountain in Figure 2.1C and the tree in Figure 2.1D - and ignore all of the competing motion elsewhere.

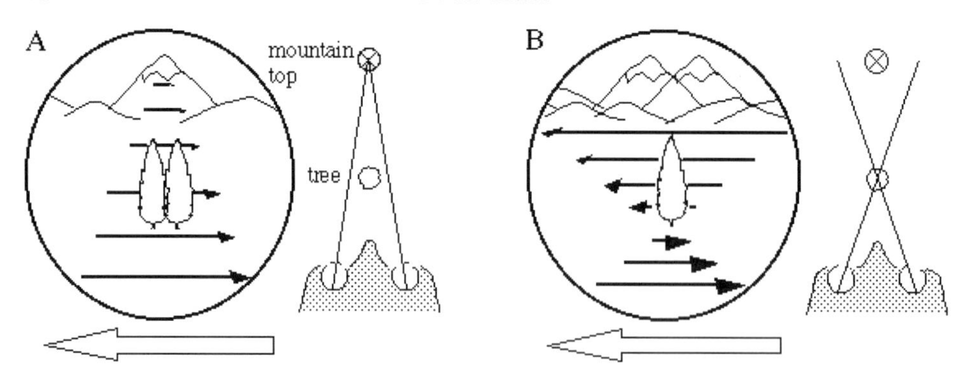

Fig. 2.2. The optic flow experienced by a translating observer (binocular viewing). (A) As in Figure 2.1C, except that, with binocular viewing, the mountain in the plane of fixation is seen as single and the nearer tree is seen double (disparate). A plan view of the observer and the two objects is shown to the right. (B) As in Figure 2.1D, except that, with binocular viewing, the tree is placed in the plane of fixation and so is seen as single whereas the distant mountain is now seen as double (disparate). Again, the plan view is shown to the right. Note that the dimensions of the eyes and their separations have been exaggerated to illustrate the disparity more clearly. In fact, disparity is much more evident with near viewing, which is also associated with the most vigorous optic flow and requires the most vigorous tracking from the observer to compensate. All of the laboratory experiments used near viewing. Reprinted with permission from Busettini, C. *et al.* (1996a). "A role for stereoscopic depth cues in the rapid visual stabilization of the eyes', *Nature*, 380: 342-345.

One way to achieve this would be to use attentional focussing mechanisms to spotlight the target of interest. Such mechanisms exist and are used by the so- called pursuit system but have the limitation that they require high-level executive decisions to select the image to be tracked and this of necessity is time consuming (Ferrera & Lisberger, 1995, 1997; Pola & Wyatt, 1993). The ocular following reflex solves this problem more expeditiously using low-level stereomechanisms that perform rapid parallel processing of binocular images, effectively sorting them on the basis of the depth plane that they occupy. This stereo algorithm, which utilizes the fact that we have two eyes with slightly differing viewpoints, is illustrated in Figure 2.2, which is a "binocular" version of the cartoons in Figure 2.1C and D. The object on which the two eyes are aligned (the mountain in Figure 2.2A or the tree in Figure 2.2B) resides in the plane of fixation and is imaged at corresponding positions on the two retinas; the object is therefore perceived as a single, fused image. In contrast, objects that are nearer or farther than the plane of fixation have images that occupy non-corresponding positions on the two retinas - they are said to have "binocular disparity" - and are seen as double (the tree in Figure 2.2A and the mountain in Figure 2.2B). Clearly, a robust algorithm for stabilizing gaze on the object(s) of regard would be to track only those objects whose images occupy corresponding positions on the two retinas: objects in the plane of fixation. Early support for this idea was the finding that optokinetic responses, which are the ocular following responses elicited by rotating a striped drum around the observer, are best for images with zero binocular disparity (Howard & Gonzalez, 1987; Howard & Simpson, 1989). However, this study examined only the closed-loop, steady-state responses, which are known to be strongly influenced by high-level processes, such as selective attention (Cheng & Outerbridge, 1975; Dubois & Collewijn, 1979; Murasugi *et al.*, 1986). More recent experiments indicate that the very earliest ocular following responses, which are generated before there has been time for such processing to influence eye movements, show a similar preference for binocular images that lack disparity (Busettini *et al.*, 1996a). Figure 2.3A shows this effect of disparity on the initial ocular following responses of a monkey to sudden movements of large-field images presented on a tangent screen facing the animal. A dichoptic viewing arrangement was used to allow the (identical) images seen by each of the two eyes to be positioned and moved independently. The very earliest responses have the usual ultra-short latency (about 55 ms) and are clearly at their most vigorous when the binocular images are in register on the screen, which is the plane of fixation (trace labelled "0" in Figure 2.3A). Responses decrement progressively as the images are presented with more and more disparity, which in effect positions them farther and farther from the plane of fixation. The disparity tuning curve for these data, based on measures of the very earliest responses, is plotted in Figure 2.3B and has a bell-shaped profile centered on zero disparity. The earliest human ocular following responses (latency <85 ms) show a very similar dependence on disparity.

The above discussion indicates that the ocular following system helps to stabilize gaze on objects of interest not by selecting a particular one but by stabilizing the image of any object that happens to lie close to the plane of fixation, an implicit assumption, therefore, being that this plane contains the objects likely to be of most interest. Note that the time-consuming process of selecting the object of interest therefore rests with the oculomotor subsystems that bring images into the plane of fixation - that is, the saccadic system working in concert with the vergence system. These latter systems redirect gaze to objects using higher-level criteria whereas ocular following relies on low-level rapid parallel filters. Thus, the general concept is of low-level reflex systems stabilizing whatever images the high-level systems happen to bring into the plane of fixation.

It has been known for some time that there are neurons in visual cortex as early as V1 that are selectively sensitive to images moving in a particular depth plane. The activation of these cells requires the images to have a specific direction of motion and binocular disparity (for review see Bishop & Pettigrew, 1986 and Poggio, 1995). Some of these neurons respond to

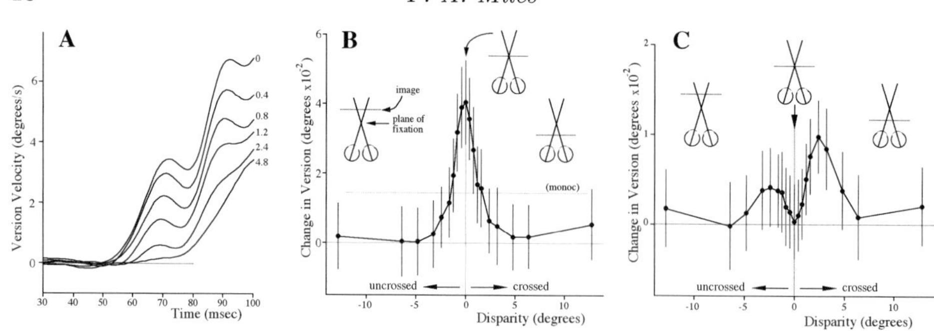

Fig. 2.3. Dependence of initial ocular following on the horizontal disparity of the moving images. (A) Mean version velocity responses of a monkey in response to downward motion (40 deg/s in all cases) when the images seen by the two eyes had crossed disparities whose magnitude (in degrees) is indicated by the numbers at the ends of the traces. Note that version is the average velocity of *both* eyes. (B) Disparity tuning curve for the ocular following responses of the same monkey. Measures based on the change in version over the time period 60-77 ms after the onset of the stimulus ramp, which was always 40 deg/s (includes the data in A). Error bars, $\pm$1SD. (C) Disparity tuning curve for a monkey whose ocular following showed a stereoanomaly for one direction of motion (rightward). The data shown are for the responses to rightward motion (80 deg/s in all cases). For leftward, upward, and downward motion, the curves were like that in B. Reprinted with permission from Busettini, C. *et al.* (1996a). "A role for stereoscopic depth cues in the rapid visual stabilization of the eyes', *Nature*, 380: 342-345.

motion within only a narrow range of depths that can lie exactly in ("tuned zero" cells), or close to ("tuned near", "tuned far" cells), the plane of fixation, whereas others respond to motion over a wider range of depths either inside ("near" cells), or beyond ("far" cells) the plane of fixation. Clearly the "tuned zero" cells are good candidates for mediating ocular following because they are selectively sensitive to images moving in the plane of fixation. However, the "tuned zero" cells in the literature all have tuning curves with half-widths much less than a degree whereas ocular following has a half-width of a degree or two. It could be that there are "tuned zero" cells with much broader tuning curves that have yet to be recorded - all recordings to date have been limited to parafoveal regions - but I think it likely that other types of "tuned" cells (with a preference for moving objects with small crossed or uncrossed disparities) also make a contribution*.

Two subjects showed extremely interesting stereoanomalies. These subjects had normal disparity tuning curves for three of the four cardinal directions of motion but, for the fourth direction, their curves exhibited a pronounced dip centered on zero disparity (Figure 2.3C). This extraordi-

* An object that does not have zero-disparity will form two images (see Figure 2.2). If the left eye's image is on the left of the fixation point and the right eye's image is on the right of the fixation point then the object is said to have an uncrossed disparity. If they are the other way around, with the left eye's image on the right and the right eye's image on the left, then they are said to have crossed disparities (see Howard & Rogers, 1995).

narily specific stereoanomaly is exactly the sort of deficit that one would expect if the subject lacked only "tuned zero" cells with a preference for motion in one particular direction. Such a seemingly cell-specific anomaly has the appearance of a naturally-occurring gene knockout. Regardless of the etiology of the deficit, its specificity lends strong support to the idea that these ocular following responses are mediated by neurons that are selective for both motion and disparity.

Although the stereoanomalies point to dependence on low-level disparity mechanisms - perhaps as early as striate cortex - there is strong evidence that ocular following derives at least some of its input from much later stages in the dorsal stream of cortex (Ungerleider & Mishkin, 1982) where motion is processed: chemical lesions in MST result in impairments of even the earliest components (Kawano *et al.*, 1997) and single unit recordings in this region have revealed many directionally selective neurons that discharge in close relation to the large-field, high-speed motion stimuli that are optimal for eliciting ocular following (Kawano *et al.*, 1994). Also, many of the neurons discharge early enough to have causal involvement. There are data indicating selectivity for binocular disparity as well as motion in neurons of MT (Maunsell & Van Essen, 1983b) and MST (Roy & Wurtz, 1990; Roy *et al.*, 1992) but stimuli optimal for ocular following were not tried in these studies.

2.4 Radial flow vergence

The previous section on ocular following dealt with the problem of stabilizing gaze when the moving observer looks off to one side. I now consider the gaze stability mechanisms that come into play when the moving observer looks in the direction of heading and so experiences the radial pattern of optic flow featured in Figure 2.1A and B. Clearly, the radial pattern of flow is associated with a change in viewing distance, necessitating that the observer converge her/his eyes (especially with near viewing) in order to keep the object of interest in the scene ahead imaged on both foveas. Recent experiments on humans (Busettini *et al.*, 1997) have indicated that radial optic flow elicits vergence eye movements at latencies that are closely comparable with the ultra-short values mentioned above for human ocular following (approximately 80ms). Centrifugal flow, which signals a forward approach and hence a decrease in the viewing distance, resulted in increased convergence. Centripetal flow, which signals the converse, resulted in decreased convergence (Busettini *et al.*, 1997). A sample vergence response profile elicited by centrifugal flow can be seen in Figure 2.4.

The clear suggestion here is that the brain is able to sense the radial pattern of flow and to infer from this that there has been a change in viewing distance. However, a characteristic of the ocular responses to these radial

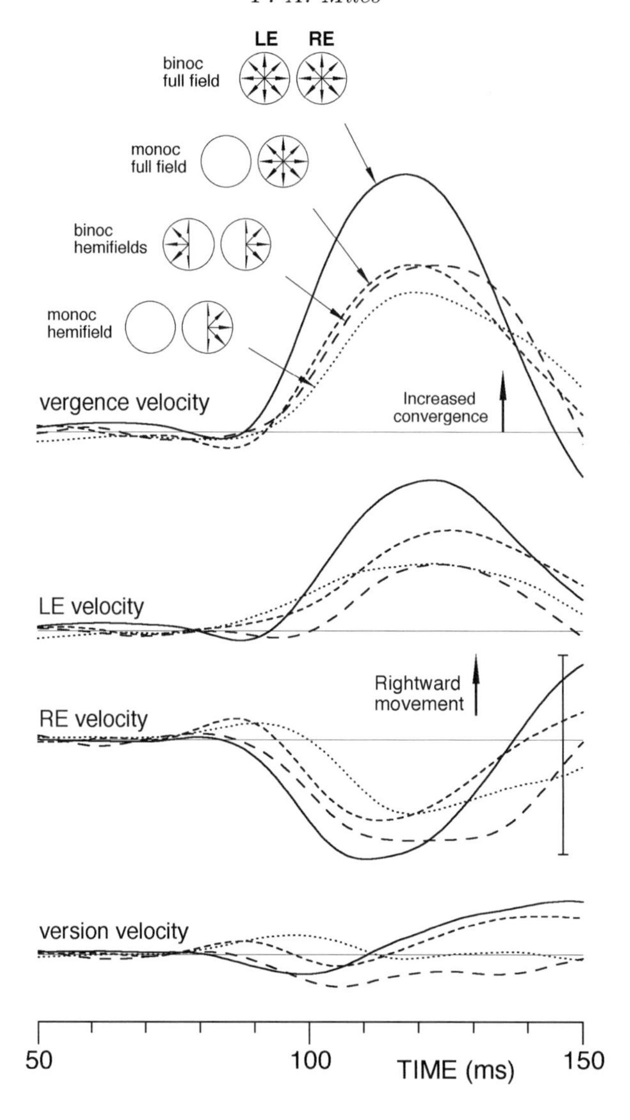

Fig. 2.4. The initial vergence eye movements elicited by radial optic flow: effect of masking off various parts of the binocular visual field (human subject). The inset cartoons indicate the extent of the masks: no mask ("binoc full field"), left eye masked ("monoc full field"), both nasal hemifields masked ("binoc hemifields"), and all but one temporal hemifield masked ("monoc hemifield"). In addition to showing the vergence velocity profiles (the difference in the velocity of the two eyes), also shown are the velocity profiles for each of the two eyes, and the version velocity (the average velocity of the two eyes). Images were random dot patterns back-projected onto a large tangent screen facing the subject. Stimuli were looming steps simulating a sudden 4% reduction in viewing distance, which was achieved by switching between two projected images, the switch occurring at time zero. Calibration bar, 2 deg/s. Reprinted from Busettini, C. *et al.* (1997). "Radial optic flow induces vergence eye movements at ultra-short latencies", *Nature*, 390: 512-515.

flow patterns is that each eye always moves in the direction of the net motion vector in the nasal hemifield, and this allows an alternative and less interesting explanation for the responses: the vergence might result from monocular tracking, in which each eye tracks only the motion that it sees and has a strong preference for motion in the nasal hemifields. For example, with centrifugal flow the net motion vector in the nasal hemifields is towards the nose and each eye moves in that direction, hence the increased convergence. That this was not the explanation was apparent from the observation that binocular vergence responses persisted, albeit weaker, when various parts of the radial flow patterns seen by the two eyes were masked off, including the whole of one eye (see traces labelled "monoc full field" in Figure 2.4), or both nasal hemifields ("binoc hemifields" in Figure 2.4), or one whole eye plus the remaining nasal hemifield so that the only parts of the patterns now visible were those seen by the right temporal hemifield ("monoc hemifield" in Figure 2.4). Note that in the last two cases each eye actually moves in the opposite direction to any net motion vector that it sees. Therefore it was concluded that the vergence responses result from a true parsing of the radial pattern of flow.

These data also imply something about the neural decoding of planar optic flow (such as that in Figure 2.1C and, to a lesser degree, Figure 2.1D). When the observer's view was limited to a single temporal hemifield there was a strong net motion vector (to the right in Figure 2.4) yet it is clear that the system still correctly attributed the flow to forward rather than to leftward motion (or rotation) of the observer because it responded with convergence rather than rightward (conjugate) ocular following. Conjugate oculomotor responses, such as ocular following, are more readily appreciated from plots of version, which is the average movement of the two eyes and is therefore insensitive to changes in vergence, than from the movements of the individual eyes. In fact, version and vergence provide an alternative, equally complete, representation of eye movements and might be more indicative of the way that eye movements are encoded in the cortical regions under consideration here. An important point here is that the system not only produces the appropriate vergence responses but avoids making inappropriate version responses despite the net motion vector to the right. The presence of vertical motion - even though there is no net vertical vector - is clearly sufficient to block the version responses.

The above discussion suggests that there are neurons or networks that act like templates or tuned filters to detect specific patterns of optic flow and generate appropriate oculomotor responses to serve the needs of visual stabilization. Again, latencies are extremely short so that the system must depend on parallel processing to arrive at an appropriate response based on the pattern of optic flow. Thus, again the system helps to stabilize gaze on objects of interest not by selecting a particular image but by sensing the global pattern of flow. Once again the general concept is of low-level reflex systems responding appropriately to whatever region of the optic flow field is brought into view by the high-level saccadic system.

There is extensive evidence that area MST in the monkey's cortex contains neurons that are selectively sensitive to radial optic flow patterns such as those now known to evoke vergence eye movements at ultra-short latencies (Saito *et al.*, 1986; Tanaka & Saito, 1989; Tanaka *et al.*, 1989; Duffy & Wurtz, 1991a, 1991b, 1995; Lagae *et al.*, 1994; Lappe *et al.*, 1996; Pekel *et al.*, 1996). In fact, MST is the first stage in this dorsal pathway at which global flow is encoded at the level of single cells: at earlier stages, such as MT, individual cells have much smaller receptive fields and encode only local motion (Van Essen *et al.*, 1981; Maunsell & Van Essen, 1983a; Albright & Desimone, 1987; Komatsu & Wurtz, 1988; Albright, 1989; Lagae *et al.*, 1994). Mention has already been made of the evidence indicating that ocular following (version) responses to planar flow in the frontoparallel plane of fixation are mediated at least in part by MST. The new observations with radial flow patterns indicate that the mechanism mediating these version responses is blocked by orthogonal motion, consistent with the finding that putative ocular following neurons in MST are suppressed by non-preferred motion. That is, motion in the opposite or orthogonal direction (Duffy & Wurtz, 1991a).

2.5 Disparity vergence

Radial optic flow is only one of several cues which indicate the forward rate of progress when the moving observer looks in the direction of heading. The possibility therefore exists that additional cues might also elicit vergence eye movements at ultra-short latencies. One such cue is the change in size of the retinal image of the objects as the observer approaches them, but it is known that this elicits convergence only at (pursuit?) latencies that are appreciably longer than those under review here (Erkelens & Regan, 1986; Cohen & Lisberger, 1996; Busettini *et al.*, 1997). Another cue, however, is very potent at generating vergence at ultra- short latencies: binocular disparity. If the observer were to move forward without converging her/his eyes adequately then the object of regard would be overtaken and repositioned inside the plane of fixation where it would be imaged at non-corresponding positions on the two retinas (so-called crossed disparity). Recent experiments have demonstrated that when random-dot patterns are viewed dichoptically and small binocular misalignments are suddenly imposed (disparity steps)*,

* In order to minimize problems due to vergence eye movements, which operate to eliminate any imposed binocular disparity, the disparity was not imposed until the onset of the movement used to elicit ocular following. Thus, the binocular stimulus had a step-ramp profile, a disconjugate step to establish the disparity and a conjugate ramp to elicit ocular following. Further, the ocular following responses were assessed by averaging the movements of the two eyes (ver-

corrective vergence eye movements are elicited at latencies of <60 ms in monkeys (Busettini *et al.*, 1996b) and <85 ms in humans (Busettini *et al.*, 1994a), values closely comparable with those for ocular following and radial flow vergence. Crossed disparity steps elicited increased convergence and uncrossed steps decreased convergence, exactly as expected of a depth-tracking servo mechanism driven by disparity. However, once more it is necessary to prove that this is truly a response to a binocularly processed visual signal and not the result of monocular tracking in which each eye merely tracks the apparent motion that it sees (towards or away from the nose). That these responses could not be the result of monocular tracking is evident from experiments in which the disparity step was confined to one eye (Busettini *et al.*, 1996b). For example, when the crossed disparity step was confined to the right eye which saw a leftward step, the result was binocular convergence in which the left eye moved rightward even though that eye had seen only a stationary pattern. The rightward movement of the left eye here is in the direction expected of a stereoscopic mechanism that responds to a binocular misalignment but is in the opposite direction to the only available motion cues - the leftward motion at the right eye.

The range of disparities over which the system behaves like a servo mechanism, that is the range over which increases in the disparity vergence error result in roughly linear increases in the vergence response, is <2 degrees. Thus, this vergence mechanism can correct only small misalignments of the two eyes, commensurate with a mechanism that performs only local stereo matches and merely attempts to bring the nearest salient images into the plane of fixation. During forward locomotion this mechanism will help to prevent images from leaving the plane of fixation. Once more, we have a mechanism that functions as a low- level automatic servo and is not involved in high-level operations like the transfer of fixation to new images in new depth planes, which requires time-consuming target selections, and (often) the decoding of large disparity errors (>10 degrees) that necessitate solution of the correspondence problem. Recent experiments (Masson *et al.*, 1997) have shown that vergence responses can also be elicited at ultra-short latencies by disparity stimuli applied to dense (50%) anticorrelated binocular patterns, in which each black dot in one eye is matched to a white dot in the other eye. Figure 2.5A shows sample mean vergence velocity profiles in response to crossed disparity stimuli applied to correlated and anticorrelated patterns. Note that the vergence responses to the anticorrelated stimuli are in the reverse direction of those to the correlated stimuli. The disparity tuning curves for these data are shown in Figure 2.5B, the curve obtained with the normal correlated patterns having a characteristic s-shape that is

sion), which meant the response measures were free of any possible vergence contaminants. Actually, vergence responses during the time window under consideration were at best very small, especially when high-speed ramps were used to elicit ocular following.

well fitted by a Gabor function. The curve for the anticorrelated data is almost a mirror image, and the cosine term for the best-fit Gabor function is phase shifted almost exactly 180 degrees. In two-alternative-forced-choice tests, subjects could readily discriminate between crossed and uncrossed disparities when applied to the correlated patterns but not when applied to the anticorrelated patterns (Masson *et al.*, 1997). This is consistent with the idea that these short-latency vergence responses derive their visual input from an early stage of cortical processing prior to the level at which depth percepts are elaborated. Actually, the large-field stimuli used in all of these disparity vergence studies contain only absolute disparity cues, which are known to be poorly perceived in depth (Erkelens & Collewijn, 1985a; 1985b; Regan *et al.*, 1986).

As mentioned above, neurons sensitive to binocular disparity have been described in various regions of the visual cortex, and these have often been invoked as the source of the error signals driving disparity vergence. However, in discussing the sensitivity of ocular following to disparity we were concerned with neurons that were selective for motion as well as disparity and had a preference for the plane of fixation (zero disparity). Now, we are presumably concerned with disparity selective neurons that have no particular motion preferences (except perhaps for motion in depth) and that discharge to nonzero disparities, thereby effectively encoding vergence error. Many such neurons have been described in visual cortex as early as V1 (see Poggio, 1995 for recent review), and in the dorsal stream, including MT (Maunsell & Van Essen, 1983b) and MST (Takemura *et al.*, 1997). At recording levels up to MT, these neurons have been classified as "near", "far", "tuned near" and "tuned far", depending on the range of disparities over which they are active. However, in MST some of the neurons that discharge in close relation to the large-field binocular stimuli used to elicit short-latency disparity vergence have disparity tuning curves that do not readily conform to any of these categories but exactly match the broad s-shaped curves characteristic of the vergence responses (such as those seen in Figure 2.5B). Once more, it seems that we have a short-latency oculomotor response that relies on signals that first occur in their entirety at the level of single cells in MST. Apropos the reversed vergence responses to anticorrelated patterns, many disparity-selective neurons in the monkey's striate cortex also respond to these patterns, despite the fact that monkeys (like humans) fail to perceive depth in them, and many of these neurons show inverted disparity tuning curves (Cumming & Parker, 1997), in accord with a local filter model of disparity selective complex cells (Ohzawa *et al.*, 1990).

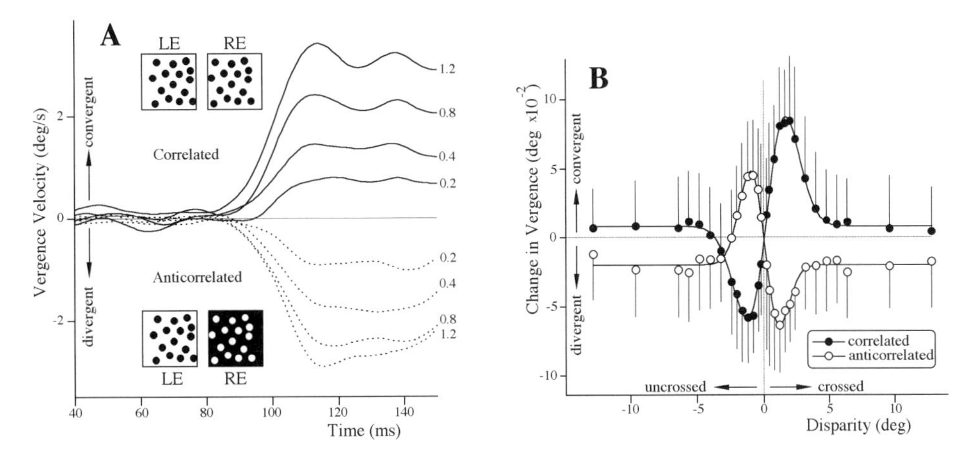

Fig. 2.5. The vergence eye movements elicited by disparity step stimuli applied to random-dot patterns using a dichoptic viewing arrangement to allow separate stimulation of each eye (human subject). (A), Mean vergence velocity responses of a monkey to crossed disparity stimuli applied at time zero to correlated (continuous line) and anticorrelated (dotted line) patterns, with stimulus magnitudes (in deg) indicated at the ends of the traces. The cartoons indicate only the general form of the patterns seen by the left (LE) and right (RE) eyes - those actually used had higher dot density (50%), each dot being 2 degrees in diameter, and the whole image extended over 80x80 degrees. (B) Plot of the mean ($\pm$1SD) changes in vergence position (over time period 60-93 ms from stimulus onset) against the disparity stimulus, with correlated (filled circles) and anticorrelated (open circles) patterns. The normal disparity tuning curves have an s-shape, the linear (servo) region being restricted to disparities $< \pm 2$ degrees. The curves are the best fitting Gabor functions and the cosine terms for the correlated and anticorrelated data differ in phase by about 180 degrees. Reprinted with permission from Masson, G. S. *et al.* (1997). "Vergence eye movements in response to binocular disparity without depth perception", *Nature*, 389: 283-286.

2.6 A family of short-latency cortical reflexes?

The three visual stablization mechanisms share a number of features - notably, ultra-short latencies and a special involvement with translational disturbances - and this has led to the suggestion that they constitute a family of reflexes (Busettini *et al.*, 1997): see Table 2.1 for a listing of their fundamental similarities and differences. All three mechanisms also show post- saccadic enhancement, whereby stimuli applied in the immediate wake of a saccadic eye movement are much more effective than the same stimuli applied a few hundred milliseconds later (Kawano and Miles 1986; Busettini *et al.*, 1994a, b; 1997). A similar enhancement occurs in the wake of a saccade-like shift of the visual scene, indicating that, in all cases, the enhancement is at least partly visual in origin and due to the visual reafference created by the saccade sweeping the image of the world across the retina. This transient priming of the three reflexes on completion of the saccade is very timely, coming when the need to re-establish retinal image

The three visual tracking mechanisms have been grouped according to the type of translation to which they are optimally responsive. With X/Y-translation, the observer moves from side-to-side or up-down, tending to produce motion in the frontoparallel plane: the plane of fixation. With Z-translation, the observer moves from front-to-back or vice versa, tending to shift objects out of the plane of fixation, producing radial optic flow and horizontal disparity.

	X/Y-TRANSLATION	Z-TRANSLATION (DEPTH)	
	OCULAR FOLLOWING	RADIAL FLOW VERGENCE	DISPARITY VERGENCE
Function	Stabilizes gaze against motion in fixation plane (tolerates position errors)	Stabilizes gaze against motion in depth (tolerates position errors)	Eliminates residual vergence errors
Input	Binocular motion in plane of fixation	Radial optic flow (monoc/binoc)	Binocular disparity (local matches)
Output (binocular)	Horizontal/vertical version	Transient horizontal vergence	Horizontal/vertical* vergence
Servo type	Velocity	Velocity	Position
Latency	<60ms monkeys <85ms humans	? ms monkeys <85ms humans	<60ms monkeys <85ms humans
Independent of perception	?	?	Yes
Post-saccadic enhancement	Yes (part visual reafference)	Yes (part visual reafference)	Yes (part visual reafference)
Adaptive gain control	Yes	?	?
Dependence on viewing distance	Yes	?	?
Neural mediation	MST + ?	MST?	MST?

* Busettini, Masson and Miles (unpublished observations).

Table 2.1. *Major features of the three visual stabilization mechanisms (from Miles, 1998)*

stability - by eliminating any residual retinal slip or binocular misalignment - is paramount. By boosting performance only transiently (with a time constant approximating that of the oculomotor plant) these control systems avoid the instability problems generally associated with excessively high gain (Miles *et al.*, 1986).

To the extent that the three mechanisms are members of a family*, one might hope to generalize from one to another. To date, only one has been shown to operate independently of perception (disparity vergence) but I would expect the same of the other two. Likewise, only ocular following has been shown to be subject to adaptive gain control, though I assume that all three reflexes could benefit and probably do. Ocular following and disparity vergence both show dependence on the viewing distance. If such dependence on proximity is also characteristic of the vergence resulting from radial flow then it might help to explain how one avoids converging one's eyes when passing through a doorway: the image of interest and the plane of fixation lie some distance beyond the doorway, resulting in a low gain that effectively vetoes a vergence response to the centrifugal flow generated by the doorframe. I suspect that the dependence on distance reflects a major involvement with near viewing because it is the retinal images of the nearest objects that are most sensitive to the observer's movements and, therefore, offer the greatest challenge to ocular alignment and stabilization. In other respects, the tracking systems might be different but still complementary. An example is the transient vergence elicited by steps of radial flow (a velocity drive) and the tonic vergence elicited by steps of disparity (a position servo).

Evidence that MST is critical for these mechanisms is currently available only for ocular following. However, it is surely not fortuitous that MST contains neurons that discharge in association with the stimuli that selectively activate each of the three mechanisms and, intriguingly, represents the first stage in the cortical processing pathway at which the adequate stimulus for each system is fully encoded at the level of individual neurons.

References

Albright, T. D. (1989). "Centrifugal directional bias in the middle temporal visual area (MT) of the macaque", *Vis. Neurosci.*, 2: 177-188.

Albright, T. D. & Desimone, R. (1987). "Local precision of visuotopic organization in the middle temporal area (MT) of the macaque", *Exp. Brain Res.*, 65: 582-592.

* This disparity vergence mechanism is in a somewhat different category from the other two short-latency tracking mechanisms insofar as its primary function is to eliminate small vergence errors, evidenced by the fact that it also operates in the vertical axis using vertical disparity, which is unrelated to depth and translation per se (Busettini, Masson & Miles, unpublished observations). While the specific involvement with vergence errors resulting from locomotion is clear, this is only a secondary function.

Bishop, P. O. & Pettigrew, J. D. (1986). "Neural mechanisms of binocular vision", *Vis. Res.*, 26: 1587-1600.
Busettini, C., Masson, G. S. & Miles, F. A. (1996a). "A role for stereoscopic depth cues in the rapid visual stabilization of the eyes', *Nature*, 380: 342-345.
Busettini, C., Masson, G. S. & F. A. Miles (1997). "Radial optic flow induces vergence eye movements at ultra-short latencies", *Nature*, 390: 512-515.
Busettini, C., Miles, F. A. & Krauzlis, R. J. (1994a). "Short-latency disparity vergence responses in humans", *Soc. Neurosci. Abstr.*, 20: 1403.
Busettini, C., Miles, F. A. & Krauzlis, R. J. (1996b). "Short-latency disparity vergence responses and their dependence on a prior saccadic eye movement", *J. Neurophysiol.*, 75: 1392-1410.
Busettini, C., Miles, F. A., Schwarz, U. & Carl, J. R. (1994b). "Human ocular responses to translation of the observer and of the scene: dependence on viewing distance", *Exp. Brain Res.*, 100: 484-494.
Cheng, M. & Outerbridge, J. S. (1975). "Optokinetic nystagmus during selective retinal stimulation", *Exp. Brain Res.*, 23: 129-139.
Cohen, G. A. & S. G. Lisberger (1996). "Motion disparity and looming cues form hierarchical inputs for smooth pursuit of targets moving in 3 dimensions", *Soc. Neurosci. Abstr.*, 22: 964.
Cumming, B. G. & Parker, A. J. (1997). "Responses of primary visual cortical neurons to binocular disparity without depth perception", *Nature*, 389: 280-283.
Dubois, M. F. W. & Collewijn, H. (1979). "Optokinetic reactions in man elicited by localized retinal stimuli", *Vis. Res.*, 19: 1105-1115.
Duffy, C. J. & Wurtz, R. H. (1991a). "Sensitivity of MST neurons to optic flow stimuli. I. A continuum of response selectivity to large-field stimuli", *J. Neurophysiol.*, 65: 1329-1345.
Duffy, C. J. & Wurtz, R. H. (1991b). "Sensitivity of MST neurons to optic flow stimuli. II. Mechanisms of response selectivity revealed by small-field stimuli", *J. Neurophysiol.*, 65: 1346-1359.
Duffy, C. J. & Wurtz, R. H. (1995). "Response of monkey MST neurons to optic flow stimuli with shifted centers of motion", *J. Neurosci.*, 15: 5192-5208.
Erkelens, C. J. & Collewijn, H. (1985a). "Motion perception during dichoptic viewing of moving random-dot stereograms", *Vis. Res.*, 25: 583-588.
Erkelens, C. J. & Collewijn, H. (1985b). "Eye movements and stereopsis during dichoptic viewing of moving random-dot stereograms", *Vis. Res.*, 25: 1689-1700.
Erkelens, C. J. & Regan, D. (1986). "Human ocular vergence movements induced by changing size and disparity", *J. Physiol.*, 379: 145-169.
Ferrera, V. P. & Lisberger, S. G. (1995). "Attention and target selection for smooth pursuit eye movements", *J. Neurosci.*, 15: 7472-7484.
Ferrera, V. P. & Lisberger, S. G. (1997). "The effect of a moving distractor on the initiation of smooth-pursuit eye movements", *Vis. Neurosci.*, 14: 323-338.
Gibson, J. J. (1950). *The Perception of the Visual World*, Houghton Mifflin, Boston, MA.
Gibson, J. J. (1966). *The Senses Considered as Perceptual Systems*, Houghton Mifflin, Boston, MA.
Howard, I. P. & Gonzalez, E. G. (1987). "Human optokinetic nystagmus in response to moving binocularly disparate stimuli", *Vis. Res.*, 27: 1807-1816.
Howard, I. P. & Rogers, B. (1995). *Binocular vision and stereopsis*, Oxford University Press, Oxford.
Howard, I. P. & Simpson, W. A. (1989). "Human optokinetic nystagmus is linked to the stereoscopic system", *Exp. Brain Res.* 78: 309-314.
Kawano, K., Inoue, Y., Takemura, A., Kitama, T. & Miles, F. A. (1997). "A cortically mediated visual stabilization mechanism with ultra-short latency in

primates", In *Parietal Lobe Contributions to Orientation in 3D Space*, P. Thier & H.-O. Karnath (Eds.), 185-199, Springer-Verlag, Heidelberg.

Kawano, K. & Miles, F. A. (1986). "Short-latency ocular following responses of monkey. II. Dependence on a prior saccadic eye movement", *J. Neurophysiol.*, 56: 1355-1380.

Kawano, K., Shidara, M., Watanabe, Y. & Yamane, S. (1994). "Neural activity in cortical area MST of alert monkey during ocular following responses", *J. Neurophysiol.*, 71: 2305- 2324.

Komatsu, H. & Wurtz, R. H. (1988). "Relation of cortical areas MT and MST to pursuit eye movements. I. Localization and visual properties of neurons", *J. Neurophysiol.*, 60: 580-603.

Lagae, L., Maes, H., Raiguel, S., Xiao, D.-K. & Orban, G. A. (1994). "Responses of macaque STS neurons to optic flow components: A comparison of areas MT and MST", *J Neurophysiol.*, 71: 1597-1626.

Lappe, M., Bremmer, F., Pekel, M., Thiele, A., & Hoffmann, K.-P. (1996). "Optic flow processing in monkey STS: A theoretical and experimental approach", *J. Neurosci.*, 16: 6265-6285.

Masson, G. S., Busettini, C., & Miles, F. A. (1997). "Vergence eye movements in response to binocular disparity without depth perception", *Nature*, 389: 283-286.

Maunsell, J. H. R. & Van Essen, D. C. (1983a). "Functional properties of neurons in middle temporal visual area of the macaque monkey. I. Selectivity for stimulus direction, speed, and orientation", *J. Neurophysiol.*, 49: 1127-1147.

Maunsell, J. H. R. & Van Essen, D. C. (1983b). "Functional properties of neurons in middle temporal visual area of the macaque monkey. II. Binocular interactions and sensitivity to binocular disparity", *J. Neurophysiol.*, 49: 1148-1167.

Miles, F. A. (1993). "The sensing of rotational and translational optic flow by the primate optokinetic system", In *Visual Motion and its Role in the Stabilization of Gaze*, F. A. Miles & J. Wallman (Eds.), 393-403, Elsevier, Amsterdam.

Miles, F. A. (1995). "The sensing of optic flow by the primate optokinetic system", In *Eye Movement Research: Mechanisms, Processes and Applications*, J. M. Findlay, R. W. Kentridge & R. Walker (Eds.), 47-62, Elsevier, Amsterdam.

Miles, F. A. (1997). "Visual stabilization of the eyes in primates", *Curr. Opin. Neurobiol.*, in press.

Miles, F. A. (1998). "The neural processing of 3-D visual information: Evidence from eye movements", *Eur. J. Neurosci.*, in press.

Miles, F. A. & Busettini, C. (1992). "Ocular compensation for self motion: visual mechanisms", In *Sensing and Controlling Motion: Vestibular and Sensorimotor Function*, B. Cohen, D. L. Tomko & F. Guedry (Eds.), 220-232, Ann. NY Acad. Sci., New York, NY..

Miles, F. A., Busettini, C., & Schwarz, U. (1992a). "Ocular responses to linear motion", *In Vestibular and Brain Stem Control of Eye, Head and Body Movements*, H. Shimazu & Y. Shinoda (Eds.),. 379-395, Springer-Verlag/Japan Scientific Societies Press, Tokyo.

Miles, F. A., Kawano, K. & Optican, L. M. (1986). "Short-latency ocular following responses of monkey. I. Dependence on temporospatial properties of the visual input", *J. Neurophysiol.*, 56: 1321-1354.

Miles, F. A., Schwarz, U. & Busettini, C. (1992b). "The decoding of optic flow by the primate optokinetic system", In *The Head-Neck Sensory-Motor System*, A. Berthoz, W. Graf, & P. P. Vidal (Eds.), 471-478, Oxford University Press, New York, NY.

Murasugi, C. M., Howard, I. P., & Ohmi, M. (1986). "Optokinetic nystagmus: the effects of stationary edges, alone and in combination with central occlusion", *Vis. Res.*, 26: 1155-1162.

Ohzawa, I., DeAngelis, G. C. & Freeman, R. D. (1990). "Stereoscopic depth discrimination in the visual cortex: neurons ideally suited as disparity detectors", *Science*, 249: 1037-1041.

Pekel, M., Lappe, M., Bremmer, F., Thiele, A. & Hoffmann, K.-P. (1996). "Neuronal responses in the motion pathway of the macaque monkey to natural optic flow stimuli", *NeuroReport*, 7: 884-888.

Poggio, G. F. (1995). "Mechanisms of stereopsis in monkey visual cortex", *Cerebral Cortex*, 5: 193-204.

Pola, J. & Wyatt, H. (1993). "The role of attention and cognitive processes", In *Visual Motion and Its Role in the Stabilization of Gaze*, F. A. Miles & J. Wallman (Eds.), 371-392, Elsevier, Amsterdam.

Regan, D., Erkelens, C. J. & Collewijn, H. (1986). "Necessary conditions for the perception of motion in depth", *Invest. Ophthalmol. Vis. Sci.*, 27: 584-597.

Roy, J.-P., Komatsu, H. & Wurtz, R. H. (1992). "Disparity sensitivity of neurons in monkey extrastriate area MST", *J. Neurosci.*, 12: 2478-2492.

Roy, J.-P. & Wurtz, R. H. (1990). "The role of disparity-sensitive cortical neurons in signalling the direction of self-motion", *Nature*, 348: 160-162.

Saito, H., Yukie, M., Tanaka, K., Hikosaka, K., Fukada, Y. & Iwai, E. (1986). "Integration of direction signals of image motion in the superior temporal sulcus of the macaque monkey", *J. Neurosci.*, 6: 145-157.

Takemura, A., Inoue, Y., Kawano, K., & Miles, F. A. (1997). "Short-latency discharges in medial superior temporal area of alert monkeys to sudden changes in the horizontal disparity", *Soc. Neurosci. Abstr.*, 23: 1557.

Tanaka, K., Fukada, Y., & Saito, H. (1989). "Underlying mechanisms of the response specificity of expansion/contraction and rotation cells in the dorsal part of the medial superior temporal area of the macaque monkey", *J. Neurophysiol.*, 62: 642-656.

Tanaka, K. & Saito, H. (1989). "Analysis of motion of the visual field by direction, expansion/contraction, and rotation cells clustered in the dorsal part of the medial superior temporal area of the macaque monkey", *J. Neurophysiol.*, 62: 626-641.

Ungerleider, L. G. & Mishkin, M. (1982). "Two cortical visual systems", In *Analysis of Visual Behavior*, D. J. Ingle, M. A. Goodale, & R. J. W. Mansfield (Eds.), 549-586, MIT Press, Cambridge, MA.

Van Essen, D. C., Maunsell, J. H. R., & Bixby, J. L. (1981). "The middle temporal visual area in the macaque: myeloarchitecture, connections, functional properties and topographic organization", *J. Comp. Neurol.*, 199: 293-326.

3

Use of horizontal disparity, vertical disparity, and eye position in slant perception

Martin S. Banks and Benjamin T. Backus

3.1 The problem of stereoscopic slant perception and the available signals

The problem of visual space perception is to determine the layout of environment from the pattern of light reaching the eyes. Although many sources of information are potentially available, an important and well-studied one is stereoscopic vision which is based on the spatial differences in the two eyes' retinal images. We have been examining how stereoscopic information is used to recover the orientation of a smooth surface. The problem is interesting because the pattern of differences in the two retinal images depends not only on the slant of the surface, but also on its location with respect to the head.

Figure 3.1 depicts the geometry for binocular viewing of a vertical planar surface. The cyclopean line of sight is the line from the midpoint between the eyes to the middle of the surface patch of interest. The objective gaze-normal surface is the plane perpendicular to the cyclopean line of sight. The slant is the angle by which the surface of interest is rotated about a vertical axis from the objective gaze-normal surface (the angle S).

What signals are available for the estimation of slant? One important signal is horizontal disparity. For a smooth surface slanted about a vertical axis, the horizontal disparity pattern can be represented locally as a horizontal size ratio (HSR), the ratio of the horizontal angles the patch subtends in the left and right eyes, respectively (α_L and α_R in Figure 3.1; Rogers & Bradshaw, 1993).* Changes in HSR produce obvious and immediate changes in perceived slant – an increase in HSR is perceived as a clockwise rotation of the surface–so this signal must be involved in slant estimation. However, HSR by itself is ambiguous. To illustrate the ambiguity, Figure 3.2 shows several surface patches in front of the observer that give rise

* Our choice of signals, such as HSR, is based on their utility for expressing disparity information in the retinal images, not on a presumption that the brain represents these quantities *per se*. Other quantities could be used (Frisby, 1984; Gårding, *et al.*, 1995; Koenderink & van Doorn, 1976; Longuet-Higgins, 1982; Mayhew & Longuet-Higgins, 1982) without loss of generality.

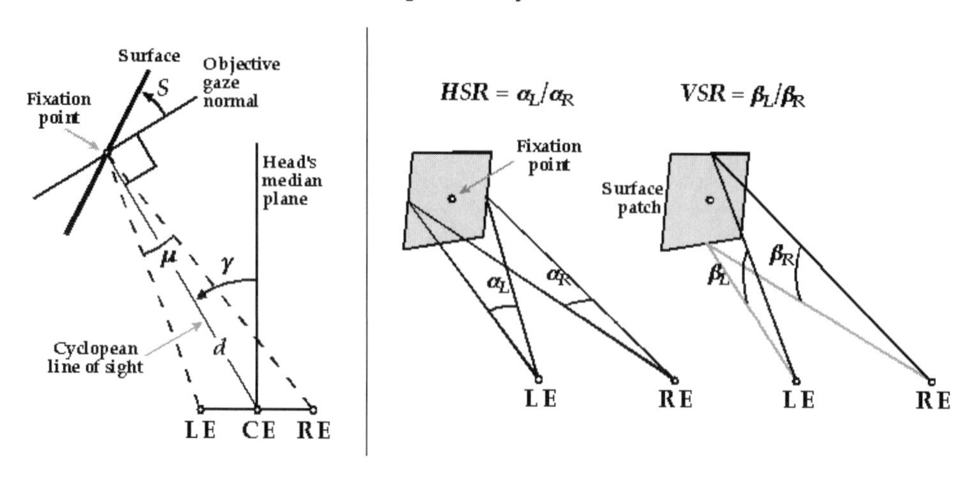

Fig. 3.1. Binocular viewing geometry. The left panel is an overhead (plan) view of the situation under consideration here. LE and RE refer to the left and right eyes. CE is the cyclopean eye, positioned at the midpoint between the left and right eyes. The head's median plane is the plane passing through the cyclopean eye and perpendicular to the interocular axis. A surface is fixated by the two eyes at the fixation point. The lines of sight from the two eyes are represented by the dashed lines. The cyclopean line of sight is the diagonal solid line. The distance d to the fixation point is measured along the cyclopean line of sight. The slant S is the angle between the surface and the objective gaze-normal plane (a plane perpendicular to the cyclopean line of sight); this angle is signed but otherwise equal to slant as defined by Stevens (1983). The angles γ and μ are the eyes' version and vergence, respectively. Positive slant (S) and positive azimuth (γ) are defined counterclockwise viewed from above. The right panel is an oblique view from behind the observer. The fixation point on the surface patch is indicated by the small white circle. Angles α_L and α_R are the horizontal angles subtended by the patch at the left and right eyes. The horizontal size ratio (HSR) is defined as α_L/α_R. Angles β_L and β_R are the vertical angles subtended by the surface patch. The vertical size ratio (VSR) is defined as β_L/β_R.

to HSRs of 1 ($\alpha_L/\alpha_R = 1$), 1.04, and 1.08. For each value of HSR, there is an infinite number of possible slants depending on the surface's location in front of the observer. Clearly the measurement of HSR alone does not allow an unambiguous estimate of the surface's orientation nor do any other descriptions of horizontal disparity (Mayhew, 1982). What other signals are used by the visual system to determine surface slant?

One potentially useful signal is vertical disparity which can be represented by the vertical size ratio (VSR), the ratio of vertical angles subtended by a surface patch in the left and right eyes (β_L and β_R in Figure 3.1). VSR varies with the location of a surface patch relative to the head, but does not vary with surface slant (Gillam & Lawergren, 1983). Figure 3.3 shows the VSR at various locations in the visual plane. Another signal, also given by vertical disparity, is the rate of change in VSR with azimuth, or $\partial VSR/\partial\gamma$.

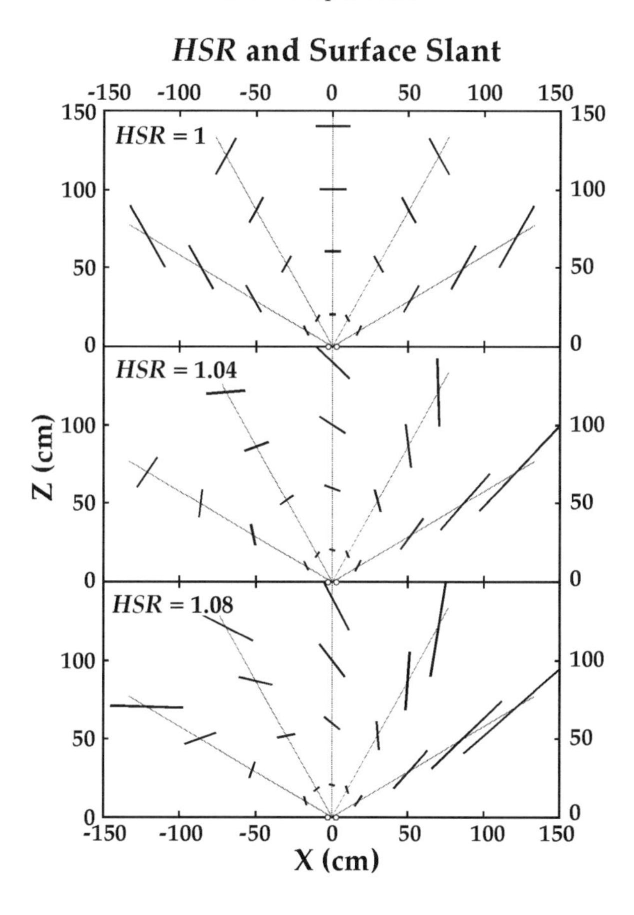

Fig. 3.2. Horizontal size ratio, position, and slant. Each panel is an overhead view of surface patches that give rise to a particular value for the horizontal size ratio (HSR). The X and Z axes are defined relative to the head and represent lateral and forward position (in cm) with respect to the cyclopean eye which is at the origin. The small circles near the origin represent the two eyes. The thin lines emanating from the origin correspond clockwise to azimuths of 60, 30, 0, -30, and -60 deg. The upper, middle, and lower panels show patches for which $HSR = 1$, 1.04, and 1.08, respectively. Notice that many surface slants are consistent with a given HSR depending on the patch's position relative to the head. For this reason, position and HSR uniquely specify slant, but HSR alone does not.

This quantity varies with distance, location, and slant (Backus *et al.*, 1998).

Other useful signals are provided by the sensed positions of the eyes. Ignoring torsion, we can represent eye position in the visual plane by two values γ and μ, which are the version and vergence angles of the eyes, respectively (see Figure 3.1). Figure 3.4 displays the values of γ and μ at various positions in front of the observer.

Finally, useful slant information can be gleaned from non-stereoscopic perspective signals such as the texture gradient created by projection onto

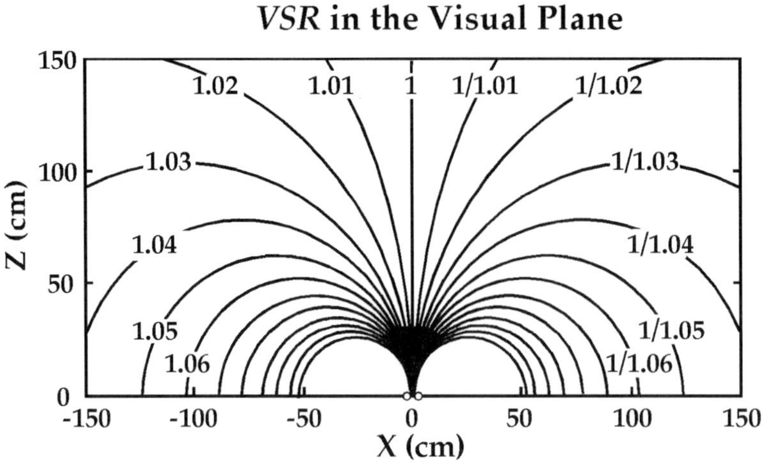

Fig. 3.3. Vertical size ratio and position in the visual plane. The figure is an overhead view through the visual plane, showing the circular contours for which the vertical size ratio (VSR) is constant. The X and Z axes represent lateral and forward position with respect to the cyclopean eye which is at the origin. The small circles near the origin represent the two eyes. Each iso-VSR contour is labeled with its VSR value. Adapted from Gillam and Lawergren (1983).

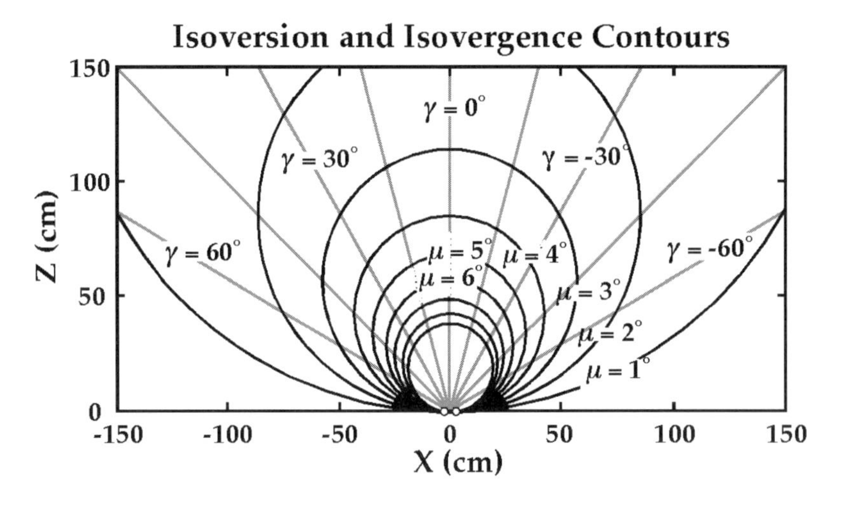

Fig. 3.4. Values of γ and μ in the visual plane. This is an overhead view showing the contours for which the eyes' vergence (μ) and version (γ) are constant. The iso-vergence contours are the black circles and the iso-version contours are the gray lines. The former are Vieth-Müller Circles. By the conventional definition of version (the average of the two eyes' rotations; Howard & Rogers, 1995), iso-version contours are Hyperbolae of Hillebrand rather than lines. However, we have defined version (γ) as the angle between the head's median plane and the cyclopean line of sight (Figure 3.1), and by this definition, the iso-version contours are lines. Together γ and μ determine the point of fixation within the visual plane. They can be estimated from extra-retinal, eye-position signals. In combination estimates of HSR, γ and μ ($\hat{HSR}$, $\hat{\gamma}$, and $\hat{\mu}$) allow an unambiguous estimate of surface slant [see equation 3.2].

the retinae of surfaces with statistically-regular textures (Buckley & Frisby, 1993; Cutting & Millard, 1984; Gillam & Ryan, 1992).

How can an unambiguous estimate of slant be obtained from various combinations of the above-mentioned signals? The slant of a vertical surface in the visual plane is related mathematically to μ, HSR, and VSR in the following way (for derivation, see Backus *et al.*, 1998):

$$S \approx -\tan^{-1}\left(\frac{1}{\mu}\ln\frac{HSR}{VSR}\right) \tag{3.1}$$

In the terminology of Gårding *et al.* (1995), μ (the vergence angle) "normalizes" the slant (scales HSR for changes due to viewing distance) and VSR "corrects" the slant specified by HSR (corrects HSR for changes due to azimuth).

Equation 3.1 shows that the eyes' vergence μ and VSR can be used together to estimate slant from HSR. From Figures 3.3 and 3.4, one can see that μ and VSR uniquely specify a location in the visual plane; therefore, the combination of μ, VSR, and HSR uniquely specifies the slant and location of the surface patch.

Other subsets of signals available to the visual system allow unambiguous estimation of slant and we summarize those in terms of three calculations (Banks & Backus, 1998): (1) slant estimation from HSR and eye position, (2) slant estimation from HSR and VSR, and (3) slant estimation from nonstereoscopic cues such as perspective.

3.1.1 Slant from HSR and eye position

Slant can in principle be estimated from HSR and sensed eye position (Ogle, 1950). VSR in equation 3.1 can be approximated by the quantity $1+\mu\tan\gamma$, yielding:

$$S \approx -\tan^{-1}\left(\frac{1}{\mu}\ln HSR - \tan\gamma\right) \tag{3.2}$$

where, as before, μ and γ are the vergence and version of the eyes, respectively. Note that the estimates of μ and γ (which we will call $\hat{\mu}$ and $\hat{\gamma}$) are derived from extra-retinal, eye-position signals*, presumably the efference copy. Figure 3.4 displays contours of constant μ and γ. Notice that μ and γ specify a location in the visual plane and, therefore, μ, γ, and HSR uniquely determine the slant and location of the surface patch. There is some evidence that sensed eye position can be used in estimating slant (Amigo, 1967; Collett, Schwarz, & Sobel, 1991; Cumming, Johnston, & Parker, 1991; Ebenholtz & Paap, 1973; Herzau & Ogle, 1937).

* This formula, using eye-in-hand position signals provides slant with respect to the cyclopean line of sight. To conver to slant relative to the head, one needs to use the version signal ($\hat{\gamma}$) again. If slant in other terms is needed, more information relating the frame-of-reference of the eye to that frame is required [see Chapter 4, Frames of Reference], ED.

3.1.2 Slant from HSR and VSR

Slant can also be estimated from retinal-image information alone; specifically, it can be estimated from HSR, VSR, and $\partial HSR/\partial\gamma$ (Gårding, *et al.*, 1995; Gillam & Lawergren, 1983; Koenderink & van Doorn, 1976; Mayhew & Longuet-Higgins, 1982). The quantity μ in equation 3.1 can be replaced by $\tilde{\mu}$:

$$S \approx -\tan^{-1}\left(\frac{1}{\tilde{\mu}}\ln\frac{HSR}{VSR}\right) \tag{3.3}$$

where

$$\tilde{\mu} \approx \frac{1}{2}\left(\frac{\partial VSR}{\partial\gamma} + \sqrt{(\partial VSR/\partial\gamma)^2 + 4\ln(VSR)\ln(HSR/VSR)}\right) \tag{3.4}$$

One cannot illustrate this method of estimating slant in the format of Figure 3.4 above, but the special case of rotating the surface until it appears to be gaze normal has a simple interpretation. For a gaze-normal plane, $S = 0$, so by equation 3.3, $HSR = VSR$ (Ogle, 1950; Howard & Rogers, 1995). Thus, an observer could in principle set a planar surface to gaze normal by adjusting its slant until $H\hat{S}R = V\hat{S}R$ (where $H\hat{S}R$ and $V\hat{S}R$ are the visual system's estimates of HSR and VSR). It is worth noting that this particular task can be done using $H\hat{S}R$ and $V\hat{S}R$ alone; errors in $\hat{\tilde{\mu}}$ [equation 3.3] should have no effect.

The visual system could estimate μ from sensed eye position ($\hat{\mu}$) and the retinal images ($\hat{\tilde{\mu}}$) in combination. Then this estimate could be used in equations 3.2 and 3.3. The distinction between estimation by HSR and eye position and by HSR and VSR would then be less strict. The fact that γ appears only in equation 3.2 and VSR only in equation 3.3 is the essential distinction between these methods.

3.1.3 Slant from non-stereoscopic cues

Useful indications of surface slant are provided by nonstereoscopic perspective cues such as the texture gradient created by projection onto the retinae of surfaces with statistically regular textures (Stevens, 1981; Buckley & Frisby, 1993; Cumming, Johnston, & Parker, 1993; Cutting & Millard, 1984). Such cues were commonly present in older stereoscopic work using real objects (Ogle, 1950; Amigo, 1967, 1972; Gillam, Chambers, & Lawergren, 1988). In more recent work employing stereoscopic computer displays, there is still generally a perspective cue that usually indicates that the surface is frontoparallel to the head (Banks & Backus, 1998). The slant specified by a given texture gradient does not vary with distance or azimuth (Backus *et al.*, 1998; Sedgwick, 1986).

3.2 Combining Signals to Obtain Slant Estimates

The slant estimates derived from these three methods generally agree under normal viewing situations. Let us call the three slant estimates $\hat{S}_{HSR,EP}$, $\hat{S}_{HSR,VSR}$, and $\hat{S}_P$. In the model presented in Figure 3.5, these estimates are combined linearly after each is given a weight based on its estimated reliability. Such a model is a "modified weak fusion" model in the terminology of Landy *et al.* (1995). In our model, the reliability is estimated from ancillary cues such as the spatial pattern, the display height, the vergence, etc. (for details, see Backus & Banks, 1998). For example, consider the various effects of increasing the viewing distance. As distance increases, there is no effect on the information carried by the perspective signal (as noted above), but the information carried by HSR is reduced because a given set of slants maps onto smaller and smaller ranges of HSR, and the information conveyed by VSR is reduced because a given set of azimuths maps onto smaller and smaller ranges of VSR. Consequently, we assume that nonstereoscopic slant estimates are weighted more heavily relative to stereoscopic slant estimates as viewing distance increases. Other factors that affect the reliability of the signals include the texture on the surface (e.g., vertically oriented textures render VSR more difficult to measure, noisy textures render perspective more difficult, and so forth), the size of the retinal image (e.g., smaller images provide a smaller area over which to estimate VSR, which should make less reliable; cf. Rogers & Bradshaw, 1995), and the presence and characteristics of other objects in the scene.

The experiments described in this chapter were designed to test whether the signals described above are used in estimating slant, and how the weights assigned to the estimates vary across viewing conditions and stimulus properties during stereoscopic slant perception.

Each of the five signals described above must be measured by the visual system before it can be used to estimate the properties of the viewed surface. Obviously, none can be measured with perfect precision, so it is of interest to know the consequences of erroneous measurements in each of the signals. An efficient estimator would use the signals that, given their uncertainties, would allow the most accurate slant estimate. The effects of signal uncertainty on slant estimation are quite dependent on the viewing situation and on the uncertainty associated with the other available signals (Backus *et al.*, 1998; Backus & Banks, 1998; Banks & Backus, 1998).

In our experiments, observers rotated a stereoscopically-rendered plane about a vertical axis until it appeared gaze normal. This is a somewhat special case because the computations involved in this task are simpler than those involved in estimating slant. By way of illustration, consider equations 3.2 and 3.3 above. The gaze-normal task requires the observer to rotate the surface until the slant S is zero. Thus, when using HSR and eye position,

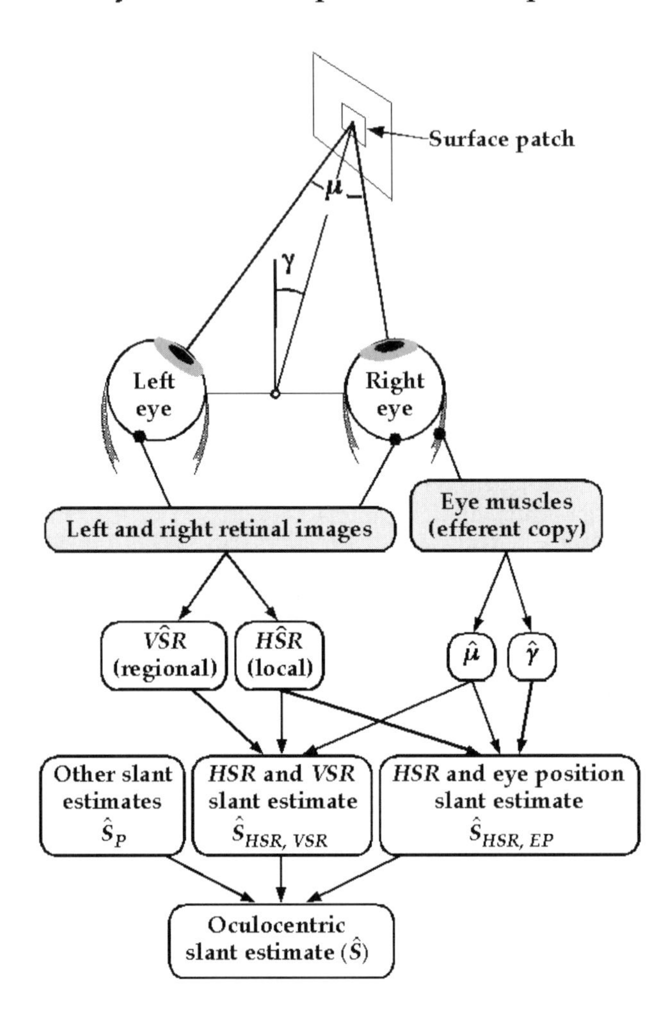

Fig. 3.5. Model of stereoscopic slant perception. The visual system measures five signals: HSR, VSR, and $\partial VSR/\partial\gamma$ (not shown) from the retinal images, and γ and μ from the eye muscles. It also measures signals based on other slant cues such as perspective. It combines these signals in three ways to estimate surface slant. HSR, VSR, and μ are used to estimate slant from HSR and VSR [equation 3.1]. The estimate of μ can be based either on eye position (shown), or on the retinal images [$\tilde{\mu}$, not shown; see equations 3.3 and 3.4]. HSR, γ, and μ are used to estimate slant from HSR and eye position [equation 3.2]. The perspective cue signal provides the third slant estimate. The various slant estimates ($\hat{S}_{HSR,EP}$, $\hat{S}_{HSR,VSR}$, and $\hat{S}_P$) are combined in a weighted average, with the weight assigned to each slant estimate dependent on the visual system's estimate of its reliability. The final slant estimate is done in oculocentric coordinates; specifically, the computations yield an estimate of surface slant relative to the cyclopean line of sight (see Figure 3.1).

$-\tan^{-1}\left(\frac{1}{\mu}\ln HSR - \tan\gamma\right) \approx 0$, which implies that $\ln HSR \approx \mu\tan\gamma$. When using HSR and VSR, $-\tan^{-1}\left(\frac{1}{\bar{\mu}}\ln\frac{HSR}{VSR}\right) \approx 0$, or simply $HSR \approx VSR$.

It has been argued that the gaze-normal task is of limited interest and that more can be learned by asking observers to estimate the slants of surfaces with non-zero slants as well (Gillam *et al.*, 1988). There are two important justifications for using the gaze-normal task to examine the use of stereoscopic signals in slant perception. First, it is unlikely that different mechanisms underlie the perception of surfaces with zero and non-zero slants. Second and most importantly, the gaze-normal task, as implemented in our experiments, allows one to isolate stereoscopic-based slant estimation from perspective-based estimation. Specifically, the residual perspective cues in our displays always indicated that the surface was gaze-normal no matter what slant was specified by the two stereoscopic means of slant estimation. Perspective cues, therefore, could not determine which of the stereoscopic means of estimation was given greater weight in a given viewing situation. For this reason, the gaze-normal task allows one to determine the relative weights of the two stereoscopic means of slant estimation without contamination from perspective cues.

3.3 Experiments on Signal Combination in Stereoscopic Slant Estimation

We conducted a number of experiments to determine the signals used in stereoscopic slant perception with particular attention to the relative weights of VSR and eye-position signals.

Apparatus

Stimuli were displayed on a haploscope consisting of two large monochrome CRT displays each seen in a mirror by one eye. Each mirror and CRT was attached to an armature that rotated about a vertical axis passing through the eye's center of rotation. The face of each CRT was always perpendicular to the line of sight from the eye to the center of the screen. A small target on each screen served as the fixation aid. As the CRTs were rotated, the eyes rotated to track the fixation aid, but the images created by the CRTs at the retinae were unchanged. Head position was fixed with a bite bar. The room was completely dark, and a black aperture occluded the frames of the monitors. Only the white dots within the display were visible.

Despite the short viewing distance of 42.5 cm, the visual locations of the dots in our displays were specified to within ~30 arcsec. This high level of spatial precision was achieved by use of two procedures: anti-aliasing and spatial calibration (Backus *et al.*, 1998).

Stimuli

The stimuli consisted of sparse random-dot (or vertical-line) displays that minimized the use of perspective cues. The dots (or lines) were placed on a simulated vertical plane; dot positions were calculated by projection through the centers of the eyes onto the virtual image planes in front of each CRT. The retinal images were therefore the same as would be created by a real object (except that dot size and intensity were constant). Consequently, the shape distortions that one frequently observes in stereoscopic displays were not present (Gillam, 1967). The two-dimensional shape of the simulated surface was elliptical with randomly varied height and width, so the outline shape provided no useful slant information.

Azimuth specified by eye position was manipulated by rotating the monitors about the pivots below the centers of the observer's eyes; such rotation did not alter the retinal images. Azimuth specified by VSR was determined by the location of the simulated surface from which the images were derived. For example, a stimulus for which eye-position azimuth is 15 deg and VSR azimuth is -15 deg, is a surface patch that produces the retinal images of a plane to the right of the head's median plane and is viewed with the monitors positioned to the left (and the eyes, therefore, rotated leftward).

Task and procedure

The observers' task was to rotate the surface about a vertical axis until it appeared perpendicular to the line of sight (Figure 3.1). Observers could perform the task easily and reliably. Discrete slant adjustments were made with key presses. A new stimulus appeared with a new set of dots after each press. Observers indicated when they were satisfied with a slant setting by pressing another key. They were given no feedback as to the accuracy of their setting.

To insure that observers' slant settings were unaffected by residual perspective cues in the stimuli, we conducted a monocular control experiment. The stimuli, procedure, and task were identical to those of the main experiments, but the observer performed the task monocularly. The standard deviations of the slant settings were generally 10 times greater than when the observer performed the task binocularly. We conclude that observers could not perform the task reliably from monocularly available cues.

Experiment 1: Natural Viewing

We first asked how well observers compensate for changes in a surface's azimuth when stereoscopic means of estimating slant agree with one another. In this natural-viewing situation, slant estimation from HSR and eye position yielded the same values as slant estimation from HSR and VSR. Non-stereoscopic cues were rendered uninformative.

The surface patch contained 300 dots. Distance specified by eye position and by VSR was 57.3 cm from the midpoint of the interocular axis. Display size varied from 22 x 32 deg to 27 x 40 deg. The azimuths specified by eye position and by VSR were always equal and were -15, -7.5, 0, 7.5 and 15 deg.

Figure 3.6 shows the results. Slant settings are plotted as a function of the eyes' version in this natural-viewing situation. The upper panel shows the data when slant relative to the objective gaze-normal plane is plotted (S in Figure 3.1). Biases in each observer's settings have been removed by translating the data vertically until the slant setting was 0 at azimuth = 0 deg. If observers compensated completely for changes in the surface's azimuth, the data would lie on the dashed horizontal line. The actual settings always show near-complete compensation.

The lower panel shows the same data plotted in terms of the natural logarithm of HSR at fixation. The solid horizontal line in the lower panel shows the $\ln(HSR)$ values if no compensation occurred and the dashed diagonal line shows the predicted values for complete compensation. Biases in each observer's settings were again removed: the data were translated vertically until $\ln(HSR)$ was 0 at version = 0 deg. These data show that observers are in fact able to set surfaces to objective gaze normal when both eye position and VSR cues are consistent with one another and non-stereoscopic cues are rendered uninformative.

Experiment 2: VSR vs. Eye Position

We next asked whether slant estimation by HSR and eye position or by HSR and VSR was the primary determinant of observers' performance in the natural-viewing situation of Experiment 1. To do so, we manipulated eye position and retinal-image information independently.

The retinal images generated by a surface at one azimuth were presented at a variety of eye positions (versions). The predicted slant settings are depicted in the upper left panel of Figure 3.7. The solid gray lines are the predictions for HSR and VSR. If settings were based on HSR and VSR, then the observed $\ln(HSR)$ values would vary with the azimuth specified by VSR; this is manifest in the separation of the three lines. Moreover, if settings were based on HSR and VSR, the observed $\ln(HSR)$ values would not vary with eye position; this is manifest in the zero slopes of

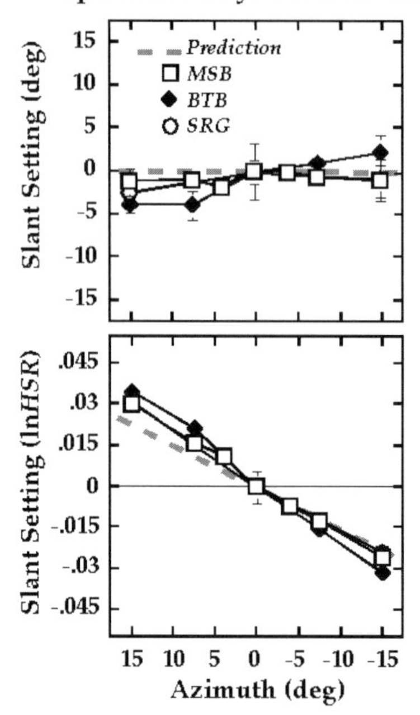

Fig. 3.6. Results of Experiment 1. In this "natural viewing" experiment, eye position and VSR specified the same azimuth. Observers adjusted the slant of a stereoscopically defined plane until it appeared gaze normal. The upper panel plots each observer's average slant settings (in deg) as a function of the azimuth of the stimulus. A value of 0 deg on the ordinate corresponds to the objective gaze-normal plane. The abscissa has positive values on the left because they correspond to leftward gaze. The dashed line represents the predicted values for complete compensation. The lower panel plots the same data, but the slant settings have been converted to values of $\ln(HSR)$ in the retinal images. The dashed gray line again represents the predictions if observers compensated completely for changes in azimuth and, therefore, set the slants to the objective gaze normal. The $\ln(HSR)$ predictions are based on an assumed interocular distance of 6.1 cm. Error bars correspond to ± 1 standard deviation. The data for each observer have been normalized to 0 deg for an azimuth of 0 deg. The values subtracted for this normalization were 3.0, -0.4, and 9.5 deg in the top panel, corresponding to -0.006, 0.001, and -0.017 in the bottom panel, for MSB, BTB, and SRG, respectively.

the prediction lines. The dashed gray line is the prediction for estimation from HSR and eye position. If settings were based on this method, then the observed $\ln(HSR)$ values would not vary with the azimuth specified by VSR (i.e., no separation between the lines), but would vary with version (i.e., non-zero slope). The stimuli were the same as those of Experiment 1 except that the azimuths specified by eye position and VSR were varied independently.

The remaining three panels of Figure 3.7 show the results, a separate panel for each observer. Biases in each observer's settings were removed by translating the data vertically until $\ln(HSR)$ was 0 for the data point

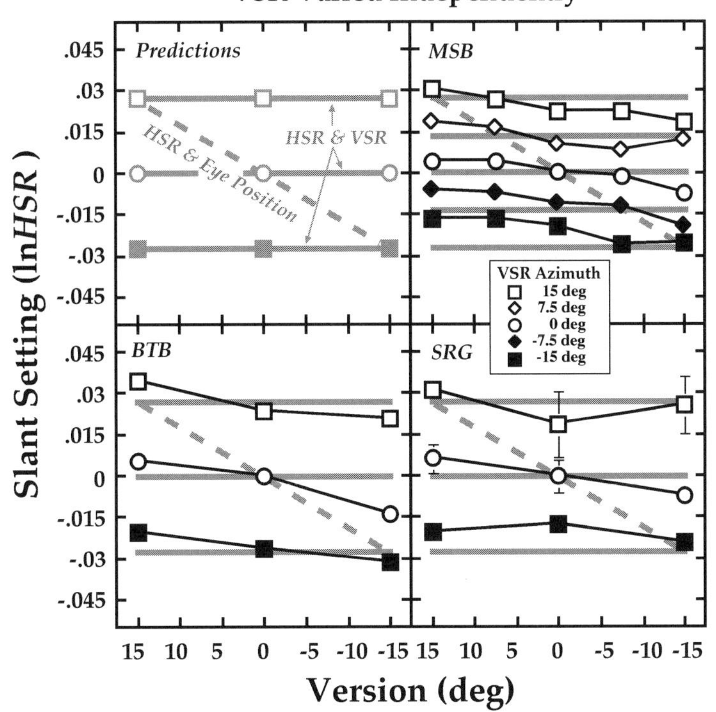

Fig. 3.7. Predictions and results for Experiment 2. In this experiment, azimuth specified by eye position and azimuth specified by VSR were independently manipulated. The average values of $\ln(HSR)$ for each observer's slant settings are plotted as a function of the eyes' version. The upper left panel shows the predictions: the dashed gray line represents the predictions for slant estimation from HSR and eye position and the solid gray lines represent the predictions for slant estimation from HSR and VSR. The other three panels plot separately the data from the three observers. The symbols represent data for different azimuths specified by VSR: the open squares are for 15 deg, open diamonds for 7.5, circles for 0, filled diamonds for -7.5, and filled squares for -15 deg. Error bars correspond to ± 1 standard deviation. Data for which azimuth specified by VSR agrees with azimuth specified by eye position are from Experiment 1. The data for each observer have been normalized to 0 deg for an azimuth of 0 deg. The values subtracted for this normalization were -0.006, 0.001, and -0.017 for MSB, BTB, and SRG, respectively.

at which version = 0 and VSR azimuth = 0. The data lie close to the horizontal lines that represent the predictions of slant estimation by HSR and VSR. Thus the data suggest that the primary determinant of perceived slant was the retinal images. However the data have a small downward slope which implies that there was a small, but consistent effect of changes in eye position.

The relative influence of slant estimation by HSR and VSR as compared with HSR and eye position can be assessed in another way. The change in $\ln(HSR)$ caused by variation in eye position or VSR can be expressed as a percentage of the change expected for complete compensation for eccentric gaze (e.g., Rogers & Bradshaw, 1993). For the effect of VSR, the average percentage was 88%. For the effect of eye position, it was 18%* The effect of VSR is about four times larger than the effect of eye position.

Experiment 3: Vertical Lines

Experiments 1 and 2 revealed little effect of actual eye position on perceived slant; instead, the visual system seemed to rely on information contained in the retinal images. In terms of underlying mechanisms, there are two ways to interpret this finding.

First, as implied by our model, the visual system might rely on the more reliable of the available signals. Specifically, with the stimuli and viewing conditions of Experiments 1 and 2, VSR may have been a more reliable indicant of the HSR correction needed than sensed eye position was. If this hypothesis is correct, then altering the measurability of VSR should reduce the visual system's reliance on VSR and increase its reliance on eye position.

Second, the visual system might always rely more on slant estimation by HSR and VSR than it does on slant estimation by HSR and eye position. If this hypothesis is correct, then changing the measurability of VSR should not change the system's reliance on sensed eye position.

In order to determine whether sensed eye position is used to correct HSR, we used the technique of Herzau and Ogle (1937) to render vertical disparities unmeasurable. Specifically, we replaced the random-dot stimulus of Experiments 1 and 2 with a series of vertical lines. Because the lines contained no vertically-separated features, vertical disparities (and $VSRs$) could not be measured.

The stimulus was a set of 12 vertical lines on a virtual plane 40 cm from the cyclopean eye. The set of lines subtended 25 x ~14 deg. The tops and bottoms of the lines were clipped at different heights by apertures close to the eyes and, consequently, VSR and $\partial VSR/\partial\gamma$ were not measurable. Azimuths of -12.5, -6.25, 0, 6.25, and 12.5 deg were presented by rotating the haploscope arms by the appropriate amounts. Observers were instructed to maintain fixation on a central marker while inspecting the stimulus. As before, they used key presses to rotate the stimulus plane until it appeared gaze normal.

The vertical lines were always the same width at the CRT, so their widths in the retinal images did not provide a reliable cue to slant. The horizontal angular subtense of the display at the cyclopean eye was constant, and the spacing of the lines was randomized. The lines were back-projected onto the

* The percentages do not add to 100% because observers did not necessarily compensate accurately.

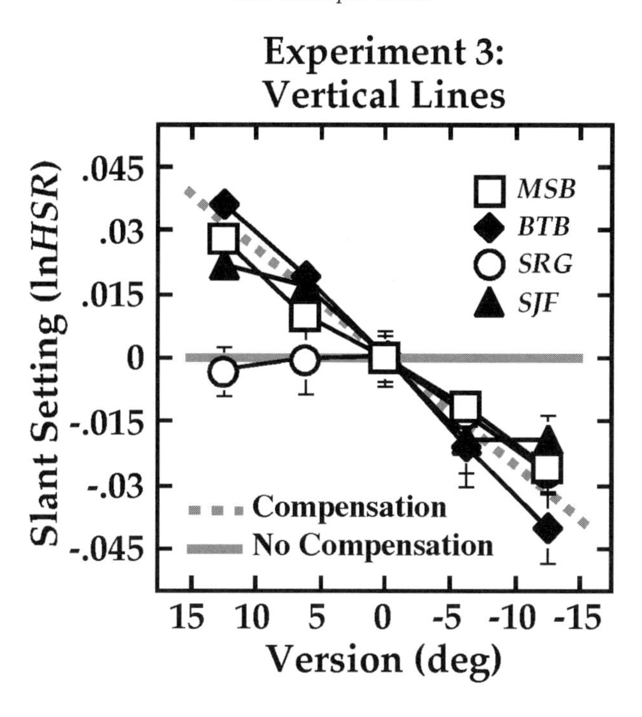

Fig. 3.8. Predictions and results for Experiment 3. In this experiment, the stimuli were vertical lines, so they contained no VSR signal. The average values of $\ln(HSR)$ for each observer's slant settings are plotted as a function of the eyes' version. The prediction for estimation from HSR and eye position (for interocular distance = 6.0 cm) is represented by the dashed gray line. If no compensation for changes in stimulus azimuth occurred, the data would lie on the horizontal solid gray line. Error bars correspond to ± 1 standard deviation. The data for the four observers are represented by the different symbols. The data for each observer have been normalized to 0 deg for an azimuth of 0 deg. The values subtracted for this normalization were 0.002, 0.004, -0.012, and -0.014 for MSB, BTB, SRG, and SJF, respectively.

surface (Banks & Backus, 1998)* and hence did not provide a cue to slant. Thus there were no reliable monocular cues to slant.

The results are shown in Figure 3.8 which plots $\ln(HSR)$ at the slant setting as a function of the eyes' version. If observers used slant estimation by HSR and eye position (and thereby took the eyes' version into account), the data should lie on the dashed diagonal line. If, on the other hand, they did not take the eyes' version into account, the data should lie on the solid horizontal line. The data for different observers are represented by different symbols. Biases in each observer's settings were removed by translating the data vertically until $\ln(HSR)$ was 0 at version = 0 deg.

The data are quite consistent with slant estimation by HSR and eye position except that SRG did not compensate for eye position in leftward

* The lines were back-projected from the cyclopean eye onto the virtual plane after rotation of the plane. With this technique, there is no useful slant information in the monocular images.

gaze. The results show that extra-retinal, eye-position signals are used to compensate for changes in azimuth when vertical disparities are rendered unmeasurable. Thus, human observers rely heavily on retinal-image information (slant estimation by HSR and VSR) when vertical disparities are measurable, but use sensed eye-position signals (slant estimation by HSR and eye position) when they are made unmeasurable.

Experiment 4: Height Manipulation

There are many other conditions that affect the magnitude and measurability of vertical disparities, so it is of interest to learn more about the conditions that favor use of vertical disparities as opposed to extra-retinal, eye-position signals.

In Experiment 4, we varied the measurability of VSR by manipulating the height of the surface patch. This manipulation does not affect VSR at a given stimulus location, but it does affect the area over which VSR can be measured and the magnitude of the largest vertical disparities in the display. Either of these factors might reduce the visual system's confidence in $V\hat{S}R$. However, this manipulation should not affect the extra-retinal signals from sensed eye position ($\hat{\mu}$ and $\hat{\gamma}$) because these signals should not depend on image size. We expect, therefore, that as display height decreases, the visual system will give relatively more weight to the estimate from HSR and eye position. When display height is zero, VSR is impossible to calculate, so there should be no effect of the azimuth specified by VSR.

Planar surface patches were presented again at a distance of 57.3 cm, as specified by eye position and by the retinal images. The width of the simulated patch varied randomly from 30–34 deg so that image width, which is maximum when the patch is gaze normal, was not a reliable cue to slant. Four stimulus heights were used: 30, 6.5, 1.3, and 0 deg. The displays for those heights contained 400, 100, 70, and 70 dots, respectively. The eyes' version and the azimuth specified by VSR were manipulated independently and took on values of -15, 0, or 15 deg.

Figure 3.9 displays the predictions and results for one observer. As before, the horizontal gray lines represent the predicted $\ln(HSR)$ values when slant judgments are based entirely on HSR and VSR; the dashed diagonal line represents the predicted $\ln(HSR)$ values when slant judgments are based entirely on HSR and eye position. Biases were removed by translating the data vertically. The slant settings for this observer were very consistent with the predictions of slant estimation by HSR and VSR when the stimulus height was 30 deg and were very consistent with the predictions of estimation by HSR and eye position when the height was 0 and 1.3 deg.

Figure 3.10 shows the effects of eye position and VSR as a percentage of the compensation required for a 30-deg change in azimuth. Percent HSR

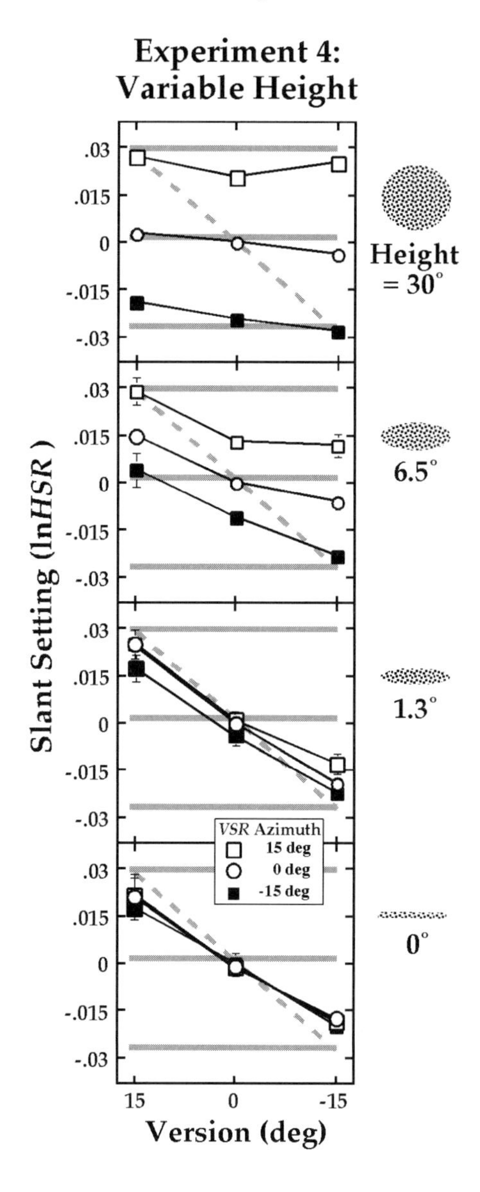

Fig. 3.9. Predictions and results for one observer in Experiment 4. In this experiment, the stimuli were similar to those in Experiment 2 (randomly positioned dots defining a plane), but the height of the stimulus was varied. Average values of $\ln(HSR)$ for the slant settings of this observer are plotted as a function of the eyes' version. The dashed gray line represents the predictions for slant estimation from HSR and eye position and the solid gray lines represent the predictions for slant estimation from HSR and VSR. The icons to the right represent the stimuli. From top to bottom, the panels show data corresponding to stimulus heights of 30, 6.5, 1.3, and 0 deg. The symbols represent data for different azimuths specified by retinal images: the open squares are for 15 deg, circles for 0, and filled squares for -15 deg. Error bars correspond to ± 1 standard deviation. The data for each stimulus height have been normalized to 0 deg for an azimuth of 0 deg. The values subtracted for this normalization were 0.004, -0.001, -0.004, -0.005, and for heights of 30, 6.5, 1.3, and 0 deg, respectively.

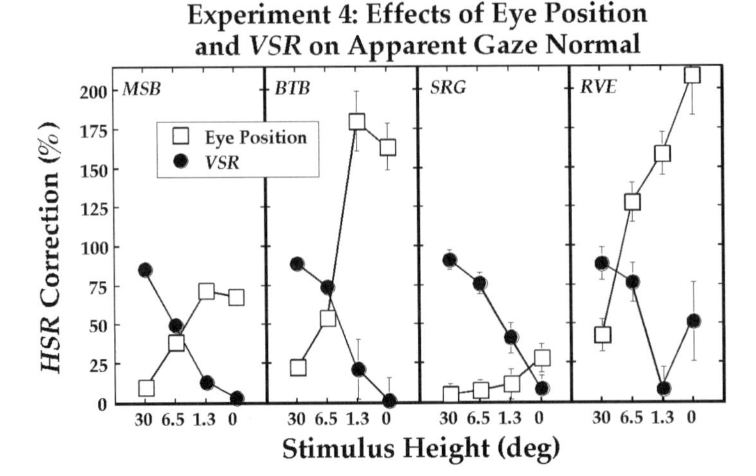

Fig. 3.10. Eye position and VSR effects in Experiment 4. The percentage change in HSR relative to that needed to compensate completely for changes in azimuth are plotted separately for the four observers; they are plotted as a function of stimulus height. Filled circles represent the percent correction due to changes in VSR and the unfilled squares represent the percent correction due to changes in eye position. The percent HSR Correction values were obtained by fitting a linear model to the data (see text for details). r^2 values for linear fits were from the tallest to shortest display 0.98, 0.92, 0.96, and 0.95 for MSB, 0.98, 0.96, 0.88, and 0.91 for BTB, 0.94, 0.91, 0.60, and 0.47 for SRG, and 0.87, 0.92, 0.91, and 0.85 for RVE. The effects of VSR and eye position were essentially additive. Error bars represent 95% confidence intervals.

correction is plotted as a function of stimulus height. When stimulus height was 30 deg, slant settings were quite consistent with slant estimation by HSR and VSR, which replicates Experiment 2. As height was decreased, however, the settings of all four observers became increasingly consistent with those predicted by HSR and eye position.

In an ancillary experiment, we asked whether the transition from VSR-based to eye-position-based estimation was determined primarily by stimulus area or by the number of visible dots. There was no discernible effect of dot number, so with the dot numbers we used, stimulus area (or height), and not dot number (or density), was the controlling variable.

The data from Experiment 4 imply that slant estimation by HSR and VSR and by HSR and eye position are both used in determining the slant of a stereoscopically defined surface. Furthermore, human observers seem to rely most on the VSR-based method when the stimulus is large and most on the sensed-eye-position method when it is small. A similar conclusion concerning perceived curvature has been drawn by Rogers and Bradshaw (1995).

In the framework of the model presented in Figure 3.5, the results of Experiments 1–4 reveal that the weights assigned to the three slant estimates vary in plausible ways with viewing condition. In particular, when vertical disparities are measurable and perspective cues are rendered uninformative (Experiment 2), $\hat{S}_{HSR,VSR}$ is given considerably more weight than $\hat{S}_{HSR,EP}$. However, when perspective cues are uninformative and vertical disparities are rendered unmeasurable (Experiment 3) or difficult to measure (Experiment 4), more weight is given to $\hat{S}_{HSR,EP}$.

Additivity of VSR and eye position effects

In the cue-conflict experiments (Experiments 2 and 4), the effects of azimuth specified by eye position and azimuth specified by VSR were roughly additive. A linear fit accounted for most of the variance in the data. The good fit of the linear model suggests that the two slant estimates are independent: that is, the estimate from HSR and VSR does not change with eye position, and the estimate from HSR and eye position does not change with VSR. It also suggests that perceived slant can in fact be modeled as a weighted average of separate slant estimates as suggested in Figure 3.5.

Our experiments involved the presentation of stimuli with properties that never arise in natural viewing. For example, some of the conditions of Experiments 2 and 4 utilized stimuli for which VSR and sensed eye position specified completely different azimuths. In such situations, observers might have access to two separate percepts, one based on VSR and the other based on eye position. If there were two percepts, one or the other would determine the response on a given trial. Alternatively, the visual system might produce one slant estimate based on a weighted combination of the estimates arising from HSR and VSR and from HSR and eye position (as suggested by the model in Figure 3.5). There are two pieces of evidence that imply that the latter is a better description. First, the appearance of the stimuli was similar whether VSR and eye position conflicted or agreed; in both cases, the stimuli appeared planar and the slant was well-specified. Second, the variability of slant settings was not markedly higher in the conflict conditions.

3.4 Geometric and Induced Effects

With a horizontal magnifier before one eye, a frontoparallel surface appears rotated about a vertical axis; this is called the *geometric effect*. With a vertical magnifier, apparent rotation is opposite in direction; this is called the *induced effect*. To restore appearance of frontoparallelism in the induced effect case, the surface must be rotated away from the magnified eye. The induced effect (Ogle, 1938, 1950) has been interesting because it was thought until recently that vertical disparities do not play an important role in surface perception.

When an observer estimates slant from HSR and VSR (instantiated by equation 3.3), the amount of rotation required to restore the appearance of gaze normal ought to be the same in magnitude, but opposite in sign, for the induced and geometric effects. Despite this clear prediction, the induced effect is typically smaller than the geometric effect at magnifications greater than 3–4%. Other theoretical formulations (Gårding *et al.*, 1995; Gillam, Chambers, & Lawergren, 1988; Gillam & Lawergren, 1983; Kaneko & Howard, 1996; Koenderink & van Doorn, 1976; Mayhew, 1982; Rogers & Bradshaw, 1995) make the same prediction as equation 3.3, so they too have been unable to explain the attenuation of the induced effect. There have, however, been some modifications to current theory that attempt to explain the attenuation.

(i) **Conflicting estimates of azimuth.** With increasing magnification, the azimuth of the fixated surface, indicated by VSR and $\partial VSR/\partial\gamma$, becomes increasingly different from extra-retinal estimates of eye position and this conflict causes an attenuation of perceived slant (Gillam *et al.*, 1988). Changes in VSR and $\partial VSR/\partial\gamma$ do not alter perceived azimuth (Ogle, 1950), so it is not clear how the proposed conflict would arise in the first place.

(ii) **Implausible stimulus location.** $VSRs$ created by vertical magnification imply azimuths and distances at which observers do not normally inspect a surface (Frisby, 1984; Gillam *et al.*, 1988; Mayhew, 1982; Ogle, 1950). This hypothesis predicts a decreasing magnification for the plateau with increasing distance and this is not observed (Gillam *et al.*, 1988; Ogle, 1938, 1950).

(iii) **Fusion failure**. The smaller induced effect could result from an inability to fuse large vertical disparities (Ogle, 1950). This hypothesis can be rejected because magnifications at plateau are the same in symmetric and asymmetric convergence (Ogle, 1940) even though vertical disparities are much larger in the latter.

(iv) **Perspective conflict.** Ogle attempted to eliminate perspective slant cues from some of his experiments by using a plane sprinkled with randomly-positioned spots (Gillam *et al.*, 1988; Ogle, 1938), but such a stimulus provides a texture gradient cue, so perspective and $HSR - VSR$ cues still specify conflicting slants. The perspective-conflict hypothesis, however, does not explain why the same conflict causes no attenuation of the geometric effect.

The conceptualization depicted in Figure 3.5 may offer insight into why the induced effect exhibits a plateau while the geometric effect does not. Under normal viewing, the three slant estimates – $\hat{S}_{HSR,EP}$, $\hat{S}_{HSR,VSR}$, and $\hat{S}_P$ – agree. Placing a meridional magnifier before one eye alters the natural relationships among HSR, eye position, VSR, and perspective cues. In the geometric effect, horizontal magnification of the right eye's image yields a decrease in HSR, but other signals are virtually unaltered. Therefore, the slant estimate from HSR and eye position ($\hat{S}_{HSR,EP}$) and from HSR and VSR ($\hat{S}_{HSR,VSR}$) are in agreement under horizontal magnification, but they both conflict with slant estimation from perspective ($\hat{S}_P$); specifically, when viewing an objectively frontoparallel surface, the two stereoscopic estimates indicate a counter-clockwise rotation of the surface while the nonstereoscopic estimate indicates that the surface is frontoparallel. In the induced effect, vertical magnification of the right eye's image yields a decrease in VSR, but the other signals are unaltered. Therefore, $\hat{S}_{HSR,VSR}$ indicates a clockwise rotation of the surface, but $\hat{S}_{HSR,EP}$ and $\hat{S}_P$ indicate that the surface is frontoparallel.

We propose that the smaller range of the induced effect relative to the geometric effect is a consequence of the differing set of slant estimates from the three means described above. In the induced effect, the two stereoscopic estimates are in conflict and slant estimation from perspective cues agrees with one of them; in the geometric effect, the stereoscopic estimates are in agreement and both conflict with perspective cues. As a consequence, the final slant estimate is affected differently by vertical and horizontal magnification. We tested this idea by varying the informativeness of perspective cues and by making slant estimation by HSR and eye position consistent and inconsistent with estimation by HSR and VSR.

Experiment 5: Effect of perspective on geometric and induced effects

We first examined the influence of perspective slant cues. Using the same apparatus as in Experiments 1-4, we measured the induced and geometric effects in the presence of strong and weak perspective cues. The stimuli were dots on a black background. The eyes were symmetrically converged at 40 cm. Observers adjusted the slant of the stimulus (rotation about a vertical axis) until it appeared gaze normal. Icons representing the weak- and strong-perspective conditions are shown in the upper panel of Figure 3.11. In the strong-perspective condition (right icon), the points in the virtual plane created a square lattice with regular spacing. When the plane was unmagnified and normal to one eye's line of sight, the lattice subtended $25 \times 25^{\circ}$ at that eye; the dots were removed within the central portion of the plane as shown (diameter of dot-free zone was 20° when the plane was gaze normal). The points were affixed to the plane before rotation so, after rotation, the square lattices projected to the two eyes in geometrically-correct fashion (dot size and brightness were constant). In the weak-perspective condition (left icon), 300 points were chosen at random from a frontoparallel disk 25° in diameter and were back-projected onto the virtual plane, *after* rotation of the plane, from the perspective of a point midway between the eyes. With this projection technique, there is no useful slant information in the monocular images. For both strong- and weak-perspective stimuli, one eye's image was then magnified horizontally or vertically in software.

The predictions of the three means of slant estimation are shown in the left panel. The magnification applied to the left or right eye is indicated by the abscissa in terms of the natural logarithm of the relevant size ratio. The predicted slant setting is represented by the ordinate in terms of $\ln(HSR)$; this quantity is the HSR created by a stimulus plane when rotated to the predicted slant setting (without incorporating the effect of magnification in the HSR value). Slant estimation from perspective predicts a slant setting of 0 ($\ln(HSR) = 0$) and, therefore, is represented by the horizontal line; of course, the perspective information is salient in the strong-perspective condition only. Slant estimation from HSR and VSR predicts the right oblique line (lower left to upper right) with horizontal magnification (geometric effect) because, according to Equation (3), the plane should be rotated by the observer by the amount that yields a value of $\ln(HSR)$ that is equal in magnitude, but opposite in sign, from the lnHSR due to the magnification alone. Slant estimation from HSR and VSR predicts the left oblique line with vertical magnification (induced effect) because according to Equation 3.3, the plane should be rotated by the observer by the amount that yields $\ln(HSR)$ equal to $\ln(VSR)$ due to the magnification. Slant estimation by HSR and eye position predicts the right oblique line with horizontal magnification, but predicts the horizontal line with vertical magnification (because the magnification in that case does not affect either HSR or eye position).

The data from two of the four observers are displayed in middle and right panels (Figure 3.11). Unfilled symbols represent settings with horizontal magnification (geometric effect) and filled symbols settings with vertical magnification (induced effect). Circles represent settings with weak perspective cues and squares settings with strong perspective cues.

The settings in the geometric-effect condition were very consistent with the predictions of equations 3.2 and 3.3 whether perspective cues were strong or weak. The other two observers exhibited similar behavior except that one showed an attenuation of the geometric effect with strong perspective. When perspective cues to slant were informative, settings in the induced-effect condition exhibited the typical plateaux at magnifications of approximately 8% ($\ln(VSR) \approx .08$). With a reduction in the salience of perspective cues, however, the induced effect was nearly identical to the geometric effect even at larger magnifications. Indeed, two observers (BTB and SJF) exhibited induced effects close to theoretical prediction (equation 3.3) up to 30% magnification ($\ln(VSR) \approx .30$). These data are consistent with an earlier observation that perceived slant increases monotonically up through large vertical magnifications (Rogers & Koenderink, 1986), but they add to it by showing that VSR alterations created by vertical magnification affect the

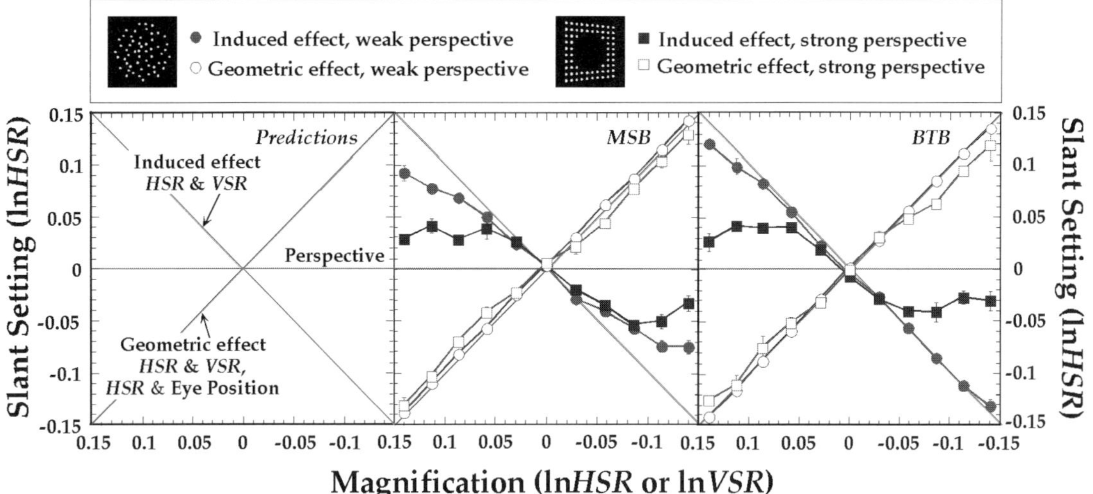

Fig. 3.11. Slant settings as a function of magnification and perspective information (Experiment 5). The stimuli were 300 dots on a black background. The eyes were symmetrically converged at 40 cm. Induced and geometric effects were measured in the presence of strong and weak perspective cues; observers adjusted the slant of the stimulus (rotation about a vertical axis) until it appeared gaze normal. The predictions are shown in the left panel and the data from two of the four observers in the middle and right panels. The abscissae are the percent magnification applied to the left or right eye expressed in terms of $\ln(HSR)$ or $\ln(VSR)$. Ordinates are the slant of the virtual plane when it appeared gaze normal (without taking into account HSR due to the magnification). Unfilled symbols represent settings with horizontal magnification (geometric effect) and filled symbols settings with vertical magnification (induced effect). Circles represent settings with weak perspective cues and squares settings with strong perspective cues. Error bars are ± 1 standard deviation.

perception of gaze normal in just the way predicted by slant estimation from HSR and VSR when perspective cues are not informative. The observation of unattenuated induced effects at large magnifications is not consistent with the hypothesis that the plateau in the induced effect is a consequence of creating implausible combinations of VSR and $\partial VSR/\partial\gamma$ (Frisby, 1984; Gillam *et al.*, 1988; Mayhew, 1982; Ogle, 1938).

Experiment 6: Effect of eye position on geometric and induced effects

Why does the manipulation of perspective cues affect the induced effect so dramatically and not the geometric effect? We propose that this difference is a consequence of the fact that slant estimation from HSR and eye position ($\hat{S}_{HSR,EP}$) agrees not with slant estimation from HSR and VSR

($\hat{S}_{HSR,VSR}$) as in the geometric effect, but rather with slant estimation by perspective cues ($\hat{S}_P$). To test this possibility, we conducted another experiment using eccentric gaze to see if the attenuation of the induced effect could be abolished by making eye position consistent with VSR. This manipulation causes $\hat{S}_{HSR,EP}$ to agree no longer with $\hat{S}_P$, but rather with $\hat{S}_{HSR,VSR}$. Similarly, we tested whether the geometric effect can be made to plateau at higher $HSRs$ when $\hat{S}_{HSR,EP}$ is made consistent with $\hat{S}_P$ rather than with $\hat{S}_{HSR,VSR}$.

By rotating the haploscope arms, the head-centric azimuth of the stimulus could be varied without altering the retinal images. The strong-perspective configuration used in Experiment 5 was used (see icon in Figure 3.12). There were four conditions: vertical and horizontal magnification in forward gaze (conventional induced and geometric effects) and vertical and horizontal magnification in eccentric gaze. In the forward-gaze conditions, the images were presented straight ahead, so the eyes'o version was 0° and vergence was $\sim 9^o$. In the eccentric-gaze conditions, the images were presented at various head-centric azimuths, so the eyes' version varied. For vertical magnification, VSR and $\partial VSR/\partial\gamma$ in the retinal images were appropriate for a plane surrounding the eccentric fixation point; stated another way, the head-centric azimuth was the one that would, in natural viewing, give rise to the VSR and $\partial VSR/\partial\gamma$ in the stimulus. For horizontal magnification, VSR and $\partial VSR/\partial\gamma$ remained appropriate for straight ahead; the head-centric azimuth was the azimuth at which a truly gaze-normal plane gives rise to an HSR value equal to the horizontal magnification in the stimulus. For gaze-normal planes, $HSR = VSR$ at fixation, so the same azimuth was used at a given magnification.

The predictions for the three means of slant estimation are represented by the horizontal and diagonal lines in each panel of Figure 3.12. Slant estimation from perspective predicts the horizontal line for all conditions. For vertical magnification (induced effect), slant estimation by HSR and VSR (equation 3.3) predicts the left oblique line for forward and eccentric gaze (because this means of estimation is unaffected by eye position *per se*). Slant estimation by HSR and eye position (equation 3.2) makes different predictions for vertical magnification depending on the gaze condition: it predicts the horizontal line for forward gaze and the left oblique in eccentric gaze. Thus, by placing the eyes in eccentric gaze, $\hat{S}_{HSR,EP}$ is made consistent with $\hat{S}_{HSR,VSR}$. For horizontal magnification, $\hat{S}_{HSR,VSR}$ would yield settings along the right oblique for forward and eccentric gaze, but $\hat{S}_{HSR,EP}$ would yield settings along the right oblique for forward gaze and the horizontal line for eccentric gaze. Hence, by placing the eyes in eccentric gaze, the estimate based on HSR and eye position is made inconsistent with the one based on HSR and VSR.

The results for two of the four observers are shown in Figure 3.12. The

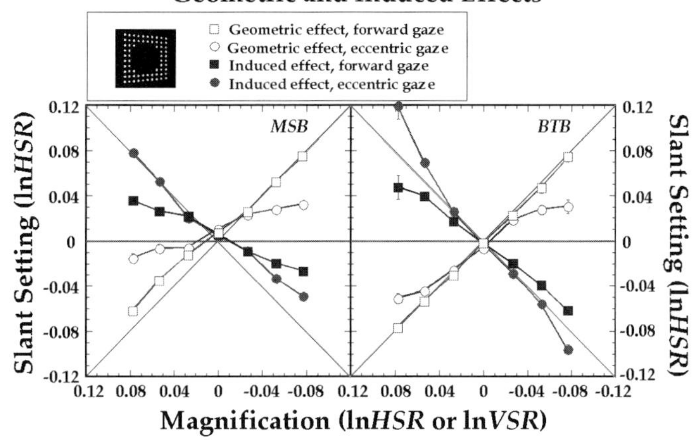

Fig. 3.12. Slant settings as a function of magnification with forward and eccentric gaze (Experiment 6). There were four conditions: vertical and horizontal magnification in forward gaze (conventional induced and geometric effects) and vertical and horizontal magnification in eccentric gaze. In the forward-gaze conditions, the images were presented straight ahead, so the eyes' version was 0° and vergence was $\sim 9^{\text{o}}$. In the eccentric-gaze conditions, the images were presented at various head-centric azimuths, so the eyes' version varied. For vertical magnification, VSR and $\partial VSR/\partial\gamma$ in the retinal images were appropriate for a plane surrounding the eccentric fixation point; stated another way, the head-centric azimuth was the one that would, in natural viewing, give rise to the VSR and $\partial VSR/\partial\gamma$ in the stimulus. For horizontal magnification, VSR and $\partial VSR/\partial\gamma$ remained appropriate for straight ahead; the head-centric azimuth was the azimuth at which a truly gaze-normal plane gives rise to an HSR value equal to the horizontal magnification in the stimulus. For gaze-normal planes, $HSR = VSR$ at fixation, so the same azimuth was used at a given magnification for horizontal and vertical magnifications. Data from two of the four observers are shown in the left and right panels. Filled and unfilled symbols represent settings for vertical and horizontal magnification, respectively. Squares and circles represent settings with the eyes in forward and eccentric gaze, respectively. Error bars are ± 1 standard deviation.

abscissae represent the unilateral magnification applied to the stimulus in terms of $\ln(HSR)$ (geometric effect) or $\ln(VSR)$ (induced effect). The ordinates represent the observers' slant setting in terms of $\ln(HSR)$ (without taking the unilateral magnification into account). Because the retinal images were the same in the forward- and eccentric-gaze conditions, any differences in slant settings between the two conditions must reflect the contribution of eye position sensed via extra-retinal signals. Unfilled and filled symbols represent the results for horizontal and vertical magnification, respectively. Squares and circles represent results for forward and eccentric gaze, respectively.

Consider first the results for vertical magnification. With forward gaze, the slant required to make the plane apparently gaze normal was significantly less than predicted by equation 3.3. However, in eccentric gaze, slant settings were larger and closer to prediction; moreover, no clear plateaux were observed. Observers reported clear phenomenological differences between the two conditions after setting the stimulus to apparent gaze normal. When the eyes were turned to the appropriate azimuth for the vertical magnification, the stimulus looked like a trapezoidal grid painted on a gaze-normal plane. When the eyes were straight ahead, settings were less certain and the stimulus did not appear so clearly planar.

Now consider the results for horizontal magnification. With forward gaze, the slant required to make the plane appear gaze normal was close to the predictions of equation 3.3. However, with eccentric gaze, the apparently gaze-normal slant was significantly lower. Thus, like the induced effect, the geometric effect can exhibit a plateau when estimation by HSR and eye position agrees with perspective rather than with estimation by HSR and VSR.

Because the retinal images were the same in the forward- and eccentric-gaze conditions, the differences in slant settings reflect the contribution of sensed eye position to the interpretation of retinal disparities.

Experiment 7: Minimizing cue conflict in the induced effect

If one minimizes the conflicts between the various means of slant estimation, we should be able to observe full induced effects up to large magnifications. We tested this hypothesis by using the weak-perspective configuration of Experiment 5 and obtaining slant settings with the eyes in forward gaze or in the eccentric position that is commensurate with the VSR in the retinal images.

The predictions are shown in Figure 3.13. With the back-projection technique, there is no useful perspective cue to slant, so the determinants of perceived slant are stereoscopic. Slant estimation by HSR and VSR predicts the left oblique for forward and eccentric gaze whereas estimation by HSR and eye position predicts the horizontal line for forward gaze and the oblique line for eccentric gaze.

The results for the two observers are also shown in Figure 3.13. In both cases, placing the eyes in eccentric gaze yielded significantly larger induced effects without plateaux. This finding demonstrates that induced-effect plateaux can be eliminated altogether when eye position is made consistent with the observed disparities and when perspective cues are made uninformative.

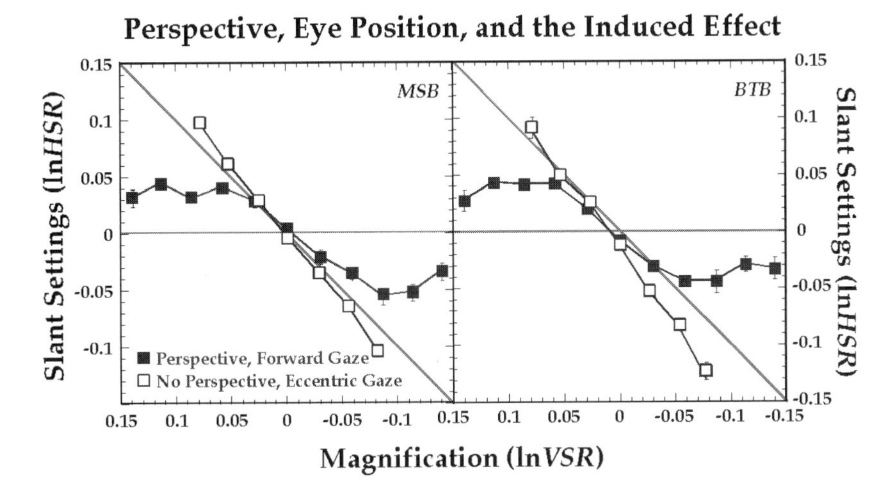

Fig. 3.13. Effects of perspective and sensed eye position on the induced effect (Experiment 7). There were two conditions: eyes in forward gaze with the strong-perspective stimulus (filled symbols) and eyes in eccentric gaze with the weak-perspective stimulus (unfilled symbols). The predictions are represented by the horizontal and diagonal lines (described in text). The panels show the data from the two observers. Slant settings in terms of $\ln(HSR)$ are plotted as a function of vertical magnification in terms of $\ln(VSR)$. Error bars represent ± 1 standard deviation.

Reconsideration of induced-effect plateau

The results of Experiments 5–7 demonstrate that the attenuation of the induced effect is due to conflicts between the three means of slant estimation described earlier. In particular, we showed that the induced effect is not attenuated once the conflicts are eliminated.

The classic induced-effect data of Ogle (1950) are typically S-shaped; see, for example, the filled symbols in Figure 3.13. It is interesting to consider why the data take this form. In the model presented in Figure 3.5, the three slant estimates-$\hat{S}_{HSR,EP}$, $\hat{S}_{HSR,VSR}$, and $\hat{S}_P$-are combined linearly after each is given a weight based on ancillary cues (such as spatial pattern and display height; for details, see Backus & Banks, 1998). Such a model is a "modified weak fusion" model (Landy *et al.*, 1995). In the classic induced effect (perspective information present and eyes in forward gaze), $\hat{S}_{HSR,EP}$ and $\hat{S}_P$ are similar to one another, and both are dissimilar from $\hat{S}_{HSR,VSR}$. A sensible strategy is to give the usual weights to all three estimates when they differ slightly, but to decrease the weight given to the discrepant estimate when it differs greatly from the other two. This is a robust estimation strategy because an estimate that differs significantly from other independent estimates (that agree with one another) is probably the least accurate estimate under normal viewing conditions. It is interesting to note that the induced-effect curve is remarkably similar in shape to the theoretical "influence curve" of a robust estimator (cf Figure 4 in Landy *et al.*, 1995). Thus, the classic S-shaped, induced-effect curve may well be the consequence of using a robust estimation strategy to estimate slant from conflicting inputs.

3.5 Discussion

Invariance of HSR and VSR signals to spatial scale

The visual system could in principle measure HSR and VSR across surface patches of a variety of sizes. What consequence, if any, is there on the estimation of HSR and VSR when the spatial scale of the measurement is varied? Consider first the measurement of HSR. Horizontal disparity is approximately a sinusoidal function of retinal eccentricity (Howard & Rogers, 1995). A sinusoid can in turn be approximated by a parabola of the form $y = ax^2 + bx + c$. Horizontal disparity for several planar surfaces is well-approximated by a parabola for retinal eccentricities out to ± 15 deg (Backus *et al.*, 1998). A property of parabolas is that chords are parallel if and only if they are centered on the same x value. As a consequence, HSR (which is nearly equal to 1 plus the slope of the chord) is little affected by differences in the size of the patch across which it is measured. Therefore, a robust estimate of HSR for a planar surface can be obtained by averaging several HSR measurements, derived from concentric patches of differing size. Naturally, measurements of HSR at small scale are required to detect local variations in surface slant.

It is more widely appreciated that VSR generally varies smoothly across the visual field (e.g., Gillam & Lawergren, 1983). For smooth surfaces, VSR varies roughly linearly with horizontal eccentricity (Gillam & Lawergren, 1983; Mayhew & Longuet-Higgins, 1982) and it varies little as the height of the stimulus patch increases from $\sim$0 to 30 deg. The visual system seems to capitalize on these facts because VSR measurements seem to be pooled across rather large portions of the visual field. For example, when stereograms are presented that contain different values of VSR, observers perceive slant specified by the average VSR across several degrees of the visual field (Kaneko & Howard, 1996, 1997; Adams, *et al.*, 1996; Stenton, Frisby, & Mayhew, 1984). For instance, Kaneko and Howard (1997) concluded that VSR is pooled across regions roughly 20 deg in diameter; variations in VSR at a finer scale have little, if any, effect on perceived slant. Thus, pooling of VSR within a large window typically yields a good estimate of VSR at the center of the window.

Recovery of slant away from fixation

Because HSR and VSR are defined relative to the visual plane, the slant equations presented earlier [equations 3.1-3.4] are valid at fixation even when the eyes are directed to the side and upward or downward. In normal viewing, the horizontal meridians of the eyes lie in or nearly in the visual plane because torsion eye movements help align the eyes' images (Rogers, 1992). With such alignment, HSR and VSR can be measured along the eyes' horizontal and vertical meridia.

Equations 3.1-3.4 can also be used to recover surface slant at any point in the visual field if the signals are defined properly. The extension to a non-fixated point in the visual plane is straight-forward. Let μ and γ be the vergence and version of the point relative to the head. They can be estimated from felt eye position and the eccentricities of the retinal images in the two eyes. HSR is defined as before. For VSR, the vertical angles β_L and β_R (Figure 3.1) must be measured along a great circle in each eye. Thus, a retinal coordinate system such as the Fick system (Howard & Rogers, 1995) allows the extension of equations 3.1-3.4 to non-fixated patches in the visual plane.

The slant estimation equations can also be extended to patches above and below the visual plane. The vertical slant axis must be defined as the line that passes through the center of the patch and is perpendicular to the epipolar plane (i.e., the plane that passes through the patch center and the optical centers of the two eyes). The objective gaze-normal plane is perpendicular to the cyclopean line to the center of the patch and contains the vertical slant axis. Thus, slant becomes the amount by which the patch is rotated about the vertical axis away from the gaze normal. The definitions of μ and γ must be generalized so that they specify location within the epipolar plane; they can again be estimated in principle from felt eye position and the eccentricities of the retinal images. To determine HSR, the horizontal angles α_L and α_R (Figure 3.1) must be measured along the great circles in each eye defined by the epipolar plane. For VSR, the vertical angles β_L and β_R must be measured along great circles perpendicular to the epipolar plane. It is not known whether the visual system does this; the point is that the slant equations presented here can in principle be generalized to nonfixated patches above and below the plane of fixation.

Oblique slants

Our discussion has focused on the estimation of surface slant about a vertical axis. Stated another way, the only tilt we considered is 0 deg (slant was varied from -90 to 90 deg). Naturally, the visual system must estimate both slant and tilt. Here we consider the utility of the slant estimation equations for tilts of different values. The ideas presented here show that even for oblique slant axes, equations 3.1-3.4 allow recovery of the component of slant about a vertical axis. We then consider the recovery of slant about other axes.

For stimuli slanted about oblique axes, equations 3.1-3.4 could in principle allow the recovery of the component of slant about the vertical axis. A fixated, planar surface intersects the visual plane in a line. The problem of estimating the component of slant about a vertical axis reduces to finding the slant of the line intersecting the visual plane. The HSR signal is determined by this line. The signals μ and γ are determined by fixation. After examining a number of viewing situations, we assert without proof that VSR is also virtually unaffected. Thus, equations 3.1-3.4 can be used to determine the slant of the line where a planar surface intersects the visual plane; this is the component of slant about a vertical axis.

To estimate the actual slant about an oblique axis, one has to measure slant about another axis besides the vertical. This could be done, for example, by estimating the slant about a horizontal axis and then combining it with an estimate of slant about a vertical axis. The horizontal axis is defined as the intersection of the visual plane and the gaze-normal plane. The slant about a vertical axis determines a line, l_1, where the surface intersects the visual plane. The slant about a horizontal axis determines a second line, l_2, where the surface intersects the plane that passes through the cyclopean eye and vertical axis. l_1 and l_2 determine the orientation of the surface in space.

Slant about a horizontal axis creates horizontal shear of one eye's image relative to the other, but horizontal shear disparity can also be caused by torsion (rotation of an eye about the visual axis). Therefore, horizontal shear disparity by itself is insufficient to specify slant about a horizontal axis. As with HSR and the estimation of slant about a vertical axis, the visual system requires more information than horizontal shear disparity to estimate slant about a horizontal axis. There are two signals that could in principle be used. First, the torsion (cyclovergence) of the eyes could be measured directly from eye position signals. An extra-retinal torsion signal does in fact exist, but its gain is rather low (Nakayama & Balliet, 1977). Vertical shear disparity is caused by cyclovergence and not by variations in surface slant, so such disparity could be used in interpreting horizontal shear disparity in a manner analogous to the use of VSR in interpreting HSR (Howard & Kaneko, 1994). It would be interesting to determine whether both signals–sensed eye position and vertical shear disparity–are used in estimating slant about a horizontal axis and to determine their relative weights in the final estimate.

Comparison with curvature findings

In Experiment 4, we observed a dramatic shift from reliance on HSR and VSR to reliance on HSR and eye position as display height was reduced. A similar effect of display size was reported for perceived curvature of stereoscopic surfaces (Rogers & Bradshaw, 1995) and for perceived depth (Bradshaw *et al.*, 1996); in both cases, observers' percepts were based more on vertical disparity as stimulus size was increased. A quantitative comparison reveals that the data reported here exhibit more reliance on vertical disparity than Rogers, Bradshaw, and colleagues observed. In their curvature study (Rogers & Bradshaw, 1995), the effect of distance specified by eye position was 30% of the complete-constancy prediction for a 38-deg display. In the work presented here, the effect of azimuth specified by eye position was only 20% in Experiments 2 and 4 for a 30-deg display. The effects of vertical disparity in Rogers and Bradshaw's curvature study and in the work presented here were 65% and 88%, respectively. The direction of these differences is *opposite* the prediction made from the difference in display size because larger displays favored vertical disparity in both studies.

This difference in reliance on vertical disparity may be the consequence of using different displays and observers or the consequence of the differing computations needed to estimate curvature and slant. Without a direct comparison of performance on curvature and slant tasks with the same stimuli and observers, one cannot decide. It is interesting nonetheless to consider how the computations required for estimating curvature differ from those in estimating slant about a vertical axis. Five signal combinations can be used to determine, stereoscopically, whether a surface is flat or curved: (1) scaling the horizontal gradient of HSR ($\partial HSR/\partial\gamma$) for distance by using an estimate of μ obtained from extra-retinal signals; (2) scaling the gradient of HSR for distance by using an estimate of μ obtained from ; (3) estimating local slants at various points on the surface using HSR and eye position, then integrating these across space; (4) doing the same, but using HSR and VSR to estimate the local slants; and (5) testing whether $HSR = kVSR^2$ throughout the stimulus. There are two means (1 and 3) that use an estimate of μ obtained from extra-retinal signals, and one (4) that might also; these three could, therefore, contribute to the observed effect of vergence. There are also three methods (2, 4, and 5) that use vertical disparity, and, therefore, could contribute to the observed effect of vertical disparity.

Use of signals in other tasks

The same signals used for estimating slant stereoscopically are needed for other computations. We have already noted that curvature and depth estimation require the same signals. Here we consider the use of μ, γ, HSR,

VSR, and $\partial VSR/\partial\gamma$ for estimating the headcentric azimuth and distance of an object.

Frisby (1984) and Gillam and Lawergren (1983) pointed out that vertical magnification of one eye's image does not yield a change in the apparent visual direction of the stimulus. This is somewhat surprising because the visual system does use vertical magnification to correct for changes in stimulus azimuth during slant estimation. Can one understand the absence of a vertical magnification effect on perceived azimuth from an analysis of the signals involved? A simple method for estimating stimulus azimuth from sensed eye position alone is:

$$Azimuth = \gamma \tag{3.5}$$

If the stimulus is not fixated, azimuth could be estimated from a combination of γ and the retinal eccentricity of the stimulus in the two eyes.

A method based on VSR is:

$$Azimuth \approx \tan^{-1}\left(\frac{\ln(VSR)}{\gamma}\right) \tag{3.6}$$

where μ is expressed in radians. Of course, μ could be determined from sensed eye position or from retinal-image information [equation 3.4]. From this, one can see why perceived azimuth might be estimated from eye position alone [equation 3.5] rather from VSR and μ [equation 3.6]. If azimuth were estimated using VSR, the estimate would be highly susceptible to errors in the estimate of μ, especially at longer viewing distances.

Now consider distance estimation using the aforementioned signals. We can simplify the situation by considering the case in which $\hat{\gamma} = 0$. In that case,

$$distance \approx I/\mu \tag{3.7}$$

where I is the interocular distance. μ can be estimated from vergence, but is also equal (in forward gaze) to $\partial VSR/\partial\gamma$ (see Backus *et al.*, 1998). Presuming that can generally be measured reliably, we predict that perceived distance will be less accurate when the surface provides no $\partial VSR/\partial\gamma$ signal (e.g., is composed of vertical lines).

3.6 Conclusion and Summary

The slant of a stereoscopic surface cannot be determined from the pattern of horizontal disparities (or HSR) alone. However, there are four other signals that, in appropriate combination with horizontal disparity, allow an unambiguous stereoscopic estimate of slant: the vergence (μ) and version (γ) of the eyes, vertical size ratio (VSR), and the gradient of VSR ($\partial VSR/\partial\gamma$).

In addition, a useful signal is provided by perspective slant cues. The determination of perceived slant can be modeled as a weighted combination of three estimates based on those signals (Figure 3.5): a perspective estimate ($\hat{S}_P$), a stereoscopic estimate based on HSR and VSR ($\hat{S}_{HSR,VSR}$), and a stereoscopic estimate based on HSR and eye position ($\hat{S}_{HSR,EP}$). The visual system must assign a weight to each slant estimate because the estimates may differ; the more reliable the estimate, the larger the weight. It is frequently not obvious *a priori* which slant estimate is most reliable in part because reliability changes with viewing distance, surface size, surface texture, eye position, and more.

In the first four experiments reported here, the conflicts between slant from HSR and VSR and from HSR and eye position were systematically manipulated while perspective cues were rendered uninformative. We found that the visual system generally gives more weight to $\hat{S}_{HSR,VSR}$ than to $\hat{S}_{HSR,EP}$ under those conditions. However, when VSR is made difficult to measure by using short stimuli or stimuli composed of vertical lines, the visual system gives more weight to sensed eye position (in particular, to the eyes' version).

In Experiments 5-7, we examined the puzzling plateaux of the induced effect by manipulating conflicts between slant from HSR and VSR, HSR and eye position, and perspective. We found that the plateaux are created by the conflicts that existed in the classic experiments and that they can be eliminated by minimizing the conflicts.

A model in which the slant percept is a linear combination of the two slant estimates accounted well for the data. An ideal slant estimator would use the signals and signal combinations that, given the information they provide and the uncertainties in their measurement, would allow the most accurate slant estimate. We do not know if the human visual system uses an ideal weighting scheme, so it will be of interest in the future to identify situations in which optimal and non-optimal schemes are used.

Acknowledgments

This work was supported by research grants from AFOSR (93NL366), NSF (DBS-9309820), and Human Frontier of Science (RG-34/96). We thank Raymond van Ee, Sharyn Gillett, and Sarah Freeman for participating as observers, Karsten Weber and Jim Crowell for assistance in software development, and Dave Rehder for assistance in constructing the haploscope.

References

Adams, W., Frisby, J. P., Buckley, D., Garding, J., Hippisley-Cox, S. D., & Porrill, J. (1996). "Pooling of vertical disparities by the human visual system", *Perception*, 25: 165-176.

Amigo, G. (1967). "The stereoscopic frame of reference in asymmetric convergence of the eyes". *Vis. Res.*, 7: 785-799.

Amigo, G. (1972). "The stereoscopic frame of reference in asymmetric convergence of the eyes: Response to 'point' stimulation of the retina", *Optica Acta*, 19: 993-1006.

Backus, B. T. & Banks, M. S. (1998). "Estimator reliability and the scaling of stereoscopic slant with distance", *Perception*, under review.

Backus, B. T., Banks, M. S., van Ee, R., & Crowell, J.A. (1998). "Horizontal and vertical disparity, eye position, and stereoscopic slant perception", *Vis. Res.*, under review.

Banks, M. S., & Backus, B. T. (1998). "Extra-retinal and perspective cues cause the small range of the induced effect", *Vis. Res.*, 38, 187-194.

Buckley, D. & Frisby, J. P. (1993). "Interaction of stereo, texture and outline cues in the shape perception of three-dimensional ridges", *Vis. Res.*, 33: 919-933.

Collett, T. S., Schwarz, U., & Sobel, E. C. (1991). "The interaction of oculomotor cues and stimulus size in stereoscopic depth constancy", *Perception*, 20: 733-754.

Cumming, B. G., Johnston, E. B., & Parker, A. J. (1991). "Vertical disparities and the perception of three-dimensional shape", *Nature*, 349: 411-413.

Cumming, B. G., Johnston, E. B., & Parker, A. J. (1993). "Effects of different texture cues on curved surfaces viewed stereoscopically", *Vis. Res.*, 33: 827-838.

Cutting, J. E. & Millard, R. T. (1984). "Three gradients and the perception of flat and curved surfaces", *J. Exp. Psych.: Gen.*, 113: 198-216.

Ebenholtz, S. M., & Paap, K. R. (1973). "The constancy of object orientation: Compensation for ocular rotation", *Perception & Psychophys.*, 14: 458-470.

Frisby, J.P. (1984). "An old illusion and a new theory of stereoscopic depth perception", *Nature*, 307: 592-593.

Gårding, J., Porrill, J., Mayhew, J. E., & Frisby, J. P. (1995). "Stereopsis, vertical disparity and relief transformations", *Vis. Res.*, 35: 703-722.

Gillam, B. (1967). "Changes in the direction of induced aniseikonic slant as a function of distance", *Vis. Res.*, 7: 777-83.

Gillam, B., Chambers, D., & Lawergren, B. (1988). "The role of vertical disparity in the scaling of stereoscopic depth perception: An empirical and theoretical study", *Percept. & Psychophys.*, 44: 473-83.

Gillam, B., & Lawergren, B. (1983). "The induced effect, vertical disparity, and stereoscopic theory", *Percept. & Psychophys.*, 34: 121-30.

Gillam, B. & Ryan, C. (1992). "Perspective, orientation disparity, and anisotropy in stereoscopic slant perception", *Percept.*, 21: 427-439.

Herzau, W., & Ogle, K. N. (1937). "Über den Grössenunterschied der Bilder beider Augen bei asymmetrischer Konvergenz und seine Bedeutung für das Zweiügige Sehen", *Albrecht von Graefes Archiv für Ophthalmologie*, 137: 327-363.

Howard, I. P., & Kaneko, H. (1994). "Relative shear disparities and the perception of surface inclination", *Vis. Res.*, 34: 2505-2517.

Howard, I. P., & Rogers, B. J. (1995). *Binocular vision and stereopsis*, Oxford University Press, New York, NY.

Kaneko, H., & Howard, I. P. (1996). "Relative size disparities and the perception of surface slant", *Vis. Res.*, 36: 1919-1930.

Kaneko, H., & Howard, I. P. (1997). "Spatial limitation of vertical-size disparity processing", *Vis. Res.*, 37: 2871-2878.

Koenderink, J. J. & van Doorn, A. J. (1976). "Geometry of binocular vision and a model for stereopsis", *Biol. Cybern.*, 21: 29-35.

Landy, M. S., Maloney, L. T., Johnston, E. B., & Young, M. (1995). "Measurement and modeling of depth cue combination: In defense of weak fusion", *Vis. Res.*, 35: 389-412.

Longuet-Higgins, H. C. (1982). "The role of the vertical dimension in stereoscopic vision", *Perception* 11: 377-386.

Mayhew, J. E. W. (1982). "The interpretation of stereo–disparity information: The computation of surface orientation and depth", *Perception*, 11: 387-403.

Mayhew, J. E. W. & Longuet-Higgins, H. C. (1982). "A computational model of binocular depth perception", *Nature*, 297: 376-378.

Nakayama, K., & Balliet, R. (1977). "Listing's law, eye position sense, and perception of the vertical", *Vis. Res.*, 17: 453-7.

Ogle, K.N. (1938). "Induced size effect. I. A new phenomenon in binocular space–perception associated with the relative sizes of the images of the two eyes", *Arch. Ophthalmol.*, 20: 604-623.

Ogle, K. N. (1940). "Induced size effect with the eyes in asymmetric convergence", *Arch. Ophthalmol.*, 23: 1023-1028.

Ogle, K. N. (1950). *Researches in binocular vision*, W.B. Saunders Co., London.

Rogers, B. J. (1992). "The perception and representation of depth and slant in stereoscopic surfaces", in G. Orban & H. Nagel (Eds.), *Artificial and Biological Vision Systems*, 241-266, Springer-Verlag, Berlin.

Rogers, B. J., & Bradshaw, M. F. (1993). "Vertical disparities, differential perspective and binocular stereopsis", *Nature*, 361: 253-5.

Rogers, B., & Bradshaw, M. (1995). "Disparity scaling and the perception of frontoparallel surfaces", *Perception*, 24: 155-179.

Sedgwick, H. (1986). "Space perception", in K. R. Boff, L. Kaufman, & J. P. Thomas (Eds.), *Handbook of perception and human performance, (Vol. I, Sensory processes and perception)*, John Wiley and Sons, New York, NY.

Stenton, S. P., Frisby, J. P., & Mayhew, J. E. (1984). "Vertical disparity pooling and the induced effect", *Nature*, 309: 622-623.

Stevens, K. A. (1981). "The information content of texture gradients", *Biol. Cybern.*, 42: 95-105.

Stevens, K. A. (1983). "Slant-tilt: The visual encoding of surface orientation", *Biol. Cybern.*, 46: 183-195.

4

Frames of Reference: with examples from driving and auditory localization

Laurence R. Harris, Daniel C. Zikovitz and Agnieszka Kopinska

Summary

All codes require a frame of reference. In this chapter we define frames of reference and describe some of the principles behind the methods that can be used to reveal which frame of reference might be used for a given task or sub-task. We describe two sets of experiments designed to reveal the frames of reference used in driving and in auditory localization.

4.1 Reference frames in the brain

When something is coded in the brain (or anywhere else for that matter), there needs to be an interpretation scheme for getting the information back out. That is, the particular code has to have a meaning. Imagine that you have a set of four switches and your job is to store the position of an object. There are two aspects to the problem: Firstly to decide what the position is represented with respect to. And secondly to decide on the relationship between the switch positions and the direction and size of the displacement from that point (possible answers are given in the appendix at the end of this chapter). This chapter addresses the first part of the problem which is called the reference frame; the second part is the coordinate system. The code cannot be decoded and used unless both of these are known.

For most tasks, there are several possible candidates for the frame of reference chosen. Most tasks in fact involve a string of conversions in which information from many sources is transformed perhaps through a series of reference frames. But at some point that the various pieces of information must be combined and this combination requires a common reference frame. Let us consider as an example, some of the processes involved in moving a pointing arm from a random starting point to touch an illuminated target (Fig 1). At some point the present and desired position (target location) of the arm have to be compared and this is only possible if all the information is in the same reference frame and in the same coordinate system.

4.1.1 Possible reference frames

Two possible schemas are outlined in the bottom part of Figure 4.1. Figure 4.1b shows the comparison being done in a reference frame defined with respect to a point fixed on the arm; in Figure 4.1c the comparison is done in the reference frame defined with respect to the head. These are not the only candidates. Alternatives include the frame of the eye (left or right?), space, some other part of the body or some other system altogether. Reference frame conversions are not the only conversions taking place: there must also be a series of conversions between coordinate systems, possibly even within one reference frame. For example the arm itself, even when it's position is coded in the frame of a fixed point on the arm, could be coded in terms of linear distances from the reference point or in terms of the angles of the joints. But the final comparison of present and desired position of the arm has to done using a common coordinate system within a common reference frame. Does it matter which frame these comparisons are done in?

4.1.2 Why does it matter which frame a comparison is done in?

Divining the criteria the brain and evolution might use in the selection of reference frames is challenging. Our arguments are most likely to be *post hoc* in assessing which benefits might accrue from the use of a particular frame after it has been identified. If all the conversions between frames are perfect and instantaneous then it does not matter in which frame a certain task is done in the sense that the outcome will not be affected.

Factors that might be involved in the selection of a reference frame include the consequences of errors, the efficiency of the computations involved and the convenience. To assess the convenience we also need to take into account what other processes might be going on at the same time. For example, if all the other body movements were done with respect to the trunk, then there might be some advantages in converting everything to this independent frame for the sake of consistency and to maintain flexibility. Although we have set off to touch the target with the right hand in the example of Figure 4.1, we might decide to kick it or head-butt it instead for some reason. Although re-converting the target position from an arm-system to a leg-system is theoretically possible, such a chain of conversions might introduce unnecessary errors or computations.

All else being equal, the choice of reference frame does not necessarily determine the behaviour. This means that we cannot necessarily tell which frame of reference is used by looking at the behaviour: the arm will reach the target using the schemes of Figure 4.1b or 4.1c or many other alternatives. How then can we design experiments to find out which frame is used?

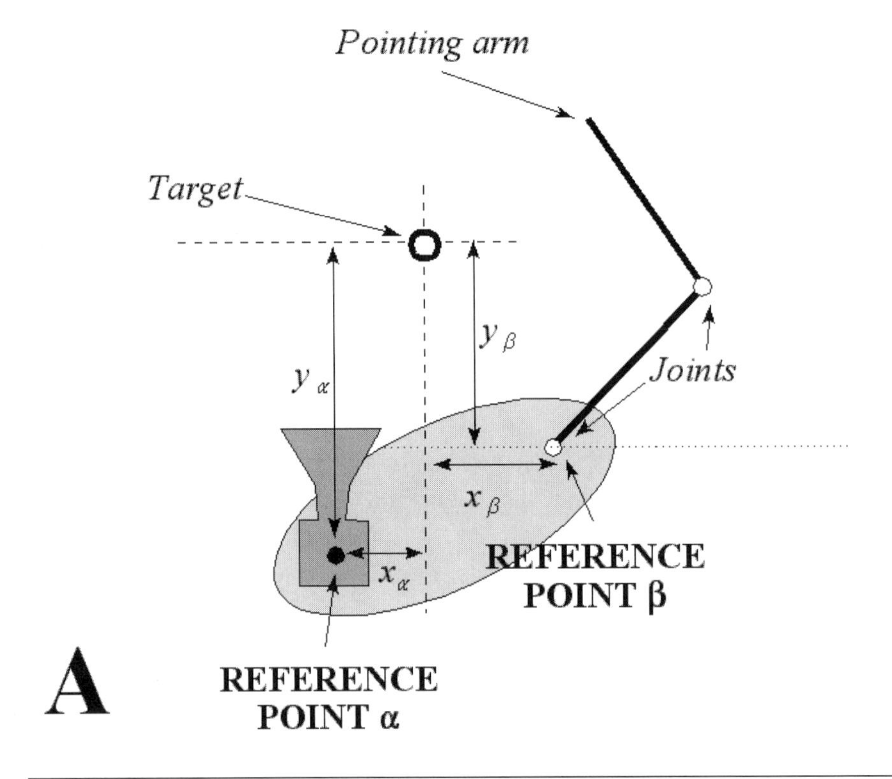

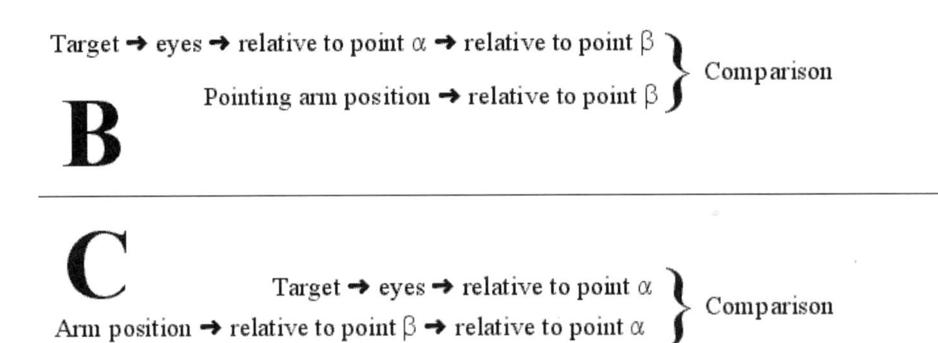

Fig. 4.1. Panel A shows a schematic representation of an arm related to reference point β and a target represented relative to point α. In order to move a pointing arm to a visually- detected target, there has to be a comparison between the present and the desired position of the arm. The comparison could be made in the reference frame of vision, relative to reference point α as shown in panel B or the arm, relative to reference point β as shown in panel C. Alternatively, both arm and target position could be converted to an independent reference frame for the comparison.

4.2 Designing experiments to find out which frame is used

Sometimes the optimum strategy for performing a task in one frame is different from that required in another. Thus it might be beneficial to maintain something as constant within a particular frame. Frames, by definition, move with respect to each other. Therefore if something is constant in one frame, it will not necessarily be in another. Experimentally varying frames and seeing if a behaviour (of organisms or cells) shifts with one or other frame can provide a clue about which frame is used. For example, if we displace the shoulder of Figure 4.1 and find that the arm made errors suggesting that the system was unaware of the shift, this could either support the scheme of Figure 4.1b (in which the conversion of target position from α to β was faulty) or the scheme of Figure 4.1c (in which the conversion of the arm position from β to α was faulty). It would suggest, however, more strongly, a third schema in which the conversion from α to β was never made and that the target location was left in an 'α' system and sent to the arm directly. Such a system must assume that α and β would normally be in register and that there would be no need for a mechanism to update shifts between them.

This technique of varying frames has been used, for example, to assess the frame of reference used for coding location in the firing pattern of superior colliculus cells. When the eyes move in the head, it is possible to see whether the visual and auditory receptive fields stay constant in a head or eye reference frame (Harris *et al.*, 1980; Jay & Sparks, 1984; Pettigrew, 1984; Hartline *et al.*, 1995; Peck *et al.*, 1995). Interestingly, it seems that some cells do one thing and some another. The receptive fields of many stay stable in neither frame which amkes these studies difficult to interpret. In the present chapter we use this technique (with clearer results) to separate the gravito-inertial frame and the visual frame as candidates for the reference frame used in driving.

Another method to identify reference frames involves adaptating some aspect of the conversion mechanism. Converting from one frame to another requires information about the relative positions of each frame. For example, converting from the retinal to the head frame needs information about where the eyes are in the head. Holding an eccentric gaze for a prolonged period will adapt this source of information and lead to errors of knowing the position of the eyes in the head (Paap & Ebenholtz, 1976). This then leads to observable and predictable errors in the pointing behaviour.

Systematic errors can sometimes arise from and thus betray, frame conversion errors. If in our Figure 4.1 example, the positions of α and β are varying very quickly then it is challenging for the system accurately to convert the position of the target measured in frame α at time 't', to its corresponding position in frame β. To do so requires knowing the relation between α and

β at time 't'. A systematic lag for example, might indicate that a conversion had taken place and reveals details of its mechanism. This technique is employed in the auditory localization experiments described in §4.4. A variant of this technique is also used by Crawford (see Crawford, 1997; Crawford's chapter, this volume) to investigate the frame of reference used in planning eye movements.

4.3 Frames of reference in driving

Casual observation of drivers and passengers shows that they tilt their heads when cornering. Why do people do this? One hypothesis is that head tilt is a response to the lateral forces created by going around a corner. Another possible reason for head tilt is that it might line the head up with some aspect of the visual demands of cornering. These different strategies would each reflect a driver's behaviour being dominated by a particular frame of reference and the attempt to optimize some aspect of processing within that frame.

A person in a car driving around a corner is exposed to two orthogonal, linear accelerations: gravity and centripetal acceleration. The combination of these two forces is equivalent and indistinguishable from a single acceleration and is called the *gravito-inertial force* (GIF). The direction of the GIF while cornering is tilted compared to the direction of gravity alone (Figure 4.2a). Our first hypothesis was that drivers and passengers might align their heads with this tilted gravito-inertial force thus indicating a dominance of a gravito-inertial reference frame and an attempt to keep the head aligned with this frame.

An alternative explanation has to do with vision. The view of the road through the windshield indicates the curvature of the road ahead. Drivers tend to look ahead at the tangent point of the road while cornering (Land & Lee, 1994; see Land's chapter this volume) and it might be useful to line up the head to keep some visual feature constant in the head frame. If riders in a vehicle were dominated by visual information corresponding to the layout of the road ahead, this might be reflected in their head orientation. Our second hypothesis was that drivers and passengers might align their heads with the visual tilt of the road.

In order to discriminate between these two hypotheses we exploited the fact that the tilt of the GIF depends on the speed of cornering whereas the road curvature (Figure 4.2b) is not affected by how fast the vehicle is moving. So by going round the same corners a number of times at different speeds we were able to vary the tilt of the GIF and the visual angle independently (Figure 4.2c) and to see with which variable the head tilt was best correlated.

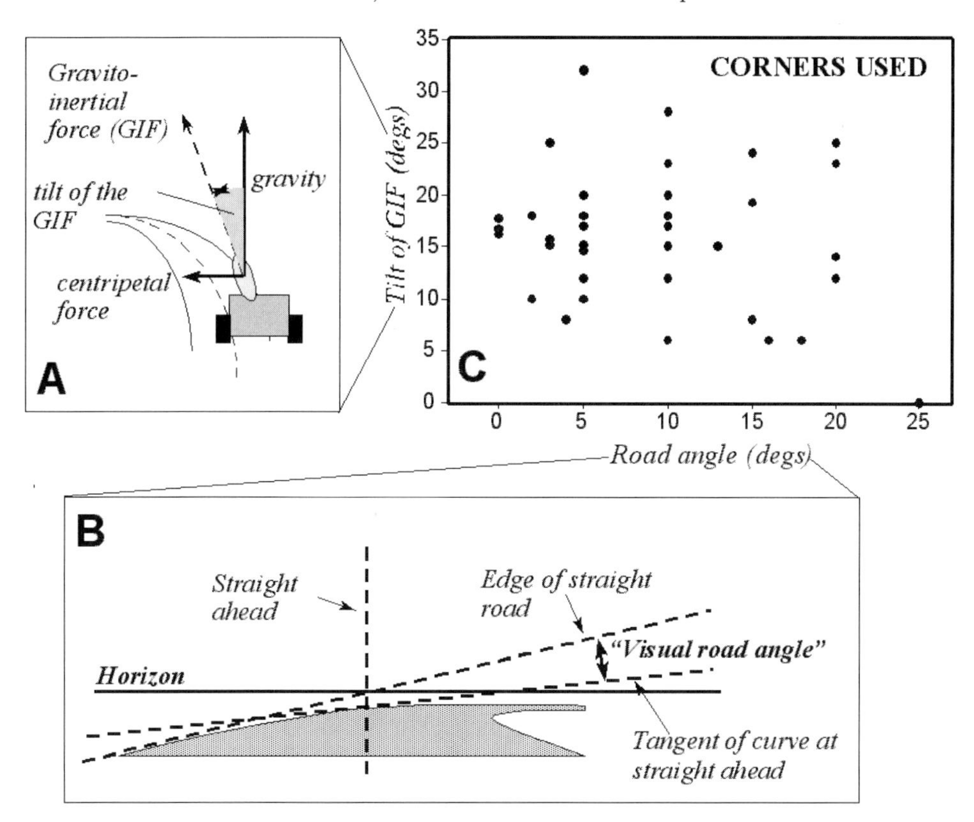

Fig. 4.2. Going around a particular corner at a given speed is associated with a tilt of the gravito-inertial force (A: see text for explanation) and a visual tilt (B: see text for explanation). Under normal conditions there is a correlation between these features because drivers tend to match their speed to the curvature of the road. In this study we de-correlated the tilt of the GIF from the visual road angle by going round the same corners at a range of different speeds (C). Each column of points in panel C corresponds going round a given corner, with a fixed visual angle, at a range of different speeds and therefore generating a range of centripetal forces and corresponding tilts of the GIF.

4.3.1 Measuring lateral accelerations

Lateral accelerations were produced by having subjects drive a normal, domestic van or car around a number of different corners of various radii at speeds of between 20 and 80 kph. The vehicle was equipped with a G-analyst (model 500; Valentine Engineering Co.) Vehicle Dynamics Monitor which is a three-axis accelerometer that measures the lateral acceleration with a sampling interval of 0.1 sec and an accuracy of 0.01g. The G-analyst stored the lateral acceleration profile digitally for later analysis. Its display screen, with a timer, was recorded on video tape that simultaneously recorded the subject's head orientation and a view of the road. This enabled the analysis of the video and the G-analyst to be synchronized.

4.3.2 Measuring the tilt of the road

The angle of the road was assessed from the video tape of the view through the front windshield of the vehicle. Video tapes were played on a freeze-frame system. A line was drawn on the video screen along the edges of the road when the road was straight. As the car went around a curve, a tangent was drawn along the far side of the road where it crossed the straight-ahead position (see Figure 4.2b). The angle this tangent made with the edge of the straight road was taken as a measure of the visual curvature of the road. This produced an arbitrary but consistent measure of the tilt of the road from 0 degs (straight ahead) to about 35 degs when the tangent was horizontal.

4.3.3 Measuring head tilt

Our subjects were twelve experienced drivers who also served as passengers. The head tilt of driver and passenger were measured from a video recorded by the same camcorder that recorded the view of the road. It was mounted firmly in the vehicle 94 cms behind the subject. In order to make measuring the orientation of their heads easier in subsequent freeze-frame analysis, drivers and passengers wore hats with several high-contrast vertical stripes. Rotation of the head without tilt produced no artefactual tilt of the lines on the hats. Tilts were measured relative to the orientation of the head before the subject went into the curve. We describe head tilt towards the centre of curvature of the corner as being 'into the corner' and tilts away from the centre as leaning 'out of the corner'. We use the convention of assigning positive numbers to tilts into the corner and negative numbers for tilts out of the corner. Clearly all our measures of tilt vary throughout a corner. In order to look for correlations between these measures we selected sections where the three variables: head tilt, road tilt and GIF tilt plateaued

4.3.4 Tilt of driver's and passenger's heads during cornering

Drivers always tilted their heads into the curve, that is in towards the centre of rotation. There was a strong correlation of the drivers' head tilt with the visual measure of road tilt (Figure 4.3a; 0.82 degs of head tilt/deg of visual angle; $r^2 = 0.86$) but none with the gravito-inertial tilt (Figure 4.3b; $r^2 = 0.001$). This is the central finding of this study and it is remarkably clear.

Passengers with eyes open generally tilted their heads out of the corner, that is in the opposite direction to drivers. Their head tilts show a negative correlation with the tilt of the GIF (Figure 4.3c; -0.72 degs of head tilt/deg of tilt of GIF; $r^2 = 0.31$) and none with vision (Figure 4.3d; $r^2 = 0.003$). There are some outliers to the regression. GIF tilts up to 20 degs were sometimes

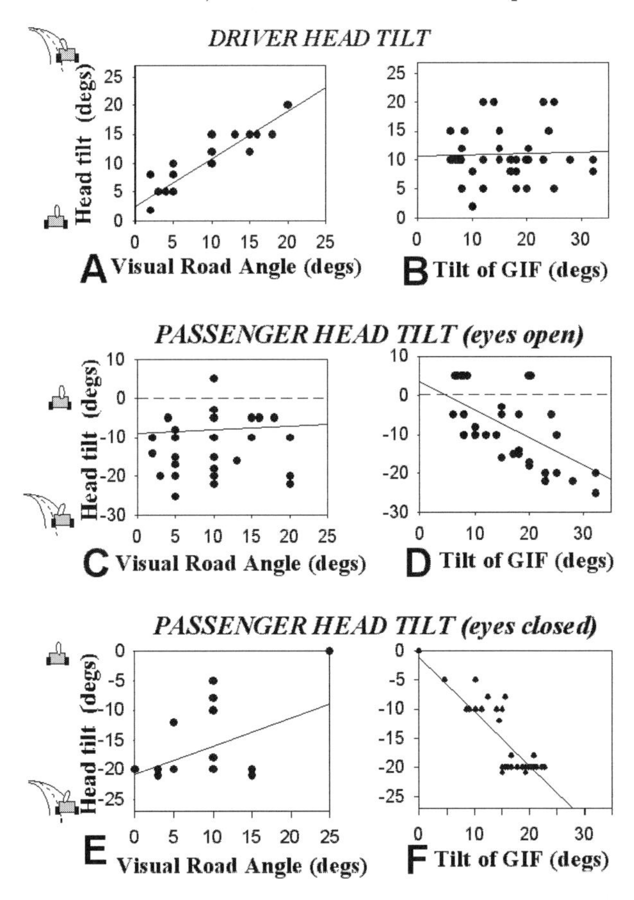

Fig. 4.3. The head tilt of drivers (panels A and B) and passengers with their eyes open (panels C and D) and closed (panels E and F) are plotted as a function of the visual tilt of the road (left panels: A, C and E) and the tilt of the GIF (right panels: B, D and F). Each data point appears in both the left and right of the pairs of panels. The solid lines are regression lines. Corners have been mirror imaged so that they all appear positive. An upward, positive slope indicates a tilt into the corner, a downward, negative slope indicates a tilt out of the corner. Both visual and GIF demands require a tilt into the corner.

accompanied by head tilts less than 5 degs. Occasionally a passenger's head tilted in the same direction as the driver's (into the curve, positive tilts in Figure 4.3c and 4.3d).

Passengers with eyes closed tilted their heads out of the corner. Their head tilts show a negative correlation with the tilt of the GIF (Figure 4.3f; -0.92 degs of head tilt/deg of tilt of GIF; $r^2 = 0.69$) and of course were not significantly correlated to the visual tilt of the road which they could not see (Figure 4.3e; $r^2 = 0.16$).

4.3.5 The frame of reference used while driving

Our results clearly show a tremendous dominance of visual information over the orientation of the gravito-inertial forces in controlling the orientation of a driver's head. The pattern of drivers' head tilts therefore strongly supports our second hypothesis and refutes our first hypothesis. Drivers tilt their heads as they go around a corner, not to counteract the centripetal force or to line up the head with a misperceived direction of gravity, but rather to line up their heads with some visual feature of the road. The pattern of passenger head tilt was not predicted by either hypothesis. Although passenger head tilt was correlated with the tilt of the GIF, it was negatively correlated. That is, there was no indication of passengers aligning their heads with the GIF, but rather they tilted the other way. This was especially true when the passenger had their eyes closed.

Although it seems to make intuitive sense that drivers might want to line up their heads with an instantaneous estimate of gravity, this would in fact not be a good strategy while driving a car. It should be remembered that the tilt of gravito-inertial force when cornering is an *illusion* and does not correspond to the orientation of anything useful except the tilt of a pendulum hung from the roof of the car! When a driver goes round a corner, the lateral accelerations can be considerable. Our drivers experienced pulls of up to 0.6g (see figure 4.2) going at modest speeds around normal suburban roads. Racing car drivers can easily pull 3g which would require a head tilt of over 70 degs! The correct strategy is neither to allow the head to move passively out under the influence of the centripetal force nor actively to line the head up with the new gravito- inertial force but to line the head up in a way that allows the driver to make maximum use of visual information relevant to the control of the car. Passengers on the other hand, especially when their eyes are closed, tend to be passively displaced by the lateral force and to pivot about their point of contact with the car.

There are other situations where it is important to line up the body with gravity in order not to fall over. Meeting the balance challenge of skiing, bob-sledding, motorbike riding or just standing up requires positioning the centre of gravity of the body on a line parallel to the gravito-inertial force that goes through the support surface. Motorbike riders are particularly interesting from the point of view of this study, therefore, as they simultaneously need to align their bodies with the gravito-inertial direction and to align their heads - if they use the same driving strategies as we have described for car drivers – with the visual properties of the road ahead. Casual inspection of motorcyclists suggests this might indeed be what they are doing. We plan to investigate head tilt in motorcyclists using the same techniques as presented in this paper in a future study.

The visual world is not altered by cornering. So what visual feature might be aligned with the head by tilting it? Lining up the head with a horizontal line drawn across the road at the point the driver is looking, requires just what we have found: a head tilt into the corner. How large a tilt is required depends how far down the road the driver is looking and the sharpness of the corner. When going very slowly, slower than normal driving speeds, we expect that the point of attention along the road becomes much closer and the head tilt required becomes negligible. Why align the head with this 'horizontal' line? Drivers need to make left/right steering decisions based on their best estimate of the road conditions ahead. The correlation of the tilt of their head with the visual road angle indicates they are working in a visual reference frame. The effect of the tilt of a drivers head is to line up this frame with the 'horizontal' of the oncoming road.

4.4 Frames of reference during auditory localization

When representing the position of a sound source, there are several possible reference frames. Candidates include the retina, the head, the body and external space. Since the ears are attached to the head, at some point the sound has to be coded in the head reference frame, but this might not be the best frame to use centrally: for example when pouncing on a sound source without taking the trouble to centre your head first. We pulled the reference frames apart by eye movements (retinal frame moving with respect to the head) or head movements (head frame moving with respect to the body) and asked subjects to make sound localization judgments. In all the experiments described here, subjects were asked to adjust a sound that was presented through head phones until it appeared to be on their head's midline. Errors reflected reference frame conversion errors.

If eye movements or eccentric eye position had any effect on sound localization this would suggest that space was represented in the brain in a retinal frame. This seems intuitively unlikely as being unnecessarily complicated. But the movement of some of the auditory receptive fields in the superior colliculus (Hartline *et al.*, 1995; Jay & Sparks, 1984; Peck *et al.*, 1995) during eye movement is compatible with such a plan. Furthermore, sound localization experiments by Lewald and Ehrenstein (1996) have indicated some movement of apparent location of sounds with eye position.

Movements of the perceived auditory midline during head movement would suggest an equally intuitively-unlikely situation, that sounds were not localized in the head frame. Errors arising from head position changes, indicate errors in knowing head position and would only affect sound localization if sounds were being subjected to a conversion requiring that information.

Notice that in both cases, accurate performance does not allow us to conclude anything about the frame of reference used. A perfect system

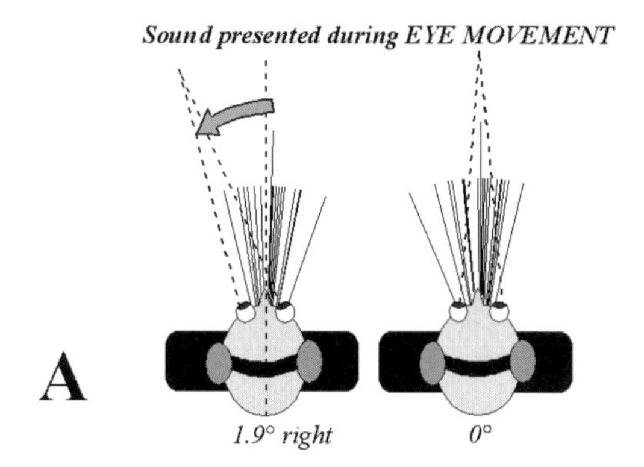

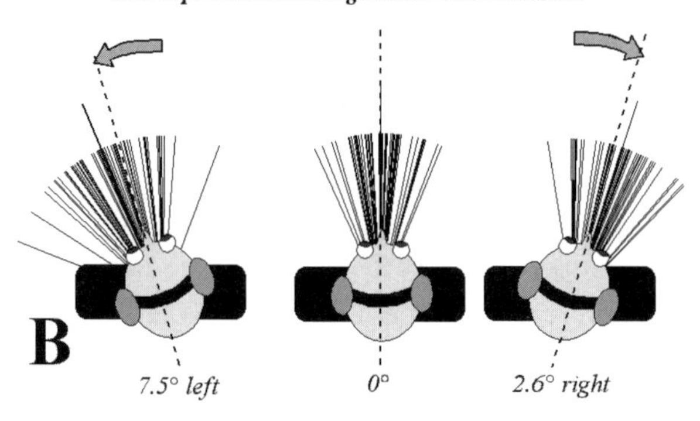

Fig. 4.4. Auditory localization during eye and head movement. Each line represents the perceived auditory midline for one subject under one condition. The average position is given by a slightly longer radiating line and as a number below the head. Panel A compares the judgments of the perceived auditory midline for sounds presented during 20 deg saccades to the left (left) with the control, eyes straight-ahead condition (right). Each subject's judgments have been normalized to their average control data. Panel B compares the judgments of the perceived auditory midline for sounds presented during a head movement to the left or right with the control, head straight-ahead condition (centre). The heads have been twisted to the left and right to indicate the turn. If the midline remained central on the head, the average would appear clustered around the dotted line that indicates the head's midline. In fact the subjective midline moves significantly ahead of this line.

seamlessly takes into account eye and head position at all times as required to arrive at the correct solution. However, we know that eye and head position information is far from perfect (e.g. Hill, 1972; Morgan, 1978; Howard & Anstis, 1974) and in any case, during the dynamic trials, we set the system the almost impossible task of keeping track of eye and head movements on the fly.

4.4.1 During eye movement

We presented sounds through head phones, carefully timed so that they occurred during saccadic eye movements. The sounds had a variable inter-aural delay and subjects were asked to estimate if the sounds were on the left or right of the midline. Eye movements were measured by an infrared technique. We honed in on the perceptual midline using a staircase technique. If auditory location is coded in a retinal frame, then the representation would have to take into account the eye position at the moment of the presentation in order to store the sound's position at the correct retinal location. We expected that the difficulty of keeping track of that location during a 20 deg saccade with a peak velocity of about 150 deg/sec would result in the frame conversion being revealed by systematic errors. An additional challenge for the system would be allowing for the auditory processing time when estimating the position of the eyes at the time of sound presentation.

In fact there were no systematic errors in estimating the perceived auditory midline during a saccadic eye movement. The position and precision of judging the perceived auditory midline did not shift during saccades (see Figure 4.4a and Harris & Lieberman, 1995) compared to judgments made while the eyes were still.

4.4.2 During head movement

We next looked to see if sounds presented during a head movement might be mislocated. This experiment exploited the same logic as the eye movement study. We presented sounds in the middle of a head movement and asked subjects to indicate if they were on the left or right of the head's midline. Surprisingly we found that the perceptual auditory midline is indeed shifted during head movements: it shifted by between 2.6 and 7.5 degs in the direction of the on-going head movement (Figure 4.4b, and Harris, Kopinska and Lieberman, 1997; Kopinska & Harris, 1998). This was rather an unexpected finding since it implies that the frame of reference for coding auditory space is not the head, but instead the body (or perhaps some other frame). That is, head-on-body information is taken into account in assessing the position of a sound in space. We are seeing here the results of errors in this processing.

4.4.3 Measuring sound localization with eccentric eye and head positions

The experiments described above were done with sounds presented during fast movements of the eye and head. We felt that keeping track of eye and head position during movement would be a challenging task that would reveal itself in errors. This was indeed the case for head position, but not eye

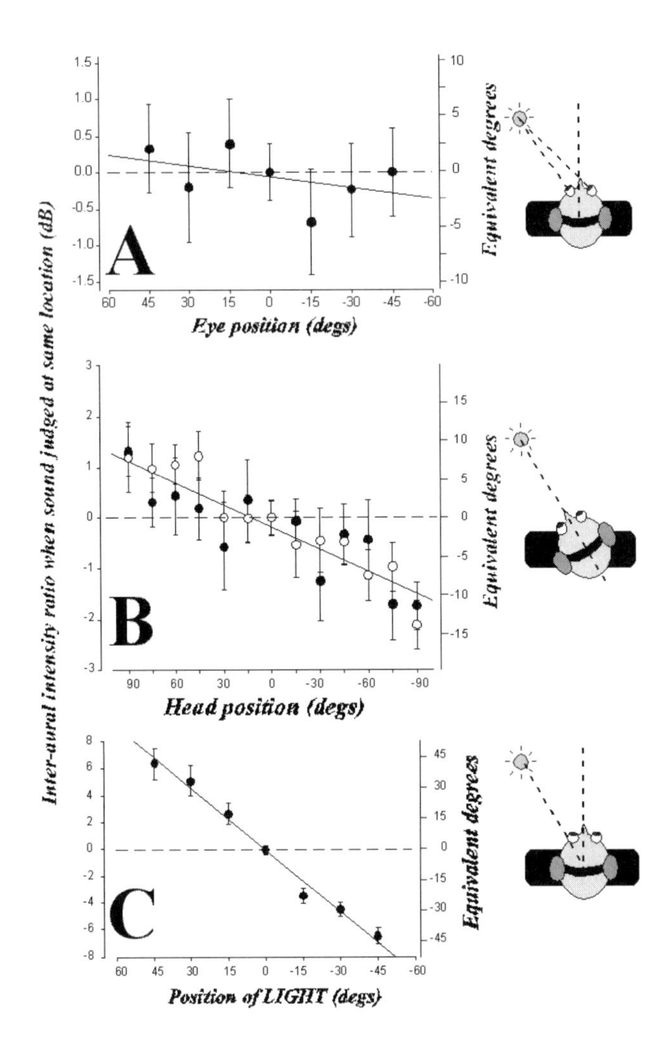

Fig. 4.5. Auditory localization during static eye and head displacement. Each point represents the mean and standard deviation of two repetitions of eight subjects. The vertical axis represents the interaural intensity difference in dB (left hand scale) and equivalent degrees (right hand scale). The equivalent degrees are estimated from the calibration graph (panel C, see text for details). Positive dB values indicate left ear louder. Positive degrees indicate left. In panel A subjects looked at lights that were arranged in a 57cm radius arc around them. Their heads pointed straight ahead (see insert to right). There was no significant shift of the midline with eye position. In panel B, subjects pointed their eyes and head at eccentric targets such that their eyes were centred in the orbits (see insert to right). Judgments of midline were shifted significantly in the direction of the head turn by 0.10 degs for each deg of head turn. Open circles indicate responses in which they were lining up the sound to the target light (ahead of the nose, see insert to right), closed circles when they were aligning with 'head's midline'. Panel C shows the calibration in which subjects adjusted the interaural intensity ratio until the sound appeared to align with targets of various eccentricities (see insert to right). This produced a calibration figure of 0.15 dB/deg.

position. Other studies, however, have claimed that simply holding the eye eccentrically can induce a shifts in the perceived location of sounds (Lewald & Ehrenstein, 1996) and certainly holding the eye eccentrically can alter the perceptual straight ahead (Hill, 1972; Morgan, 1978). Such a shift could be interpreted as indicating a transfer of auditory location information from a head to a retinal frame since it would seem to arrise an error in the processing of eye position in the head. Eye-in-head information is only needed by a sound localization system if it converts information from the head frame to the eye frame. However, it seemed unlikely that a system might make errors during the relatively simple static estimate of eye position and yet be able to keep perfect track of the eye in head during fast movements! On the other hand, our dynamic experiments with the head suggested that we might find some shifts not with eye position but with eccentric head position.

Consequently we repeated the experiments under static conditions. Eye position (range= $\pm 45^{\circ}$) was controlled by having the subject look at a target. Head position (range= $\pm 90^{\circ}$) was controlled by asking subjects to point to a target with a laser that was mounted on their hat. Having arranged their eyes and heads in space in this way, subjects then indicated the perceived location of their auditory midline by adjusting the interaural intensity ratio of a 1kHz, 15 msec sound presented repeatedly through headphones until the sound appeared to be on their head's midline. Calibration was done by asking subjects to adjust the interaural intensity of sounds until they appeared aligned with lights of various eccentricities. This provided a calibration figure of 0.15dB interaural intensity difference/deg of target eccentricity (Figure 4.5c).

We found no consistent shift with eccentric eye position (0.03 deg of shift/deg of eye movement, or 0.0045dB/deg, $r^2 = 0.20$; Figure 4.5a cf. Lewald & Ehrenstein, 1996). But there were very significant ($p<0.0001$) and consistent shifts with head position (0.1 deg of shift/deg of head movement, or 0.015dB/deg, $r^2 = 0.9$; Figure 4.5b). As in the dynamic experiments of Figure 4.4, the shift was in the same direction as the head displacement.

4.4.4 Conclusions about the frame of reference for auditory localization

Systematic shifts of the auditory midline with head position indicate that (i) sound localization takes head position into account but does so inaccurately and (ii) sounds are not perceptually localized in a head frame. It seems intuitive that a system that uses head-on-body information would make errors when asked to perform while the head was moving quickly on the body (dynamic experiments, Figure 4.4). But why would a system that incorporated a head-to-body frame conversion make errors when the head was statically displaced? Our results suggest that if the head is displaced left by 40 degs, then a sound emanating from 44 degs left (4 degs to the left of the nose) will be judged as coming from straight ahead. And a sound actually coming from the direction of the nose will be judged as coming from the right. It is as if the system assumes that when the head is displaced 40 degs, a target at 44 degs corresponds to 'straight ahead'. This error suggests that the head-to-body conversion process assumes (and does not check) that there is only a partial contribution of the head to an eccentric gaze position.

To check on the feasibility of this suggestion we (i) compared the instructions 'line the sound up with the light' (letting the system allow for slop between head and target position; open circles in Figure 4.5b) with 'line the sound up with your midline' (closed circles in Figure 4.5b) and (ii) measured the relative amount of eye and head displacement used to achieve an eccentric gaze. Lining up with the light produced a bigger effect for most eccentricities (Figure 4.5b) suggesting that the system was not expecting the head to be pointing directly at an eccentric target. Our measurements of actual head and eye positions during eccentric gaze showed that the eyes generally contributed 11% of the gaze and the head 89%.

We conclude that the central mechanism of spatial localization of sounds probably utilizes the body frame. We cannot rule out an external-space frame and are planning further experiments to investigate this possibility.

4.5 Overall conclusions about the use of frames of reference by the brain

These sample experiments on driving and auditory localization illustrate that the fundamental questions of frames of reference can be usefully approached by simple behavioural tasks. We hope that our results will have theoretical consequences in understanding the coding of various aspects of the world by the brain. We hope also that these studies will draw attention to the problems of selection and distinction of reference frames. They might also be useful in identifying steps involved in the specific processes of driving and sound localization. They might also find application for example in devising strategies for high-performance driving and for presenting accurately-localizable sounds to a moving subject.

APPENDIX 1

ANSWER to the four switch puzzle given in the introduction:

It is a trick question. The important point is that no solution codes the reference frame: that has to be decided beforehand! Notice that if the coder and the decoder use different reference frames, predictable but disastrous errors would occur. There is no right answer, but here are some possibilities.

(i) There are 2 positions for 1 switch, 4 combinations for 2, 8 combinations for 3 and 16 for 4. So you could divide the area into 16 zones and identify each with a combination.

(ii) You could use 2 for direction (allowing 4 directions) and 2 for distance (allowing 4 distances). This might be easiest to remember when it came to reading back the switches later.

Acknowledgments

We would like to acknowledge the support of Natural Science and Engineering Research Council of Canada OGP0046271 and Centre for Research in Earth and Space Technologies CRESTech.

References

Crawford, J. D. (1997). "Geometric transformations in the visual-motor interface for saccades", in. M. Fetter, T.Haslwanter, H. Misslisch, &D. Tweed (Eds.), "Three-dimensional kinematics of eye, head and limb movements", 85-99, Harwood Academic Publishers; The Netherlands.

Harris, L. R., Blakemore, C. & Donaghy, M. (1980). "Integration of Visual and Auditory Space in the Mammalian Superior Colliculus", *Nature*, 288: 56-59.

Harris, L. R., Kopinska, A. & Lieberman, L. (1997). "Auditory localization during eye and head movement", *Neurosci. Abstr.*, 23: 261.11.

Harris, L. R. & Lieberman, L. (1996)."Auditory stimulus detection is not suppressed during saccadic eye movements", *Perception*, 25: 999-1004.

Hartline, P. H., Vimal, R. L. P., King, A. J., Kurylo, D. D., & Northmore, D. P. M. (1995). "Effects of eye position on auditory target localization and neural representation of space in the superior colliculus of cats", *Exp. Brain Res*, 104: 402-408.

Hill, A. L. (1972). "Directional constancy", *Percept. & Psychophys.*, 11: 175-178.

Howard, I. P. & Anstis, T. (1974). "Muscular and joint-receptor components in postural persistence", *J. Exp. Psych.*, 103: 167-170.

Jay, M. F. & Sparks, D. L. (1984). "Auditory receptive fields in primate superior colliculus shift with changes in eye position", *Nature*, 309: 345-347.

Kopinska, A. & Harris, L. R. (1998). "Mapping auditory onto visual space: the effect of eye and head position", *Invest. Ophthalmol. and Vis. Sci.*,. 39:(in press).

Land, M. F. & Lee, D. N. (1994). "Where do we look when we steer", *Nature*, 369: 742-744.

Lewald, J. & Ehrenstein, W. H. (1996). "The effect of eye position on auditory lateralization", *Exp. Brain Res.*, 108: 473-485.

Morgan, C. L. (1978). "Constancy of egocentric visual direction", *Percept. & Psychophys.*, 23: 61-63.

Paap, K. R., & Ebenholtz, S. M. (1978). "Perceptual consequence of potentiation in the extraocular muscles: An alternative explanation for adaptation to wedge prisms", *J. Exp. Psychol.*, 2: 457-468.

Peck, C. K., Baro, J. A. & Warder, S. M. (1995). "Effects of eye position on saccadic eye-movements and on the neuronal responses to auditory and visual-stimuli in cat superior colliculus", *Exp. Brain Res.*, 103: 227-242.

Pettigrew, J. D. (1984). "Mobile maps in the brain", *Nature*, 309: 307-308.

5

On the role of time in brain computation

Dana H. Ballard, Garbis Salgian, Rajesh Rao and Andrew McCallum

Abstract

Ultimately computation in the brain must be reduced to the firing patterns of its 10^{10} nerve cells. Each of these cells communicates with stereotypical voltage spikes of one millisecond duration. However in order to synthesize complex behaviors, the brain must organize these firing patterns into temporal hierarchies. We are very used to spatial hierarchies in the brain owing to the preponderance of different anatomical structures at different scales. Herein we develop some of the arguments, first advanced by Alan Newell, that the brain must use temporal hierarchies in computation.

5.1 Introduction

Communicating in physical systems over long distances has costs in time and space.* There is great incentive to organize computation so that most of it is done by communicating *locally*, but in any complex system some long-distance communication will be needed. The only way around this problem is to organize such systems hierarchically. This approach allows one to tailor effects so that they can be reliably accessed at the correct temporal and spatial scale. One can have systems for which local effects can be of great consequence at long scales. Such systems are termed *chaotic*.† But these systems cannot be easily utilized in computation. The bottom line is that for any physical system to be manageable, it must be organized hierarchically.

5.1.1 Constraints of time and space

Hierarchical systems have a fundamental constraint that stems from the organization of space itself. Whenever a system is constructed of units that

* The ideas in this section were originally developed by Allen Newell in his book *Unified Theories of Cognition* (Cambridge, MA: Harvard University Press, 1990).

† Gregory L. Baker and Jerry P. Gollub, *Chaotic Dynamics: An Introduction* (New York: Cambridge University Press, 1990).

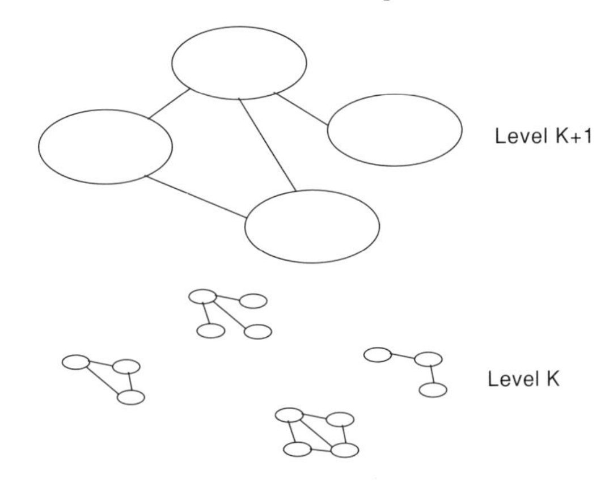

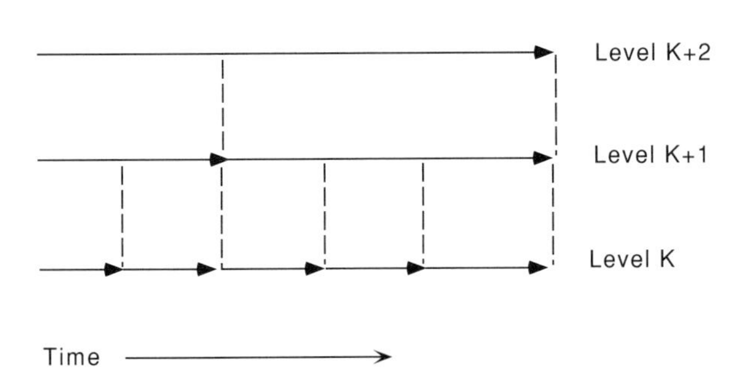

Fig. 5.1. (Top) Spatial scaling in hierarchical systems. Composite units each take the sum of the space of their components. Thus more abstract circuits necessarily take up more space than their components. (Adapted from Newell, 1990.) (Bottom) Temporal scaling in hierarchical systems. Composite units each take the sum of the time of their components. Thus more abstract operations are necessarily slower than their component operations. (Adapted from Newell, 1990.)

are composed of simpler primitives, the more abstract primitives are necessarily larger and slower. The effects of spatial scale are shown in Figure 5.1. Suppose the 0th level of a system is constructed of components of size C that may be thought of as primitives. Then the next level in the system that uses K of these will take up space KC. Moving up the hierarchy, the nth level will cost K^nC.

Newell's scaling relationships are very well fit by brain anatomy. Using $K = 10$ allows the description of structures observed across six orders of magnitude, as shown in Table 5.1.

Time scales in a similar way, as shown in Figure 5.1, but the reason time may not scale geometrically is that temporal costs can be ameliorated with special algorithms. These usually use special connectivity so as to push temporal costs into the spatial domain. For example it is almost certain

Spatial Scale	**Primitive**
1 m	CNS
10 cm	Systems
1 cm	Cortical Maps
1 mm	Networks
100 μm	Neurons
1 μm	Synapses

Table 5.1. *The organization of brain anatomy into structures at different spatial scales (after Churchland and Sejnowski, 1989).*

that the brain uses parallel connections between neurons to compute over the retinotopic array in parallel. This is an example of using spatially distributed hardware to make computations go faster. However to interpret these results, the overall system must be encapsulated in some way. When this is done the temporal scaling constraint can be applied. Thus instead of a cost of tM for level n, the actual cost may be $T^n M$, where $T = f(t)$ is the temporal scaling owing to the special algorithm. The point is that it is still an exponential growth function. The computation of more abstract entities must slow down at exponential rates.

We have just seen that there are fundamental spatiotemporal constraints on hierarchical systems. In a hierarchical system, the more abstract components run slower and take up more space at geometric rates. In the rest of this paper we concentrate on the temporal hiearchy. Our point will be that the brain is best understood in terms of separate functions that operate at separate time scales.

5.1.2 Cognitive hierarchies

The temporal constraints on levels in a hierarchical system can be interpreted in terms of biology.* The most fundamental constraint for the human brain is the communication system between neurons. Neurons communicate by sending electrical spikes that take about one millisecond to generate. As a result, the circuitry that uses these spikes for computation has to run slower than this rate. Let us assume that about ten operations are composed at each level. Then local cortical circuitry will require 10 ms. These operations are in turn composed for the fastest "deliberate act." A primitive deliberate act is then 100 ms. Measurements of perception show that such acts take about 50-100 ms, so the 100 ms estimate is in the ballpark. The next level is the physical act. The shortest such act is an eye movement which

* Allen Newell's *Unified Theories of Cognition* again.

Though the cost of developing algorithms may be amortized over a much longer period, ranging from years for developmental learning to centuries for genetic learning, the timescale for their execution is tightly constrained to lie within a range of 100 ms to 10 seconds.

Temporal Scale	**Primitive**	**Example**
10 sec	Complex task	Moving in speed chess
2 sec	Simple task	Saying a sentence
300 ms	Physical act	Moving the eyes
80 ms	Memory act	Noticing a stimulus
10 ms	Neural circuit	
2 ms	Neuron spike	

Table 5.2. *The organization of human computation into temporal bands (after Newell, 1990).*

typically takes on theorder of 300 milliseconds. At the next level we have a primitive task such as the uttering of a short sentence. Composing these results is a complex task. A good example of a complex task is a chess move. Speed chess is played at about 10 seconds per move.† Table 5.2 shows these relations.

It is important to understand the fundamental sense behind these scaling relationships. An acquaintance who was a chess fanatic and also very good with circuits bought a chess computer and turned up its clock by a factor of two. The computer still functioned and played faster. However he was lucky. Designers of computers are limited by the switching times of the transistors. There must be enough time for the transistor to switch states before it is polled. If it is polled too soon, when the transistor is in an uncertain state, catastrophes would result. This notion of catastrophe can serve as an important guide to our intuition of behaviors organized along temporal scales. If one has less than the requisite ten seconds per move, one's game falls apart. Similarly for the next two scales. One can only talk so fast, and the neural circuitry that moves the eyes must have sufficient time to compute their target.

In the rest of the paper we attempt to enflesh the reality of these timescales by describing modeling efforts that encapsulate struture at the different timescales.

† Another reason that the modeling methods here are different than traditional symbol manipulation in AI is that the timescales are much shorter, too short to form a symbol.

5.2 The cortex as an 80ms memory

There are many results that suggest that the cortical memory has a fundamental time constant in the range of 50-100 milliseconds. To describe a few:

(i) Subjects can readily name letters presented sequentially on a CRT display. However the naming can be blocked by a one Tesla magnetic field applied to the cortex 50 ms after letter onset.
(ii) The response of cortical cells can be modulated by stimuli presented outside of their classical receptive fields. However Zipser, *et al.* (1996) have shown that this modulation takes about 100ms to appear, implying it is a product of cortical memory feedback.
(iii) Subjects viewing a moving bar and an aligned flashed bar simultaneously perceive the former as advanced in position by the time traversed in 50 ms as shown by Nijhawan (1994).

The fundamental problem that the cortical memory must solve is that of self-calibration. It must continuously encode particular memories and have a metric for recording the progress of encoding of such memories. Neurons in the visual cortex encode information in a highly selective fashion, typically responding only to a limited range of input stimuli such as moving bars of a particular orientation, velocity, and direction of movement (Hubel & Wiesel, 1962, 1965). The first model proposed to explain the structure of this code was by Barlow (1987). Barlow's idea was that the job of the cortex was to decorrelate the incoming signal to make it more useful for subsequent processing. Since then very specific models have been studied for example (Dobbins *et al.*, 1987; Hancock *et al.*, 1992). focusing mainly on geniculocortical/corticocortical feed-forward connections. More recently very specific properties of striate cortex have been reproduced with a generative system by Olshausen and Field (1996). This theory uses a sparseness function that strikes a compromise between decorrelation and redundancy in the resultant neural encoding. The Olhausen and Field theory is extremely important advance in that it can reproduce the distribution of the response properties of simple cells in striate cortex. However it leaves much to be done.

(i) It does not extend readily to complex cells.
(ii) It does not address the complete circuitry of cortex including a) hierarchies and b) feed-forward and feedback connections.
(iii) It stops short of explaining why biology picked that particular metric.

These difficulties can be addressed by situating the choice of a metric within the Minimum Description Length Principle (MDL). The MDL principle is a breakthrough in the understanding of theories as it provides a way of making different theories commensurate. The way this is done is to trade-off the accuracy of the theory against its size. The accuracy of a theory can be measured in terms of the errors or *residuals* of the theory's predictions. The size of the theory can be measured directly in terms of its length. Both these quantities can be measured in terms of information bits.

The MDL principle can be seen as directly related to cortical memories, in that the cortex can be seen as the embodiment of a theory about itself. It grounds this theory by trying to predict its input. The goodness of the theory is measured in terms of the accuracy of the predictions balanced by the cost of making them. This cost, or in other words the length of the theory, is measured in terms of the parameters of the cortical memory. A natural choice for these parameters are the synapses and firing rates of cortical cells. We still have to understand the motivation for a particular metric, as there are many ways of encoding the cost of the cortex as a theory, but the MDL principle leads to a new understanding of the role for such a metric: it measures the cost of the cortex as a theory.

Our model is based on a particular form of the MDL principle. Its central feature is that it contains functional roles for both feedback and feed-forward connections. We posit that the cortex encodes memories by trying to predict its input. In the hierarchy, feedback signals carry top-down predictions of lower level inputs from higher, more abstract areas. The feed-forward connections convey the residuals (MacKay, 1956; Barlow, 1994; Mumford, 1994) (or differences) between the current input and its prediction from the higher area. The residuals provide an error signal that can be used to adjust the synapses that are the basis for the underlying memories. The computationally attractive idea of encoding differences between an input signal and its prediction from a preexisting internal model is captured succinctly by the statistical construct of the *Kalman filter* (Kalman, 1960; Maybeck, 1979) from classical control theory. The Kalman filter model can be used both to adjust the neural firing rates in the course of making a specific prediction, and, on a longer time scale, to adjust the synapses. The combined prediction and synaptic learning scheme generalizes a number of previous estimation/learning methods (Williams, 1985; Daugman, 1988; Oja, 1989; Pece, 1992; Olshausen & Field, 1995, Harpur & Prager, 1995).

5.2.1 The model in outline

Assume that an image, denoted by a vector $\mathbf{I}$ of n pixels, can be represented as a linear combination of a set of k basis vectors $U_1, U_2, \ldots, U_k$. These basis vectors can be written in matrix form as:

$$\mathbf{I} = U\mathbf{r} \tag{5.1}$$

where U is the $n \times k$ matrix whose columns consist of the basis vectors U_j and $\mathbf{r}$ is the $k \times 1$ vector consisting of coefficients r_j.

The goal is to estimate the coefficients $\mathbf{r}$ for a given image and, on a longer time scale, learn appropriate basis vectors in U. A standard approach is to define a least-squared error criterion of the form:

$$E_1 = (\mathbf{I} - U\mathbf{r})^T(\mathbf{I} - U\mathbf{r}) \tag{5.2}$$

Note that E is simply the sum of squared pixel-wise errors between the input $\mathbf{I}$ and the image reconstruction $U\mathbf{r}$. Estimates for U and $\mathbf{r}$ can be obtained by minimizing E.

Unfortunately, in many cases, minimization of a least-squares optimization function such as E without additional constraints generates solutions that are far from being adequate descriptors of the true input generation process. For example, a popular solution to the least-squares minimization criterion E is principal component analysis (PCA). PCA optimizes E by finding a set of mutually orthogonal basis vectors that are aligned with the directions of maximum variance in the input data, but does not take into account the cost of representing model paprameters. To take into account the prior distributions of the parameters $\mathbf{r}$ and U, one can minimize an optimization criterion of the form:

$$E = E_1 + \mathbf{r}^T L\mathbf{r} + \lambda||U||^2 \tag{5.3}$$

where $\mathbf{r}^T L\mathbf{r}$ and $g(U)$ are terms related to the prior distributions of the parameters $\mathbf{r}$ and U. Strictly speaking, these regularization functions are not as good an approximation as those of Olshausen & Field (1996) who used functions of the form $f(x) = \log(1 + x^2)$, $f(x) = |x|$ and $f(x) = -e^{-x^2}$ to encourage sparseness in $\mathbf{r}$. However our simpler functions readily extend to interconnected networks.

For a given set of basis vectors U, we can minimize E with respect to $\mathbf{r}$ using gradient descent (Churchland & Sejnowski, 1992) to obtain the following dynamics for estimating object identity and transformation:

$$\dot{\mathbf{r}} = -\frac{k_1}{2}\frac{\partial E}{\partial \mathbf{r}} = k_1 U^T(\mathbf{I}(\mathbf{x}) - U\mathbf{r}) - k_1 L\mathbf{r} \tag{5.4}$$

where $\dot{\mathbf{r}}$ represents the temporal derivatives of $\mathbf{r}$, T denotes matrix transpose, and k_1 is a positive time constant of the dynamics that determine the rate of descent towards the minima of E. Thus, given a transformed image $\mathbf{I}(\mathbf{x})$, one needs to compute the residual error between the input $\mathbf{I}(\mathbf{x})$ and its prediction $U\mathbf{r}$ which was made using the internal model given by U.

For specific object vectors $\mathbf{r}$, one can minimize E with respect to the object basis matrix U to obtain the following learning rule for adapting the synaptic efficacies represented by the U matrix:

$$\dot{U} = -\tfrac{c_1}{2}\tfrac{\partial E}{\partial U} = c_1(\mathbf{I}(\mathbf{x}) - U\mathbf{r})\mathbf{r}^T - c_1\gamma U \tag{5.5}$$

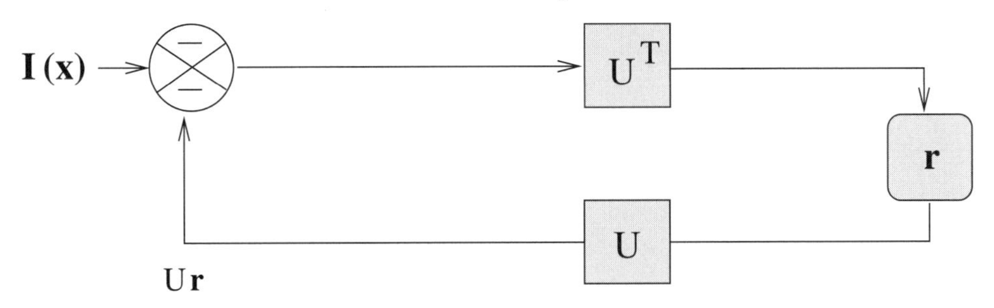

Fig. 5.2. The basic circuitry of the predictive model can be modularized into separate feed-forward and feedback pathways. The learning rule results in the projective(feedback) synapses being the algebraic transpose of the receptive(feed-forward) synapses. the feed-forward pathway just carries the residual error between the predictions and the input.

where $\dot{U}$ represents temporal derivatives, c_1 is a positive time constant that determines the learning rate.

5.2.2 Extending the model to hierarchies

In the development of the system equations, it was helpful to think of the input as being in terms of an image, but there is no fundamental reason for this. Thus it can be just as practical to think of the responses $\mathbf{r}$ as a kind of "image" also. In that case there can be virtue in encoding their responses: the overall structure of the cortical machinery as measured in terms of the bits needed to encode synapses and responses might be reduced. Thus if $\mathbf{r}_1$ is the number of responses of cortical area 1 which shares connections with area 2 higher in the hierarchy, one can use the responses of the second area to try and predict those of the first area in the same way as was done to encode the original image. To do this, approximate the responses $\mathbf{r}_1$ by

$$\mathbf{r}_1 = U_2\mathbf{r}_2 \qquad (5.6)$$

The mathematical treatment is identical to that of the original encoding, Equations 5.4 and 5.7, however one additional problem is to specify the feed-forward synapses, as now the requirement is for a complete circuit. it turns out that the feed-forward synapses can be trained with a variant of the feedback rule.

$$\dot{W} \quad = c_1(\mathbf{I}(\mathbf{x}) - U\mathbf{r})\mathbf{r}^T - c_1\gamma W \qquad (5.7)$$

As shown by (Williams, 1985), such a rule results a solution $U = W^T$.

The ability to model hierarchies allows one to make contact with experimental data. The results can be surprising. For example we show in simulation that modulation in a neuron beyond the classical receptive field

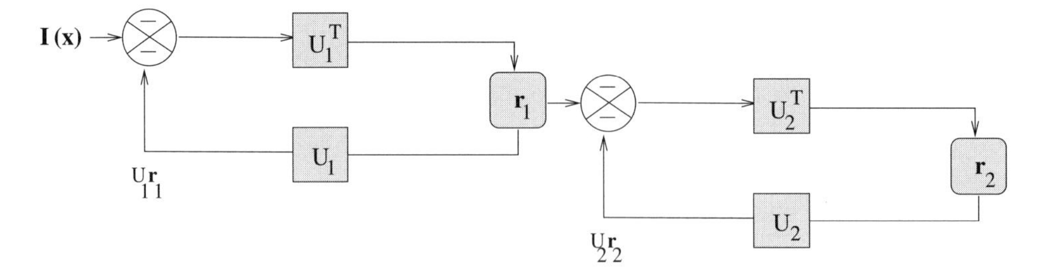

Fig. 5.3. Using the module concept, it is simple to visualize the effect of hierarchical connections. The responses at one level in the hierarchy are further encoded by the subsequent level.

can be explained in terms of the MDL principle. Consider the effect of end-stopping observed in cortical area V1. A neuron that responds to a bar stimulus can have that response attenuated if the bar extends beyond the classical receptive field. We show in simulation that this effect can be seen as a network effect. Neurons in a simulated V2 attempt to predict the response of neurons in V1, and thus that neuron need only send the residual. As the bar extends beyond a V1 neuron's receptive field, they are able to do this more effectively, resulting in lower residuals.

Since the equations are all linear, one might wonder why hierarchies in this instance are a virtue. The reason for skepticism is the well-known result that linear hierarchical networks can be collapsed into a single-level equivalent network. However there are a number of virtues in preserving the hierarchy. The foremost is that it turns out that even though connection between areas are linear, it is possible to build the hierarchy in such a way that *within* the area, the responses can be non-linear. This will be demonstrated when we tackle the issue of time in the predictive process.

In order to validate the model, we trained a two-level hierarchical network on a sample of natural images (Figure 5.2A). By considering interactions between top-down and bottom-up signals as captured by the residuals in the network, we were able to model complex receptive field (RF) properties analogous to those observed in the cortex. We consider here the classical phenomenon of *endstopping*: neurons that respond vigorously to a bar have diminished responses as the bar extends beyond the classical RF (Hubel & Wiesel, 1965). Endstopping has previously been characterized as a feed-forward effect caused by lateral inhibition from neighboring cells (Hubel & Wiesel, 1965; Dobbins, Zucker & Cynader, 1987), but results from the model suggest that end-inhibition in a lower cortical area such as V1 may be explained more generally as occurring due to predictive feedback from a higher area such as V2. As shown in Figure 5.5, the typical response of a model neuron (solid line) drops off sharply as a test bar extends beyond the classical RF. The attenuation in response can be attributed to diminishing

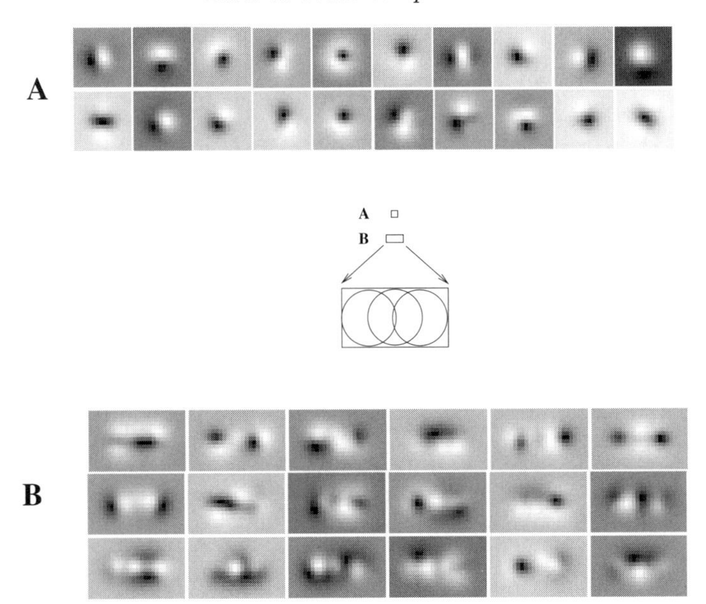

Fig. 5.4. A two-level network. The second level has larger receptive fields that encompass three smaller receptive fields as shown diagrammatically. The lower set of receptive fields shows the code used by the second level in the hierarchy.

level 1 residuals in the hierarchical predictor network caused by better predictions from the higher level as stimulus length is incrementally increased up to the size of the larger RF of the level 2 model neurons (see Figure 5.4 for an example). Such an explanation is supported by the RF profiles of level 2 neurons (Figure 5.2C), some of which appear to be tuned towards long line segments. Further confirmation of the role of feedback in endstopping was obtained by disabling the feedback connections from level 2 to 1, which eliminated endstopping in most of the level 1 model neurons (Figure 5.5A and 5.5B).

Though not explicitly trained on bars, the model exhibited strong endstopped responses as a consequence of the statistical structure of natural images as captured by the MDL-based optimization function and its hierarchical Kalman filter implementation. The endstopped model responses were strongly influenced by top-down feedback. End-inhibition was quantified as the percentage difference between peak response and average plateau response for lengths greater than 18 pixels: (peak - plateau)/peak$\times$100. Model neurons were classified into 10 categories according to their degree of end-inhibition, with 100% inhibition denoting a plateau response of zero to long bars. As shown, disabling the feedback connections eliminated endstopping (defined as greater than 50% inhibition) in 84% of the layer 2+3 model neurons.

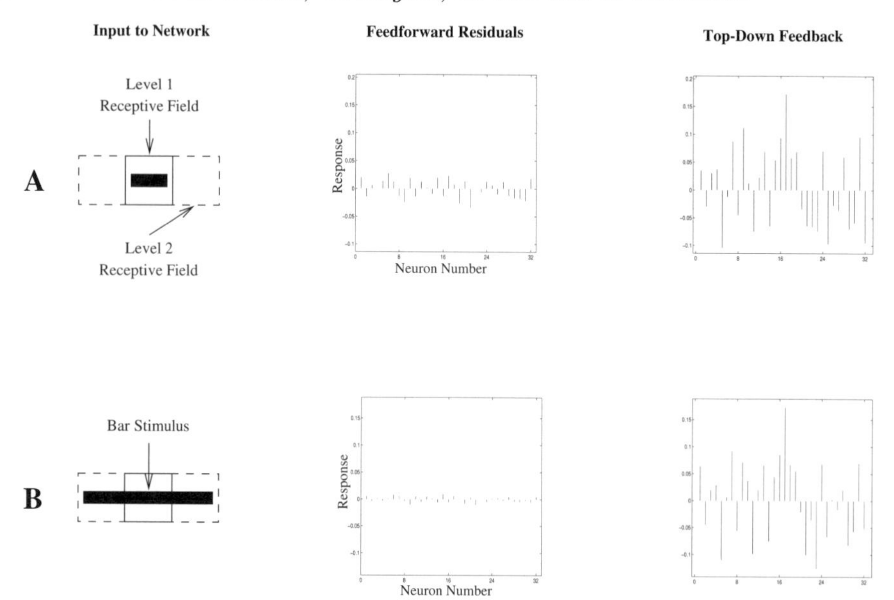

Fig. 5.5. The endstopping effect is illustrated by this network simulation. Neurons trained on natural images are also sensitive to bar patterns and thus respond to the bar pattern when provided as input. (A) In the hierarchical network the level two model neurons attempt to predict the response of the level one model neurons to a short bar but have only limited context. (B) When the bar is longer the responses of flanking level one model neurons provides information that alters the prediction of level two. Thus the feed-forward response of the central level one model neuron is reduced.

5.3 Visual routines

The 80 ms timescale allows the cortex to organize itself as a memory, but to *use* this memory, memory states must be organized into larger functions. The structure of such functions takes more time as memory operations typically have to be composed for some encompassing purpose. As little as a decade ago, it was widely accepted that the visual world could be completely segmented into constituent parts in the memory prior to analysis. However the complexity of vision's initial segmentation can easily be unbounded for all practical purposes (Tsotsos, 1989), so that the goal of determining a complete segmentation of an individual scene in real time is impractical. Thus to meet the demands of ongoing vision, the focus has shifted to a more piece-wise and on-line analysis of the scene, wherein just the products needed for behavior are computed as needed.

The on-line analysis of a time-varying scene cannot, in practise, be done with a single special-purpose program since there is still the problem of dealing with the goals of the scene analysis which can be many and varied. A solution to this is to have a library of *visual routines*. The concept of visual routines was first developed by Kosslyn (1973) but the rationale for their use was elaborated by Ullman (1984). The central idea is to have a collection of routines that represent different kinds of basic image processing sub-functions. These then can be composed to subserve more elaborate goal-directed programs. The crucial compositional capability allows the visual routines to span the huge space of different task goals.

Although the idea of visual routines is compelling, and they have been used in several simulations (Chapman, 1991; Reece, 1992), so far they have been used in image analysis only in a few restricted circumstances (Johnson *et al.*, 1994). Dickmanns *et al.* (1990) mentions among the advantages of his spatio-temporal approach "intelligent nonuniform image analysis ... allowing to concentrate limited computer resources to areas of interest". Burt defined smart sensing as the "selective, task oriented, gathering of information from the visual world" (Burt, 1988). The framework gives general guidelines for limiting visual searches, but it does not present specific methods for choosing the regions of interest for further analysis. Here we illustrate such functions with the central example of deploying gaze in a visual search task.

Human vision relies extensively on the ability to make saccadic eye movements. These rapid eye movements, which are made at the rate of about two or three per second, orient the high-acuity foveal region of the eye over targets of interest in a visual scene. The characteristic properties of saccadic eye movements (or saccades) have been well studied, and the neural circuitry underlying the execution of the movement comparatively well understood (see, eg Viviani, 1990).

Starting from Yarbus' classical work (Yarbus, 1967), a number of studies have suggested that gaze changes are most often directed according to the ongoing cognitive demands of the task at hand. The task-specific use of gaze is best understood for reading text (O'Regan, 1990) where the eyes fixate almost every word, sometimes skipping over small function words. In addition, it is known that saccade size during reading is modulated according to the specific nature of the pattern recognition task at hand (Kowler & Anton, 1987). Tasks requiring same/different judgments of complex patterns also elicit characteristic saccades (Just & Carpenter, 1976). In chess, it has been shown that saccades are used to assess the current situation on the board in the course of making a decision to move, but the exact information that is being represented is not yet known (Chase & Simon, 1973). In a task involving the copying of a model block pattern on a board, subjects have been shown to employ fixations for accessing crucial information during different stages of the copying task (Ballard, Hayhoe, & Pelz, 1995, Ballard *et al.*, 1996). In natural language processing, there is recent evidence that fixations reflect the instantaneous parsing of a spoken sentence in the current visual context (Tanenhaus *et al.*, 1995). The role of gaze has been studied by Land & Furneaux (1997) in a variety of other visuo-motor tasks such as driving (see Land's chapter, this volume), music reading and playing ping pong. In each case, gaze was found to play a central *functional* role, closely linked to the ongoing task demands.

Given these factors, what possible computational mechanisms underlie the control of gaze? What properties of the pre-saccadic peripheral stimuli are used in the target selection process? A model of these mechanisms is becoming increasingly important in the light of the central nature of gaze control in visual cognition, and also in the context of the growing numbers of neurophysiological studies that reveal task-specific attentional effects on neurons involved in the visuo-spatial machinery (eg Duhamel, Colby, & Goldberg, 1992). To support task-directed vision, such a model must meet at least the following four criteria:

(i) **Generality**: Any proposed mechanism for targeting parts of an image must have broad generality since saccadic targets can vary greatly according to the requirements of the current task, and must necessarily use only partial information about the target.

(ii) **Speed**: Targets must be computed quickly in order to model observed human performance. Using millisecond neural circuitry, the targets for the next fixation need to be computed in approximately 80-100 milliseconds, allowing barely one pass through the cortex (Thorpe & Imbert, 1989; Oram & Perrett, 1992).

(iii) **Resolution**: The computation of the target must use spatial scales that are available extrafoveally. Approximately 6 cycles per degree are available.

(iv) **Spatial Memory**: Since the retinotopic relationship of targets to the point of gaze changes with each saccadic eye movement, the gaze control mechanism must include some form of spatial memory for operations such as remembering, inhibiting, and other forms of book-keeping associated with saccadic targets.

5.3.1 Iconic representations

The representation used in our model is guided by the first three of the four criteria listed in the introduction: it should be able to represent general targets in arbitrary natural scenes, it must be computed quickly, and it should use low resolution information. These criteria are met by employing low resolution iconic representations of targets and scenes that can be extracted directly from the optic array. This allows general portions of a scene to be represented in a pre-categorical format without requiring any elaborate segmentation. This is an essential property, since the information required for such complex operations is frequently the goal of the eye movement itself.

R	G	B	
0.3333	0.3333	0.3333	R+G+B
0.5000	-0.5000	0.0000	R-G
-0.2500	-0.2500	0.5000	B-Y

(a)

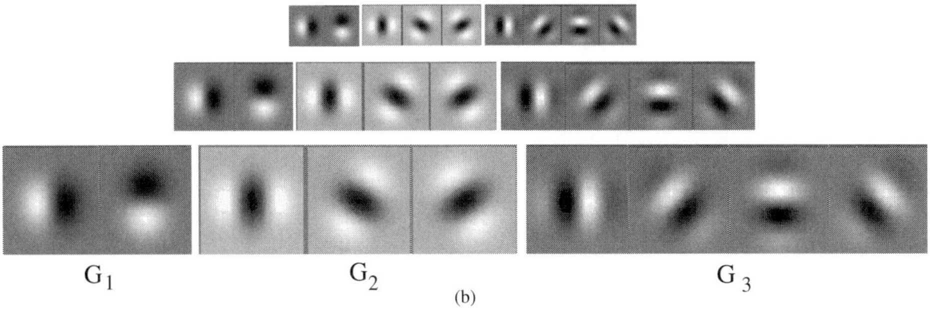

G_1 G_2 G_3

(b)

Fig. 5.6. **Spatiochromatic Basis Functions**. Motivation for these basis functions comes from statistical characterizations of natural image stimuli (Derrico & Buchsbaum, 1991; Hancock *et al.*, 1992; Olshausen & Field, 1995; Rao & Ballard, 1997; Bell & Sejnowski, 1996). (a) shows the weights assigned to the three input color channels, generating a single achromatic channel (R+G+B) and two color-opponent channels (R-G and B-Y). (b) shows the nine oriented spatial filters at three octave-separated scales for each of the three channels in (a) (bright regions denote positive magnitude while darker regions denote negative magnitude). At each scale, these nine filters are comprised of two first-order derivatives (G_1) of a 2D photometric Gaussian, three second-order derivatives (G_2), and four third-order derivatives (G_3). Thus, there are three color channels, three scales per channel, and nine spatial filters per scale, for a total of 81 filter responses characterizing each location in the image. These 81 spatiochromatic measurements at a given image location can be regarded as a photometric signature of the local image region centered at that location.

The computation of saccadic target coordinates is accomplished by correlating the iconic memory of the target with the iconic representation of the current optic array. A correlation peak indicates the most likely location of the target in the current image, allowing a saccade to be executed to that location.

It would be prohibitively expensive to encode icons in their literal form of raw images, since the memory needed would then scale with the size and number of icons. A more efficient alternative is to encode the icons using their responses to a set of *spatio-chromatic basis functions*. For example, color can be represented by a single achromatic and two color opponent channels (Figure 5.6a). Along each of these channels, a local image patch can be characterized using a zeroth order Gaussian G_0 and nine of its oriented derivatives (Figure 5.6b) (Freeman & Adelson, 1991).

Figure 5.7 shows the filter-based representations at three different loca-

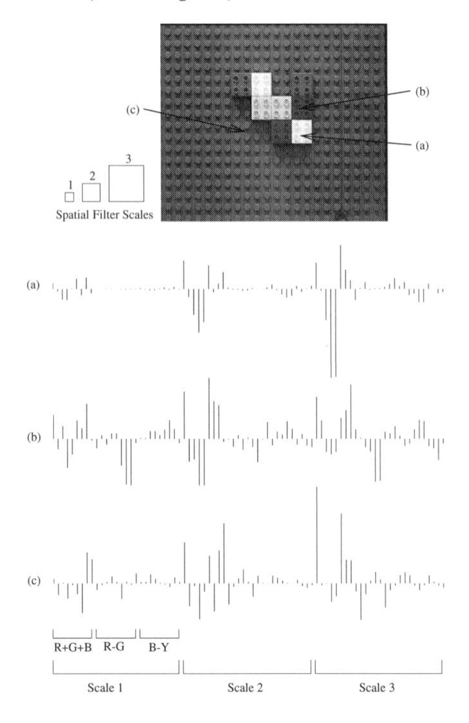

Fig. 5.7. **Using Spatiochromatic Filters to Extract Task-Dependent Properties**. A portion of the blocks image used in the copying task is shown at the top. The three scales at which the filters of Figure 5.6 were applied to the image are shown on the left. (a) The filter responses for a location on a white block. Each individual filter, when convolved with the local image intensities near the given image location, results in one measurement, for a total of 81 measurements per image location. Positive responses in the vector are represented as an upward bar above the horizontal, negative responses as a downward bar below the horizontal. As expected, the vector for the white block has many low responses due to the color-opponent channel coding. (b) The filter response vector for a location on a red block. (c) The filter response vector for a location in the green background.

tions on a display of Duplo blocks. Despite the similarities in the display, the responses of the three points are very different from each other.

The use of multiple scales is crucial to the visual search model. In particular, the larger the number of scales, the greater the perspicuity of the representation as depicted in Figure 5.8, which shows the frequency distribution of correlations between all points in a dining table image (like the one in Figure 5.10a) and a fixed target point in the same image.

Figure 5.9 illustrates the simple correlation algorithm in a search task. Gaze, as denoted by the cross-hairs, is first directed to a given scene location as shown in (a). At that point the filter responses are memorized. Next, at some point in the course of the rest of the behavior, it may be desirous to return to the original location from a distal point. The targeting algorithm is used to correlate the memorized features with the current retinotopic

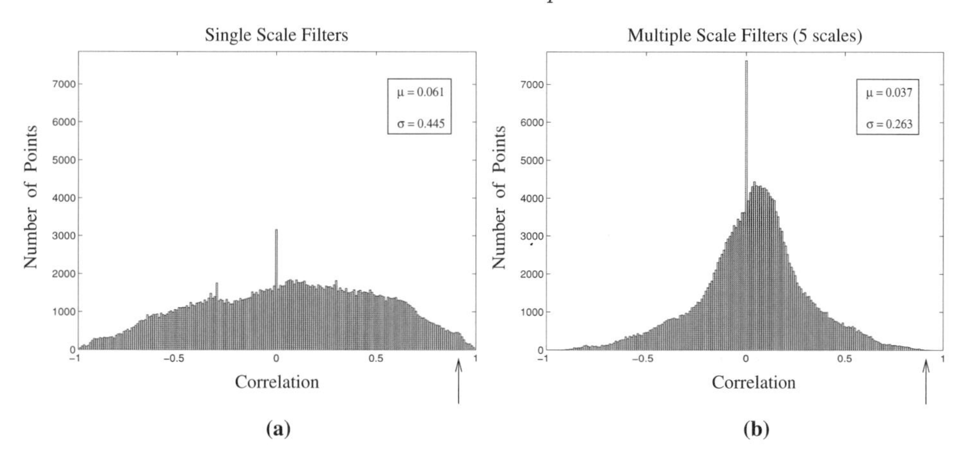

Fig. 5.8. **The Effect of Scale.** The distribution of distances (in terms of correlations) between the filter response vector for a selected target point in the dining table scene and all other points in the scene is shown for single scale response vectors (a) and multiple scale vectors (b). Using responses from multiple scales (five in this case) results in greater perspicuity and a sharper peak near 0.0. The most important feature of these plots appears at the extreme right hand side. Only one point (the target point) has a correlation greater than 0.94 (demarcated by an arrow) in the multiple scale case (b) whereas 936 candidate points fall in this category in the single scale case (a).

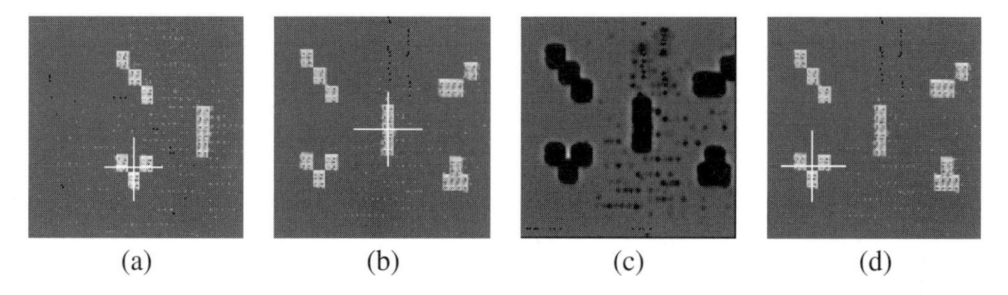

Fig. 5.9. **Visual Search Using Spatial Filter Responses**. The simplest form of the visual search model is based on winner-take-all correlation matching. (a) At a given location, the filter responses are remembered. (b) Next, gaze is transfered to another point. The search problem is to find the original location in this new view. (c) The saliency map, showing the highest correlation value (brightest point) at the original location. (d) Gaze is transfered back to that location.

image, resulting in a saliency map as shown in (c). Note that the coordinate system of the saliency map can also be interpreted in terms of "saccadic motor error." Thus, the saliency peak can be used to drive the oculomotor command for returning the eyes to the original target.

5.3.2 Fixation patterns in visual search

To compare the model's performance with human search and targeting behavior we used the data from eye movements in a visual search task described

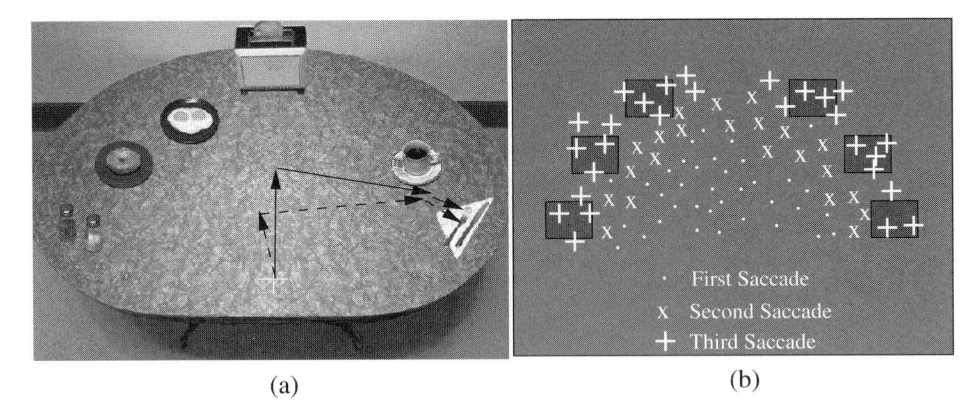

Fig. 5.10. **Eye Movements in the Visual Search Task**. Measurements from actual human data show marked differences from the simple winner-take-all model. (a) shows the typical pattern of multiple saccades (shown here for two different subjects) elicited during the course of searching for the object composed of the fork and knife. The initial fixation point is denoted by '+'. (b) depicts a summary of such movements over many target-present search trials as a function of the six possible locations of a target object on the table.

in (Zelinsky *et al.*, 1996). In this experiment, fixation patterns were observed in a simple search paradigm using natural images. The task was designed to evoke natural eye movement performance while maintaining control over some of the stimulus properties. Subjects were asked to fixate a point near the bottom of a 12×16 degree display. They were given a one second presentation of an image containing a single object (e.g. a tool) at the fixation point, defining the search target. This was followed approximately one second later by a scene that filled the display and contained one, three, or five objects (e.g. various tools) on an appropriate realistic background (e.g. work bench). The objects themselves subtended about $2°$ of visual angle. The subjects were asked to indicate (by pressing a button), as quickly and accurately as possible, whether the previewed object was among the group of one to five objects in the subsequent view.

The typical eye movements elicited in this particular task are shown in Figure 5.10a. The surprising result was that rather than a single movement to the location of the memorized target, several saccades are typical, with each successive saccade moving closer to the target location (Figure 5.10b). This "skipping" of the saccades in this search paradigm proved to be an extraordinarily robust finding, occurring in almost all trials across all four subjects (Zelinsky *et al.*, 1996). The model as described cannot account for these fixation patterns, since the winner-take-all strategy means that only a single target is computed. However, if this assumption is relaxed, multiple fixations can be fairly easily accounted for within the context of the model.

Multiple "skipping" saccades to a target will occur if the computation of the saliency map has three properties:

(i) Saliency map computation was slower than the time needed to make an eye movement. This would imply that eye movements are made to target locations as determined by the *current* state of the saliency map, rather than waiting until the final state has been computed.
(ii) The saliency map was computed using the larger spatial scale filters first, adding saliency information from successively finer scales as the search process evolves over time. This might naturally occur as a consequence of the absence of the high spatial frequency information when the target lies in the peripheral retina.
(iii) The most likely target location was computed using a weighted averaging scheme. In conjunction with (i) and (ii) above, this would imply that early eye movements are directed to "center-of-gravity" locations since only coarse scale information regarding the objects and the background is available at the early stages of the search, thereby biasing the weighted averaging model towards the center of the scene.

The new targeting model was implemented on a pipeline image processor, the Datacube MV200, which can compute convolutions at frame rates (30/sec). Figure 5.11 shows the saliency maps for this image after each of three iterations, with the middle and highest frequencies included in (b) and (c) respectively. Part (d) of the figure shows the sequence of fixations generated by the model for this image, together with those from a human subject. The target (composed of the fork and the knife) was the same in both cases.

5.4 Programs

The functions described in the previous section can be thought of on a still longer timescale, such as a few seconds, as actions in a program that directs behavior. The mathematics that is useful for describing such programs is based on Markov models. Markov models allow the description of processes in the world, but they are passive. They cannot model an agent's actions that change the world. That task requires an extension of the model to incorporate the different actions available to the agent. This extension is termed a *Markov decision process (MDP)*. A Markov decision process differs from a Hidden Markov Model (HMM) in that actions, though they are stochastic, are chosen by the user.

MDPs incorporate the bookkeeping needed to keep track of user-controlled actions. Modeling such actions introduces three factors:

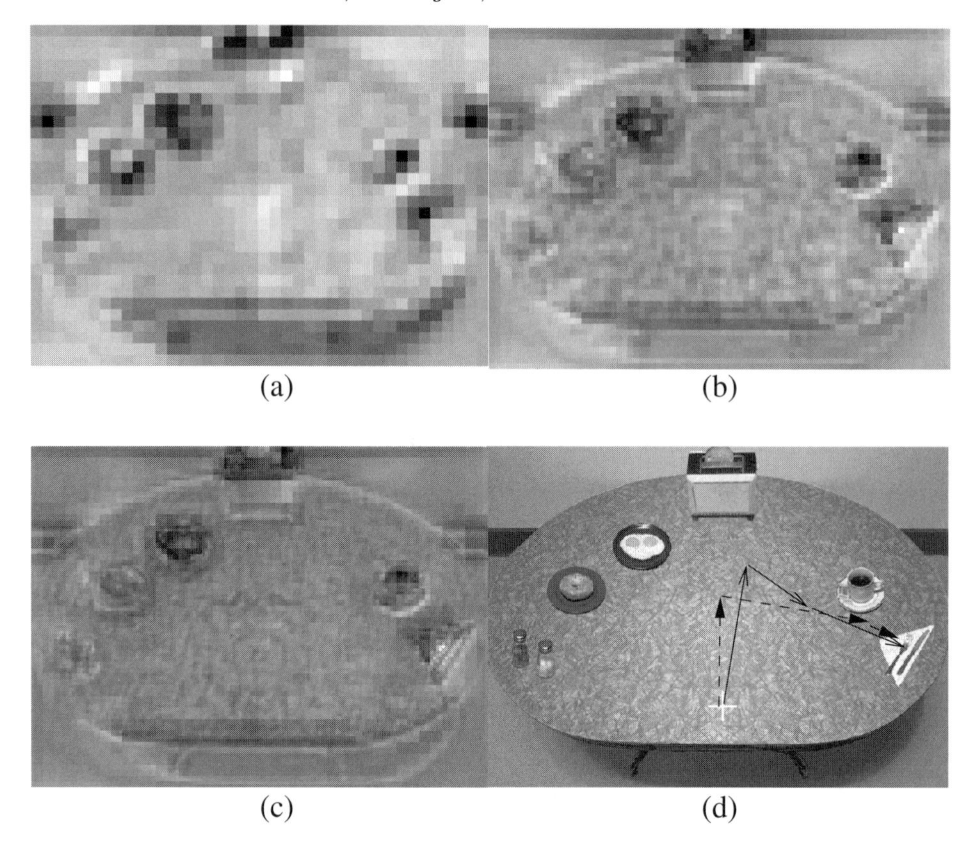

Fig. 5.11. **Illustration of Coarse-to-Fine Saccadic Targeting.** The saliency map $S(x, y)$ after the inclusion of the largest (a), intermediate (b), and smallest scale (c) as given by filter response distances to the prototype (the fork and knife); the brightest points are the closest matches. (d) shows the predicted eye movements as determined by the weighted population averaging scheme. For comparison, saccades from a human subject are given by the dotted arrows.

(i) The model is unknown but must be discovered. The equations of motion are only learned by applying forces and observing their consequences.

(ii) The utility of an action must also be learned by trying it out and seeing what happens.

(iii) The success or failure of the control policy is not known immediately, but only after the pole is horizontal or the cart hits the stop (failure), or when the pole is vertical (success).

The mathematics to handle delayed rewards is termed *reinforcement learning*. As emphasized in the introduction to this section, delayed rewards are a fundamental property of the world, and dealing with them successfully is vital to survival. Reinforcement learning does so in a direct way, creating a discrete state space that is a model for the world and then associating

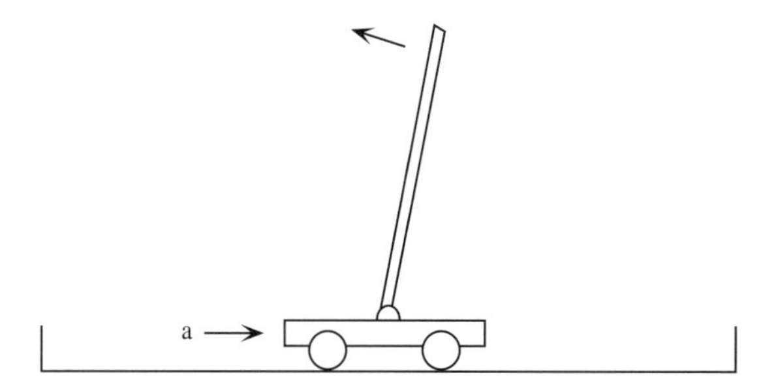

Fig. 5.12. The cart and inverted pendulum problem. A cart moves on a one-dimensional track. Attached to the cart is a pendulum that is to be balanced. The control of the cart and pendulum is in the form of a discrete set of horizontal impulses $u = \pm a$ that are applied to the cart.

a *utility* for each state that reflects the possibility of getting a reward by taking actions from that state.

5.4.1 Learning state information from temporal sequences

The key element in an MDP is the definition of the state space. One way to define such a space is to rely on the sequences of observations encountered in real-world experience. When faced with making a decision, interrogate a library of previous experiences in which similar situations came up, and make the consensus decision. This method is shown in Algorithm 1. The library is built by recording the agent's experience as a sequence of states

$$\boldsymbol{x}_i, i = 1, \ldots, t \tag{5.8}$$

where t is the current time. Associated with each state, is the triple o_t, a_t, r_t denoting, respectively, the observation in state $\boldsymbol{x}_t$, the action taken, and the reward received. In the earlier formulation, there are several actions possible from each state, and each has an associated Q-value. Here, the principal data structure is the record of what actually happened with one action per state. Thus there is only one Q-value, q_i, per state.

To understand how the algorithm works, consider the analogy with a k-nearest neighbor strategy, as shown in Figure 5.13. In that strategy, to decide what to do in a given state, the k nearest neighbors in state space are interrogated. The states are represented abstractly as points in a hypothetical state space. Associated with each such point will be a policy. In the figure $k = 3$, and the three closest neighbors are indicated with shading. The action taken is a function of their policies; for example, the most frequent action may be taken. The k-nearest sequence strategy works similarly. The agent's history is examined for the k most similar temporal state trajectories. Here again $k = 3$, and the three most similar sequences are indicated

Learning in a Geometric Space

k-nearest neighbor, $k = 3$

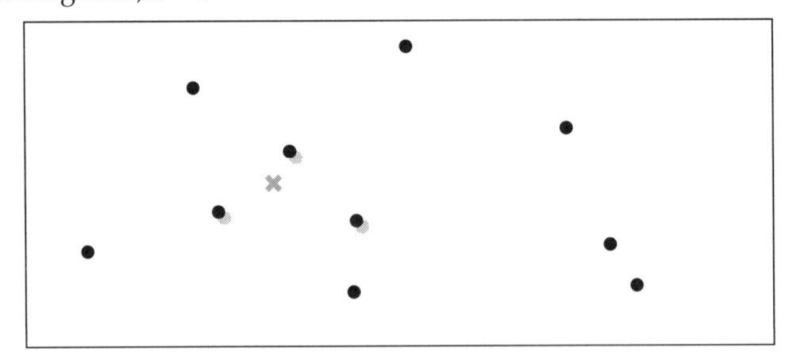

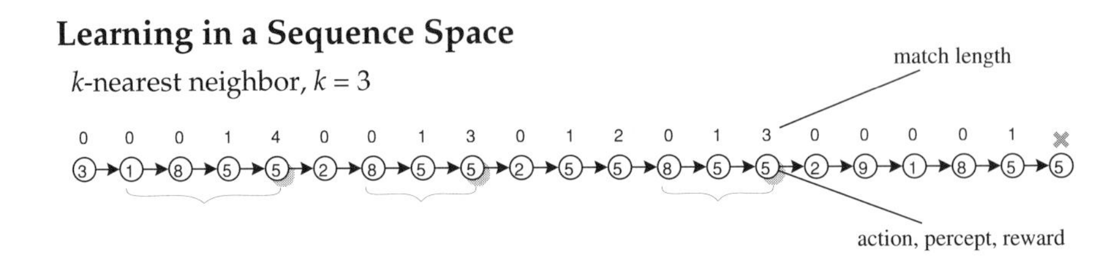

Fig. 5.13. Using k-nearest sequence memory for state identification. (*upper*) In k-nearest neighbor the action to take in a given situation is determined from the actions stored in similar states, where similarity is determined from the state space encoding. (*lower*) In k-nearest sequence a record of the agent's history is kept. To determine the action to take at the current time, as indicated by the "x," the k closest sequences are determined from the history, and their actions are used to choose the current action. (From McCallum, 1995a.)

with brackets. The chosen policy is a function of the policies recommended by the leading state in those sequences.

As shown in Figure 5.14, the k-nearest sequence algorithm can lead to dramatic performance improvements.

5.4.2 Distinguishing the value of states

The idea behind using temporal sequences to make decisions is simple on the surface: if there is a set of previous histories that resulted in a common action, then take that action this time as well. However, there is a subtlety to consider: how long should the sequence be? It could be the case that sequences that are short have higher utility than longer sequences, or vice versa. Guessing the sequence length based only on matches between the state action values in the absence of utility can lead to suboptimal decisions. A way around this difficulty is to let utility guide the selection of the sequence lengths. The goal is to partition the set of temporal sequences into subsets that have the same utility.

One way of creating these partitions is to use a tree to store the temporal

Algorithm 1 Nearest Sequence Memory

(i) For each of the current actions, examine the history, and find the k most similar sequences that end with that action.
(ii) Each of these sequences will end in a state with an action. Each state votes for that action.
(iii) Compute the new Q-value by averaging the values for the successful voters.
(iv) The action with the most votes may be selected, or a random action (for exploration) may be selected.
(v) Execute the action, recording the new state, together with its observation and reward.
(vi) Update the new Q-values.

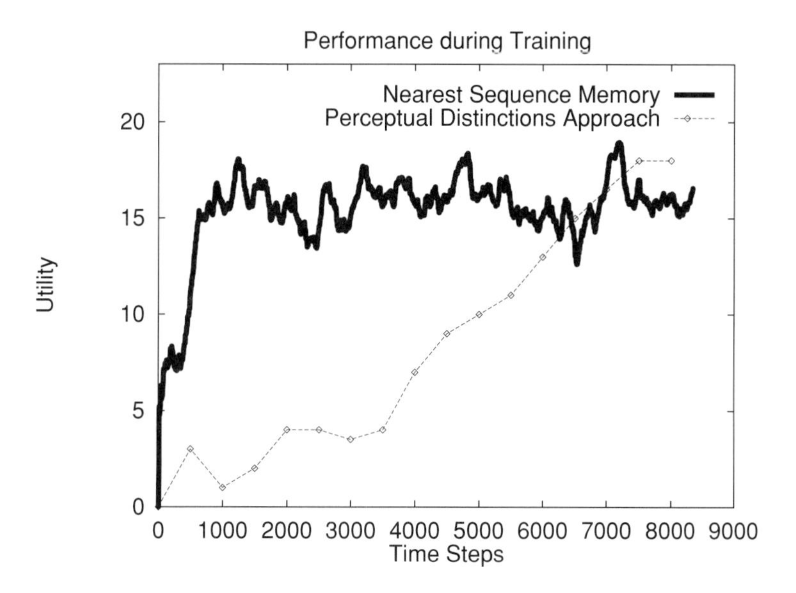

Fig. 5.14. The k-nearest sequence algorithm (Algorithm 1) can lead to significant speedups. Here it is compared to the Chrisman algorithm (Chrisman, 1992) for resolving aliasing by adding perceptual bits. The graph shows roughly an order of magnitude in the time to converge to a high-utility policy. (After McCallum, 1995a.)

sequences (McCallum, 1995a). The tree consists of alternating levels, starting with the most recent observations, then the actions that led to those observations, next the observations that were obtained in the penultimate states, and so on. A representative tree for a hypothetical problem that has two possible actions and two possible observations is shown in Figure 5.15. The tree has two parts, one upon which the current policy is going to be based and another that stores longer sequences that are potential extensions to the current policy. The latter part is termed a *fringe* and is represented by the dotted lines in the figure.

A sequence of observations and actions accesses the current policy. For example, given the sequence `0 b 1`, the current best action is stored at the leaf of the tree indicated by the arrow in Figure 5.15.

The fringe is used to check that longer sequences might not result in decisions that are better still. To do this checking one can test the value of a sequence against the penultimate sequence of shorter length. If the longer sequence makes useful distinctions, then that sequence is used as a baseline, and even longer sequences are represented in the fringe. This procedure is followed when taking an action generates a new observation of the state, resulting in a new temporal sequence which is then stored in the tree. For example, given the sequence

```
0 b 1 a
```

a test would be made to see if the tree should be extended to make the distinction between that and

```
0 b 1 b
```

or whether the current partition

```
0 b 1
```

should be kept as is.

The sequences are being used to define the state space incrementally. Each node in a leaf of the tree represents a state in the state space, and the agent is learning a policy that when executed will put it in another state that will also be a leaf node. The information in fringe nodes represents potential refinements of the state space. Thus the tree also guides the updating of the policy *values*.

To make these ideas more specific requires some definitions:

- An *instance* at time t is a four-tuple consisting of the previous instance, the action taken from that instance, and the observation and the reward received as a result of the action; that is,

$$T_t = < T_{t-1}, u_{t-1}, o_t, r_t > \tag{5.9}$$

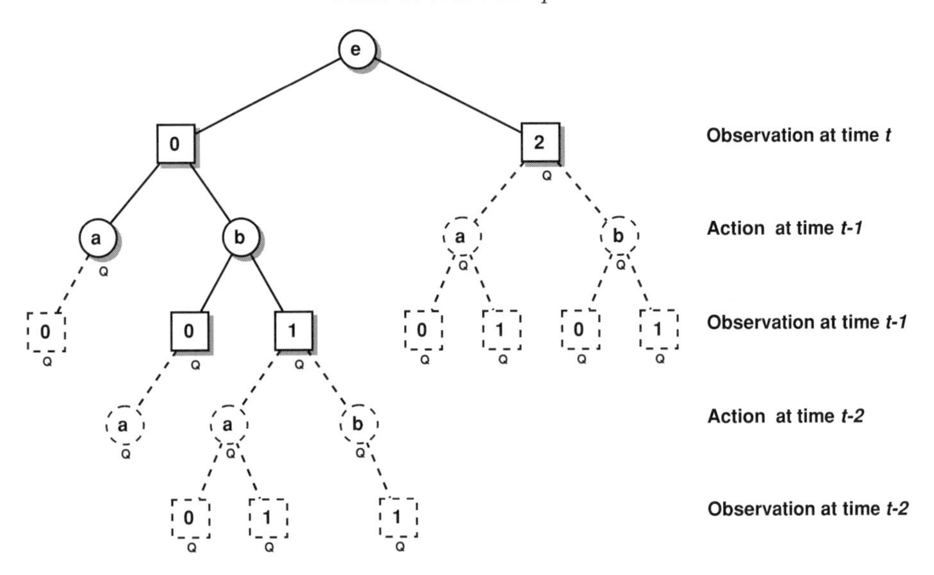

Fig. 5.15. The tree data structure for indexing into temporal sequences. Percepts are indicated by integers, actions by letters. The fringe nodes are drawn in dashed lines. Nodes labeled with a Q are nodes that hold Q-values. (After McCallum, 1995b.)

Stringing together the set of instances results in the temporal sequence used in the previous section.

- A *state* of the agent's model is represented as a leaf node of the *history tree*. This is explicitly denoted by referring to the transition that is stored in the leaf; that is,

$$\boldsymbol{x} = L(T) \tag{5.10}$$

- The set of all instances that is associated with the leaf $\boldsymbol{x}$ is referred to by

$$\mathcal{T}(\boldsymbol{x}) \tag{5.11}$$

The ultimate goal of this version of reinforcement learning is of course to calculate a policy that specifies the action to be taken in each state. To rate the value of different actions, the Q-value iteration algorithm is used, with the updating determined by the appropriate partitions as captured by the tree. That is,

$$Q(\boldsymbol{x}, \boldsymbol{u}) \leftarrow R(\boldsymbol{x}, \boldsymbol{u}) + \gamma P(\boldsymbol{x}'|\boldsymbol{x}, \boldsymbol{u}) U(\boldsymbol{x}') \tag{5.12}$$

where $R(\boldsymbol{x}, \boldsymbol{u})$ can be estimated as the average amount of reward obtained when action $\boldsymbol{u}$ was taken from state $\boldsymbol{x}$,

$$R(\boldsymbol{x}, \boldsymbol{u}) = \frac{\sum_{T_i \in \mathcal{T}(\boldsymbol{x}, \boldsymbol{u})} r_i}{|\mathcal{T}(\boldsymbol{x}, \boldsymbol{u})|} \tag{5.13}$$

and the probability $P(\boldsymbol{x}'|\boldsymbol{x}, \boldsymbol{u})$ can be similarly estimated as the number of

Algorithm 2 Utile Distinction Memory

(i) Initialize the history tree so that it is initially empty.
(ii) Make a move in the environment, and record the result in the instance chain.
(iii) Add the instance to the history tree.
(iv) Do one step of value iteration.
(v) Test the fringe of the tree using the Kolmogorov-Smirnoff test.
(vi) Choose the next action such that

$$\boldsymbol{u}_{t+1} = \text{argmax } Q[L(T_t), \boldsymbol{u}]$$

or choose a random action and explore.

---

times taking $\boldsymbol{u}$ from $\boldsymbol{x}$ resulted in state $\boldsymbol{x}'$ in the partition divided by the total number of times $\boldsymbol{u}$ was chosen, or

$$P(\boldsymbol{x}'|\boldsymbol{x}, \boldsymbol{u}) = \frac{|\forall T_i \in \mathcal{T}(\boldsymbol{x}, \boldsymbol{u}) \text{ s.t. } L(T_{i+1}) = \boldsymbol{x}'|}{|\mathcal{T}(\boldsymbol{x}, \boldsymbol{u})|} \tag{5.14}$$

Example: Hallway Navigation The use of a tree to index sequences of the same utility is illustrated with a simple example of searching a hallway to find a reward. In this case, as shown in Figure 5.16, the reward is the state labeled G at a central location. The agent has actions for moving along the compass directions of north, south, east, and west. What makes the problem partially observable is that the sensory information does not uniquely identify a state but only provides information as to local barriers in each of the four directions. Thus the code on the squares in the figure indicates the local surround. Positions that have the same configuration of surrounding walls get the same code. The agent receives a reward of 5.0 for reaching the goal, -1.0 for bumping into a wall, and -0.1 otherwise. For this example, the discount factor γ is 0.9, and the exploration probability is a constant 0.1.

Algorithm 2 is applied to this case, and the resultant tree that is built is shown in Figure 5.17. You can see that where the decision of which direction to go is unambiguous, the tree is shallow, but where the situation is ambiguous, the tree is deep. As a case in point, consider the hallways that have a code of 10. The decision for what direction to take next is very ambiguous but can be resolved with a local history of the agent's recent perceptions and actions. The figure shows that the tree accomplishes this

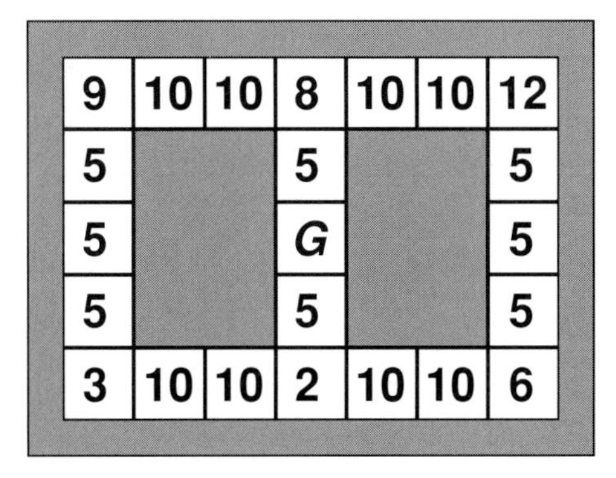

Fig. 5.16. A hallway navigation task with limited sensors. World locations are labeled with integers that encode the four bits of perception. (After McCallum, 1995b.)

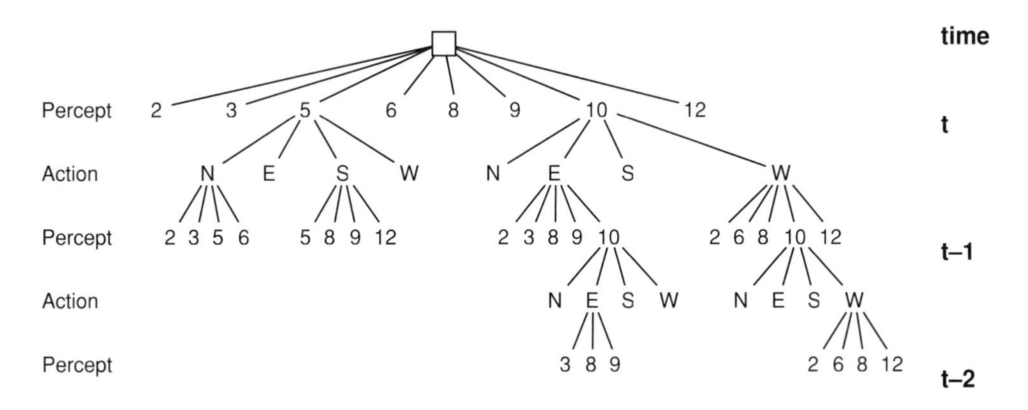

Fig. 5.17. A tree for navigating through the environment in the maze task. Deeper branches of the tree correspond to longer, more detailed memories. Memory was only created where needed to solve the task at hand. (After McCallum, 1995b.)

purpose. This algorithm is an instance of the use of *decision trees* in learning and is only representative of a larger set of learning algorithms that use such trees as a memory structure*. The main reason for its prominence here is its use of reinforcement learning.

5.5 Systems

Having put the elements of behavior together, we can now turn to the integration of such elements in a complete system. We will illustrate this

* See also B. D. Ripley, *Pattern Recognition and Neural Networks* (New York: Cambridge University Press, 1996); J. R. Quinlan and R. L. Rivest, "Inferring Decision Trees Using the Minimum Description Length Principle," *Information and Computation* 80 (1989):227–48; and J. R. Quinlan, "The Minimum Description Length Principle and Categorical Theories," in W. W. Cohen and H. Hirsh, eds., *Proceedings of the 11th International Machine Learning Conference* (San Francisco, CA: Morgan Kaufmann, 1994).

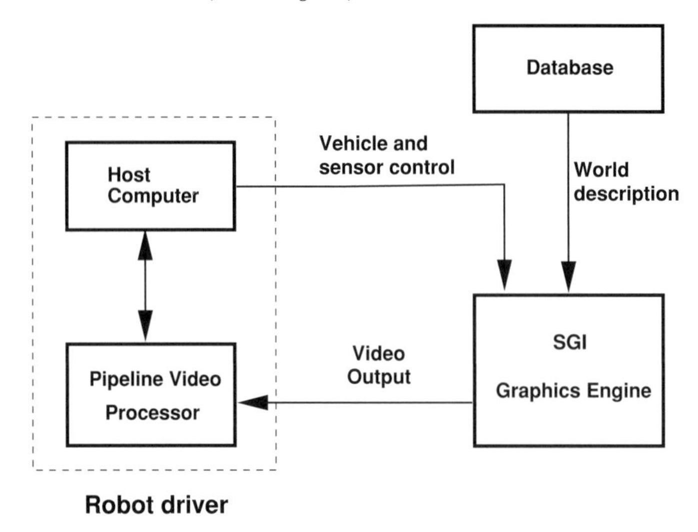

Fig. 5.18. The graphical output of the simulator is sent to the real-time image processing hardware (Datacube color digitizer and MV200 processing board) connected to a host computer. The host analyzes the incoming images and sends back to the simulator controls for the vehicle and virtual camera.

integration with the example of driving a car. The system is developed in a unique platform. The view from a car driving in a simulated world is fed into a Datacube pipeline video processor. Terzopoulos & Rabie (1995) pioneered the use of simulated images in his animat vision architecture. However, the processing is carried out in software only, and it is not clear how the algorithms would transfer into the real world. In our case, the graphical output from the simulator is sent to a separate subsystem (host computer with pipeline video processor) where the images are analyzed in real-time and commands are sent back to the simulator.

Visual routines are scheduled to meet the temporary task demands of individual driving sub-problems such as stopping at lights and traffic signs. The output of the visual routines is used to control the car which in turn affects the subsequent images. The images are generated by an SGI Onyx Infinite Reality engine which uses a model of a small town and the car (figure 5.18). The integration of photo-realistic simulation and real-time image processing represents a proof of concept for a new system design which allows testing computer vision algorithms under controllable conditions. In addition to the simulations, the routines are also tested on similar images generated by driving in the real world to assure the generalizability of the simulation.

5.5.1 Perceptual and control hierarchy

A hierarchy of perception and control levels that forms the framework for the driving program is presented in figure 5.19. While the picture is definitely

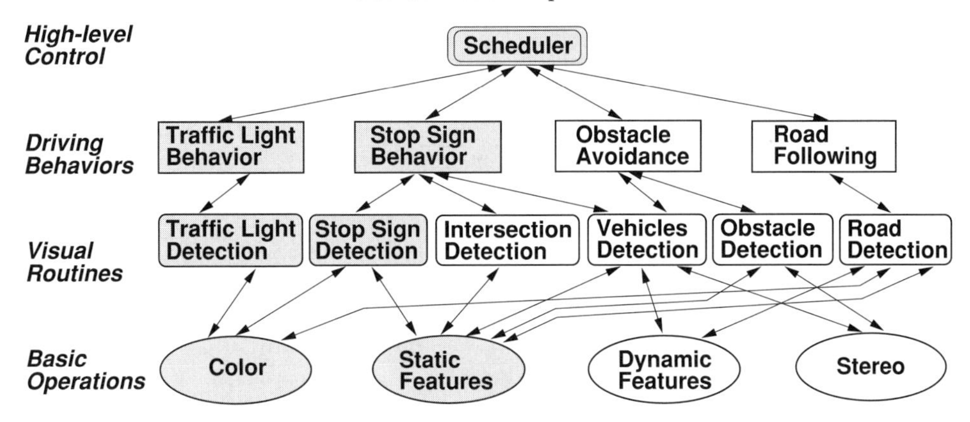

Fig. 5.19. Hierarchy of perceptual and control modules.

not exhaustive, all the relevant components are represented. At the top a *scheduler* selects from a set of task-specific *behaviors* the one that should be activated at any given moment. The behaviors use *visual routines* to gather the information they need and act accordingly. Finally, the visual routines are composed from an alphabet of *basic operations* (similar to Ullman's proposal, 1984).

To illustrate how modules on different levels in the hierarchy interact, consider the case when the scheduler activates the stop sign behavior. In order to determine whether there is a stop sign within certain range, the stop sign detection visual routine is invoked. The routine in its turn uses several basic operations to determine if a stop sign is present. For instance, it uses color to reduce the image area that needs to be analyzed. If there are any red blobs, they are further verified to see if they represent stop signs by checking distinctive image features. Finally, the routine returns to its caller with an answer (stop sign found or not).

If no stop sign was detected, that information is passed to the scheduler which can then decide what behavior to activate next. On the other hand, if a stop sign was found, the agent has to stop at the intersection to check for traffic. For that, it needs to know where the intersection is, so the intersection detection routine will be activated. It can use static image features such as lines and corners to determine where in the image the intersection is located. At the behavior level this information can be used for visual servoing until the intersection is reached.

The shaded modules in figure 5.19 are the ones that have been implemented so far. One may argue that we should have started with the road following behavior, since a real vehicle will need it to be able to move. Road following has been intensely studied for more than a decade (Dickmanns *et al.*, 1990; Pomerleau, 1995; Land, this volume) and it was successfully demonstrated at high speeds and over extended distances. Therefore we decided not to duplicate these efforts initially and instead to take advantage of the simulated environment. In our experiments the car is moving on a predefined track and the driving program controls the acceleration (the gas and break pedals).

The remainder of this section briefly describes the modules that we have implemented, starting from the lowest level in the hierarchy. For more details, see Salgian & Ballard (1998).

5.5.2 Basic operations

At the lowest level in the hierarchy are basic operations. These are simple low-level functions which can be used in one or more of the higher level task-specific visual routines. So far we have implemented color blob extraction and image feature extraction primitives.

The implementation uses special real-time image processing hardware, namely two Datacube boards. One is a color digitizer (Digicolor) and the other is the main processing board (MV200) with six dual-port memory surfaces and processing elements for convolution, linear and non-linear arithmetic operations, and a warper module. The Datacube boards are connected to a host computer (Sun Sparc) via a VME bus which allows image transfers from memory surfaces to host memory for further processing.

Color. The role of the *color primitive* is to detect blobs of a given color. An incoming color image is digitized in the Hue, Saturation, Value space. Colors are defined as regions in the hue-saturation sub-space and a lookup table is programmed to output a color value for every hue-saturation input pair. A binary map corresponding to the given color is further extracted and analyzed using a blob labeling algorithm. The end result is a list of bounding rectangles for the blobs of that color.

Static features. The role of this primitive is to detect objects of a specific appearance. It uses the responses of multiple filters at different spatial scales, as described in §5.3.1

Searching for an object in an image is realized by comparing the index of a suitable point on the model with the index for every image location. The first step is to store the index (response vector) $\mathbf{r^m}$ for the chosen point on the model object. To search for that object in a new image the response $\mathbf{r^i}$ at every image point is compared to $\mathbf{r^m}$ and the one that minimizes the distance $d_{im} = ||\mathbf{r^i} - \mathbf{r^m}||$ is selected, provided that d_{im} is below some threshold.

We use five spatial scales and five filters (first and second order Gaussian derivatives), so it takes 25 convolutions to determine whether the desired features are present in the image or not.

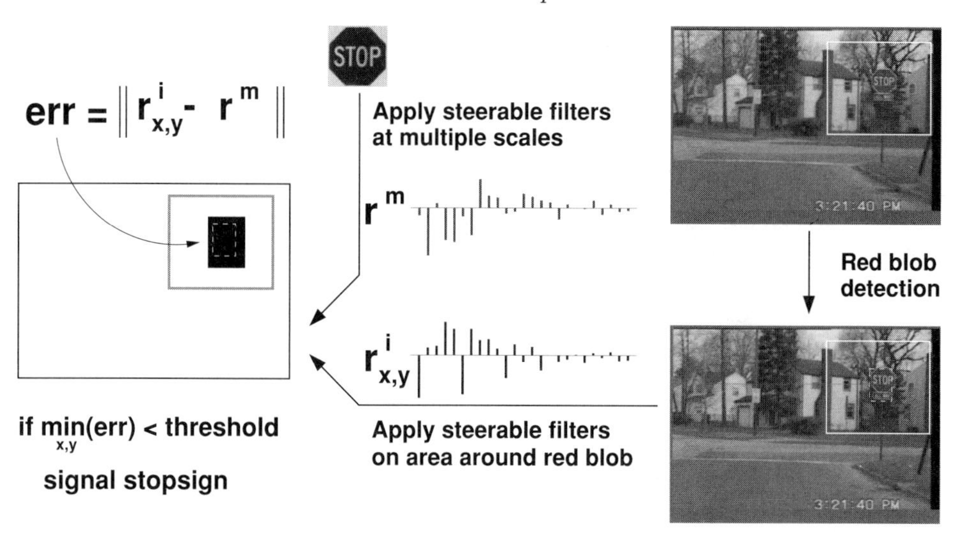

Fig. 5.20. Stop sign detection routine.

5.5.3 Visual routines

Basic operations can be combined into more complex, task-specific routines. Since the routines are task-specific, they can make use of high level information such as a geometric road model or known ego-motion to limit the region of the image that needs to be analyzed, which leads to reduced processing time. We have implemented two such routines, for stop light and stop sign detection.

Stop light detection. The stop light detection routine is an application of the color blob detection primitive to a restricted part of the image. Specifically, it searches for red blobs in the upper part of the image. If two red blobs are found within the search area (the corresponding bounding rectangles are marked with thick white lines in Figure 5.21), then a stop light is signaled. Currently, the search window is fixed a priori. Once we have a road detection routine, we will use that information to adjust the position and size of the window dynamically.

Stop sign detection. The area searched for stop signs is the one on the right side of the road (the green rectangle in the right side of every image in figure 5.20). The first step is to detect red blobs in this area, which are candidates for stop signs. Since other red objects can appear in this region (such as billboards, brick walls, etc.) the color test alone is not enough for detecting the stop signs, being used just as a "focus of attention" mechanism to further limit the image area that is analyzed.

Once a red blob is detected, it is further analyzed by comparing the steerable filter responses for the area with the previously stored responses for a stop sign. If the error (difference) is below some predetermined threshold, a stop sign is reported.

Fig. 5.21. Stop light (left) and stop sign (right) detection from simulated sequences (up) and real video (down). Overlayed at the bottom of every image is the relevant information for the current state (from left to right): time for the current cycle and current state name. For stop sign detection: minimum error (distance) between the filter responses for the current analyzed region (around the red rectangle) and the stored responses; if error is below a threshold then signal a stop sign.

The routines have been tested both in the simulated environment and on real video sequences shot with a hand-held camera from a car. Sample results are presented in in Figure 5.21.

5.5.4 Driving behaviors

The visual routines are highly context dependent, and therefore need an enveloping construct to interpret their results. For example, the stop light detector assumes the lights are in a certain position when gaze is straight ahead, thus the stop light behavior has to enforce this constraint. To do this, the behaviors are implemented as finite state machines, presented in figure 5.22.

Traffic light behavior. The initial state is "Look for stop lights", in which the traffic light detection routine is activated. If no red light is detected the behavior returns immediately. When a red light is detected, the vehicle is instructed to stop and the state changes to "Wait for green light" in which the red light detector is executed. When the light changes to green, the routine will return "*No red light*" at which time the vehicle starts moving again and the behavior completes.

Stop sign behavior. In the "Look for stop signs" state the stop sign detection routine is activated. If no sign is detected the behavior returns immediately. When a stop sign is detected, the agent needs to stop at the intersection. Since we don't have an intersection detector yet, once the stop

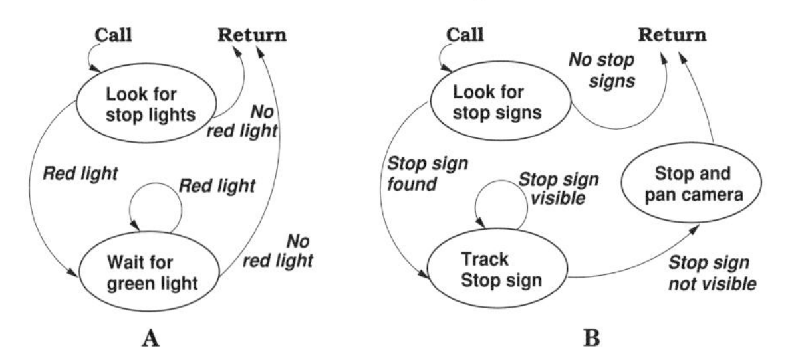

Fig. 5.22. Finite state machines used for two driving behaviors **A** traffic light behavior and **B** stop sign behavior. The conditions that trigger a state transition are labeled on the arcs (in italics).

sign is detected, the state changes to "Track stop sign" in which the vehicle moves forward while tracking the sign. When the sign is no longer visible, a new state is entered in which the agent stops and pans the camera left and right.

5.5.5 Scheduling

Given a set of behaviors, and a limited processing capacity, the next problem to address is how to schedule them in order to ensure that the right behavior is active at the right time. This issue has been addressed by other researchers and several solutions have been proposed: inference trees (Reece, 1992), and more recently, distributed architectures with centralized arbitration (Rosenblatt, 1997; Sukthankar, 1997).

We have just started to experiment with different alternatives for the scheduler design. Currently our principal method is to alternate between the two existing behaviors (traffic light and stop sign), but there are important subsidiary considerations. One is that the performance of difficult or important routines can be improved by scheduling them more frequently. Another is that the performance of such routines can be further improved by altering the behavior, for example by slowing down. The effect of different scheduling policies is addressed in the experiments section.

5.5.6 Experiments

Experiments illustrate the performance of the stop sign behavior. In particular, they show the effect of two parameters, the scheduling frequency and vehicle speed. Intuitively, the more often the stop sign routine is scheduled the higher the recognition rate should be. Similarly, for a fixed scheduling rate, the recognition rate should be higher for lower vehicle speed. However, having a quantitative measure of how the performance degrades is important for the scheduler design.

In these experiments the vehicle is moving on a straight path along a

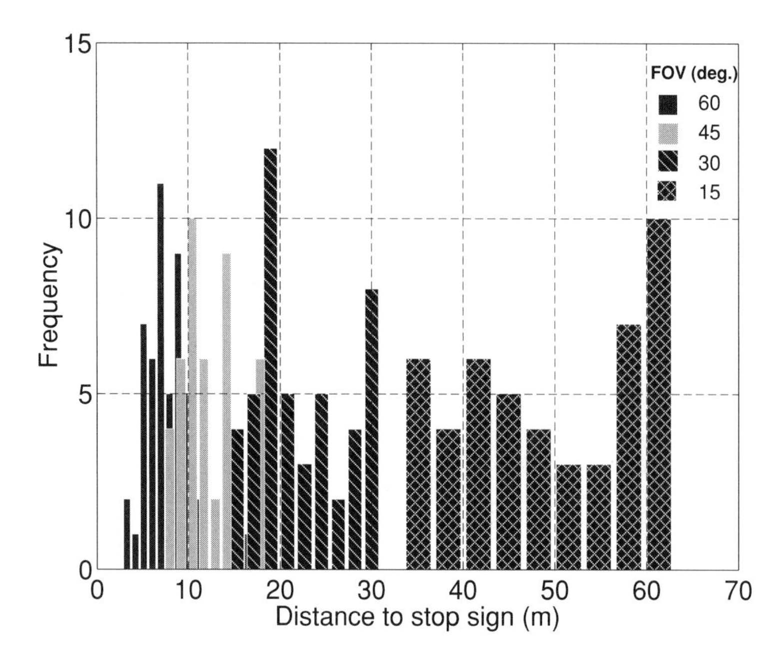

Fig. 5.23. Frequency histogram of the distance at which the stop sign is first detected, for different fields of view (FOV). The field of view in degrees corresponding to each shading density is indicated at the top of the figure.

road, approaching an intersection with a stop sign on the right side. We first estimated the influence of the camera field of view (FOV) on the distance at which the stop sign is first detected. Based on these results (figure 5.23), we selected a FOV of 30 degrees, in order to be able to detect the stop signs early enough.

The path is covered 50 times for every combination of parameter values and we count how many times the stop sign is detected. The hit ratio represents the normalized success rate (1 means that all the encountered signs were detected). We varied the scheduling by changing the "timeout" limit for the two behaviors, ie the number of cycles after which the other behavior is activated if nothing is detected. We repeated the experiment for different scheduling policies and for different vehicle speeds.

The results are illustrated in figure 5.24A. As expected, the hit ratio for the stop sign increases when the stop sign routine is scheduled more often and when the vehicle speed is lower. The one-dimensional slice shows how the number of successful cases decreases as the speed increases, but the degradation is less severe if the stop sign routine is scheduled more often (higher SS/SL ratio).

The results, while expected, show that the recognition of the stop sign is significantly better at low speeds. Therefore, we modified the stop sign

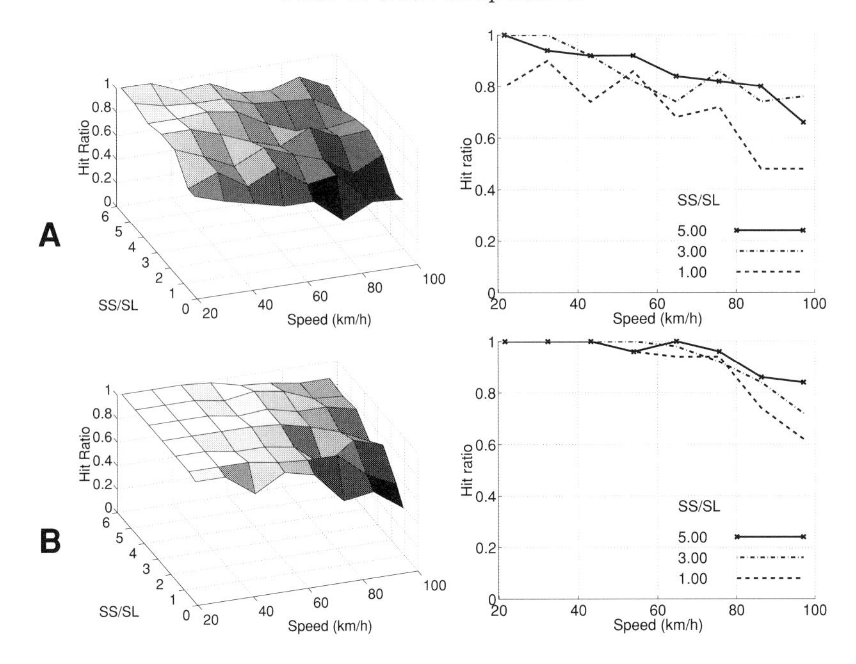

Fig. 5.24. Hit ratio for different vehicle speeds and different scheduling rates for the two routines (Stop Sign and Stop Light), **A** at constant speed and **B** with the vehicle slowing down when red blobs are detected.

behavior such that as soon as a red blob is detected the vehicle slows down. After passing the stop sign the vehicle accelerates until the previous speed is attained.

Figure 5.24B presents the results for this modified behavior. In this graph the values on the horizontal axis represent the vehicle speed before slowing down. The graph shows that perfect performance can be extended up to 40 km/h and that performance is much less sensitive to changes in the scheduling ratio.

5.6 Conclusion

To summarize the discussion of hierarchies, there are several important points.

The brain does not have time to construct complete models of the world, but can compensate for this with the ability to construct task-dependent models quickly. A key element in such models is determining the correct state space of the task, but this can be learned through multiple trials.

Many effects can be explained in terms of specific interactions within hierarchies. For example, skipping eye movements occur when the time to compute a target is longer than the time to initiate an eye-movement.

The abstract analysis of hierarchical systems finds a reality in the construction of the brain. The evidence is overwhelming that the brain is organized hierarchically. However the crucial question is "How are there levels composed for behavior?" This is a question of great difficulty owing to the complexity of the brain. To address this question we have describe our modeling efforts which are based on the premise that the different functions of the brain can be isolated in different temporal bands, as suggested by Newell. We have focused on three such bands.

(i) **Memories** In order to isolate a stimulus, the input must be processed by the cortical memory. Our estimate for the time to do this is 80 milliseconds. In our simulation, the well-known phenomenum of "endstopping" is shown to be a natural product of the memory dynamics.

(ii) **Routines** Multiple memory accesses can be composed to compute specific objects which in turn serve as focal points for behavior. The example used herein is that of gaze targeting, but many other examples could have been used. The point is that within about 300 milliseconds, the brain can compose extremely complcated functions of individual memory states.

(iii) **Programs** At still longer timescales the brain has time to integrate multiple sensori-motor actions into a program that accomplishes a specific behavioral goal. Such programs are naturally stored as hidden Markov models. Our simulation herein addresses how these models can be constructed from scratch to encode just the right amount of state information. But the larger point is that at the timescale of about 2-3 seconds, the brain has time to access vast libraries of such programs to direct behavior.

As an example of a complete small system, we have used the example of driving. The driving example illustrates that hierarchical models can be developed based on the principles described herein that carry out complex tasks.

References

Ballard, D., Hayhoe, M., & Pelz, J. (1995). "Memory representations in natural tasks", *J. Cog. Neurosci.*, 7:66–80.

Ballard, D., Hayhoe, M., Pook, P., & Rao, R. (1997). "Deictic codes for the embodiment of cognition", *Behav. Brain Sci.*. In press.

Barlow, H. (1994). What is the computational goal of the neocortex? In C. Koch and J. Davis (Eds.), *Large-Scale Neuronal Theories of the Brain*, pages 1–22, MIT Press, Cambridge, MA.

Barlow, H. (1987). "Learning receptive fields", In *Proceedings of the IEEE Int. Conf. on Neural Networks*, 115–121.

Bell, A. & Sejnowski, T. (1997). "The 'independent components' of natural scenes are edge filters", *Vis. Res.* (in press).

Burt, P. (1988). "smart sensing" in machine vision. pages 1–30.

Chapman, D. (1991). *Vision, Instruction and Action.* MIT Press, Cambridge, MA.
Chase, W. & Simon, H. (1973). "Perception in chess". *Cog. Psych.*, 4: 55–81.
Chrisman, L. (1992). "Reinforcement learning with perceptual aliasing: the perceptual distinctions approach", Proc. 10th National Conference on Artificial Intelligence, 183-188, San Jose, CA.
Churchland, P. & Sejnowski, T. (1992). *The Computational Brain.* MIT Press, Cambridge, MA.
Daugman, J. (1988). Complete discrete 2-D Gabor transforms by neural networks for image analysis and compression. *IEEE Trans. Acoustics, Speech, and Signal Proc.*, 36: 1169–1179.
Derrico, J. & Buchsbaum, G. (1991). "A computational model of spatiochromatic image coding in early vision", *J. Vis. Commun. Image Rep.*, 2: 31–38.
Dickmanns, E. D., Mysliwetz, B. D., & Christians, T. (1990). "An integrated spatio-temporal approach to automatic visual guidance of autonomous vehicles", *IEEE Trans. on Sys., Man, and Cybern.*, 20(6):1273–1284.
Dobbins, A., Zucker, S., & Cynader, M. (1987). "Endstopped neurons in the visual cortex as a substrate for calculating curvature", *Nature*, 329: 438–441.
Duhamel, J.-R., Colby, C., & Goldberg, M. (1992). "The updating of the representation of visual space in parietal cortex by intended eye movements" *Science*, 255: 90–92.
Field, D. (1994). "What is the goal of sensory coding?" *Neural Computation*, 6: 559–601.
Foldiak, P. (1990). "Forming sparse representations by local anti-Hebbian learning", *Biol. Cybern.*, 64: 165–170.
Freeman, W. & Adelson, E. (1991). "The design and use of steerable filters", *IEEE Trans. Pattern Analysis and Machine Intelligence*, 13: 891–906.
Hancock, P., Baddeley, R., & Smith, L. (1992). "The principal components of natural images", *Network*, 3: 61–70.
Harpur, G. & Prager, R. (1996). "Development of low-entropy coding in a recurrent network", *Network*, 7: 277–284.
Hubel, D. & Wiesel, T. (1962). "Receptive fields, binocular interaction, and functional architecture in the cat's visual cortex", *J. Physiol. (Lond.)*, 160: 106–154.
Hubel, D. & Wiesel, T. (1965). "Receptive fields and functional architecture in two non-striate visual areas (18 and 19) of the cat", *J. Neurophysiol.*, 28: 229–289.
Johnson, M. P., Maes, P., & Darrell, T. (1994). "Evolving visual routines", In *Artificial Life IV Conference.*
Just, M. & Carpenter, P. (1976). "Eye fixations and cognitive processes", *Cog. Psych.*, 8: 441–480.
Kalman, R. (1960). "A new approach to linear filtering and prediction theory", *Trans. ASME J. Basic Eng.*, 82: 35–45.
Kanerva, P. (1988). *Sparse Distributed Memory.* Bradford Books, Cambridge, MA.
Kosslyn, S. M. (1973). "Scanning visual images: Some structural implications", *Percept. Psychophys.*, 14: 90–94.
Kowler, E. & Anton, S. (1987). "Reading twisted text: Implications for the role of saccades", *Vis. Res.*, 27: 45–60.
Land, M. & Furneaux, S. (1997). "The knowledge base of the oculomotor system", In *Proceedings of the Royal Society Conference on Knowledge-Based Vision.*
MacKay, D. (1956). "The epistemological problem for automata" In *Automata Studies*, 235–251. Princeton University Press, Princeton, NJ.
Maybeck, P. (1979). *Stochastic Models, Estimation, and Control (Vols. I and II)*, Academic Press, New York, NY.
McCallum, A. K. (1995a). *Reinforcement learning with selective perception and hidden state*, Ph.D. thesis, Computer Science Department, University of Rochester, Rochester, NY.

McCallum, A. K. (1995b). "Meuronal architectures for pattern-theoretic problems", in *Large-Scale Neuronal Theories of the Brain*, C. Koch, & J. Davis (Eds.), 125-152, MIT Press, Cambridge, MA.

Mumford, D. (1994). "Neuronal architectures for pattern-theoretic problems", In *Large-Scale Neuronal Theories of the Brain*, C. Koch and J. Davis (Eds.), 125–152. MIT Press, Cambridge, MA.

Nijhawan, R. (1994). "Motion extrapolation in catching", *Nature*, 370:256.

Oja, E. (1989). "Neural networks, principal components, and subspaces", *Int. J. of Neural Sys.*, 1: 61–68.

Olshausen, B. & Field, D. (1996). "Emergence of simple-cell receptive field properties by learning a sparse code for natural images", *Nature*, 381: 607–609.

Oram, M. & Perrett, D. (1992). "Time course of neural responses discriminating different views of the face and head", *J. Neurophysiol.*, 68: 70–84.

O'Regan, J. (1990). "Eye movements and reading", In *Eye Movements and Their Role in Visual and Cognitive Processes*, E. Kowler (Ed.), 455–477 Elsevier, New York, NY.

Pece, A. (1992). "Redundancy reduction of a Gabor representation: a possible computational role for feedback from primary visual cortex to lateral geniculate nucleus", In *Artificial Neural Networks 2*, I. Aleksander & J. Taylor (Eds.), 865–868, Elsevier Science, Amsterdam.

Pomerleau, D. (1995). "Ralph: rapidly adapting lateral position handler", In *Proceedings of the Intelligent Vehicles '95 Symposium*, 506–511, New York, NY.

Rao, R. & Ballard, D. (1995). "An active vision architecture based on iconic representations", *Art. Intel, (Special Issue on Vision)*, 78: 461–505.

Rao, R. & Ballard, D. (1997). "Dynamic model of visual recognition predicts neural response properties in the visual cortex", *Neural Comp.*, 9: 721–763.

Reece, D. A. (1992). "Selective perception for robot driving", Technical Report CMU-CS-92-139, Carnegie Mellon University.

Rosenblatt, J. K. (1997). *DAMN: A Distributed Architecture for Mobile Navigation.* PhD thesis, The Robotics Institute, CMU. CMU-RI-TR-97-01.

Salgian, G. & Ballard, D. (1998). "Visual routines for autonomous driving", In *Proceedings of the 6th International Conference on Computer Vision (ICCV-98)*, Bombay, India.

Sukthankar, R. (1997). *Situation Awarness for Tactical Driving.* PhD thesis, Robotics Institute, CMU, Pittsburg, PA 15213. CMU-RI-TR-97-08.

Tanenhaus, M., Spivey-Knowlton, M., Eberhard, K., & Sedivy, J. (1995). "Integration of visual and linguistic information in spoken language comprehension", *Science*, 268: 632–634.

Terzopoulos, D. & Rabie, T. F. (1995). "Animat vision: Active vision in artificial animals", In *ICCV-95*, 801–808.

Thorpe, S. & Imbert, M. (1989). "Biological contraints on connectionist modelling", In *Connectionism in Perspective*, R. Pfeifer, Z. Schreter, F. Fogelman-Soulie, and L. Steels (Eds.), 63–92, Elsevier, Amsterdam.

Tsotsos, J. (1989). "The complexity of perceptual search tasks", In *Eleventh International Joint Conference on Artificial Intelligence.*

Ullman, S. (1984). "Visual routines", *Cognition*, 18: 97–160.

Viviani, P. (1990). "Eye movements in visual search: Cognitive, perceptual, and motor control aspects", in E. Kowler (Ed.), *Eye Movements and their Role in Visual and Cognitive Processes. Reviews of Oculomotor Research V4*, 353-383, Elsevier.

Williams, R. (1985). "Feature discovery through error-correction learning", Technical Report 8501, Institute for Cognitive Science, University of California at San Diego.

Yarbus, A. (1967). *Eye Movements and Vision.* Plenum Press, New York, NY.

Zelinsky, G., Rao, R., Hayhoe, M. & Ballard, D. (1997). "Eye movements reveal the spatio-temporal dynamics of visual search", *Psychological Science.* in press.

Zipser, K., Lamme, V., & Schiller, P. (1996). "Contextual modulation in primary visual cortex", *J. Neurosci.*, 16: 7376-7389.

6

Effects of orbital pulleys on the control of eye rotations

Lance M. Optican and Christian Quaia

Abstract

Rotations of a rigid body, such as the eyeball, around arbitrary axes in three dimensions are not commutative. That is, the final position depends on the order in which the rotations around the various fixed axes are carried out. Thus, Tweed and colleagues (1987; 1994) proposed that a non-commutative controller in the brain is required to move the eyes. In contrast, Schnabolk & Raphan (1994a) have argued that a non-commutative controller is not required because the eyes are controlled by muscle torques, which are commutative vectors. However, the eye does not rotate freely, like a rigid body, but rather is constrained by orbital tissues and the extraocular muscles. A most peculiar feature of the orbit is the trochlea, a pulley which changes the direction of action of the superior oblique muscle. Miller, Demer and colleagues (1989; 1993) have shown that there are also pulleys on other eye muscles. As proposed by Miller and colleagues (1993; 1997), the presence of these pulleys makes the direction of the torque applied by the muscles to the eyeball a function of eye orientation. Raphan (1997) has recently proposed a new model of the ocular plant incorporating these pulleys. With the proper constraints on the location of the pulleys, we now demonstrate quantitatively that the oculomotor plant appears essentially commutative to the neural system controlling it. This is possible because the signals neurally encoded are the orientation of the eye and a signal that is very close to its derivative. Consequently, in a pulley model, the neural controller is nearly commutative and the control of eye rotations is greatly simplified.

6.1 Introduction

One of the most remarkable anatomical features of the ocular plant is the trochlear pulley. The tendon of the superior oblique goes to the front of the orbit, through a tendonous pulley attached to the nasal side of the orbit, and returns to insert on the globe at a point behind the pulley. This allows the superior oblique muscle to exert a downward and outward torque on the globe. Now, perhaps even more remarkably, the work of Miller, Demer and colleagues (Miller, 1989; Miller *et al.*, 1993; Miller & Demer, 1997) has revealed that in fact all of the extraocular muscles with the exception of the inferior oblique go through specialized regions of Tenon's capsule that

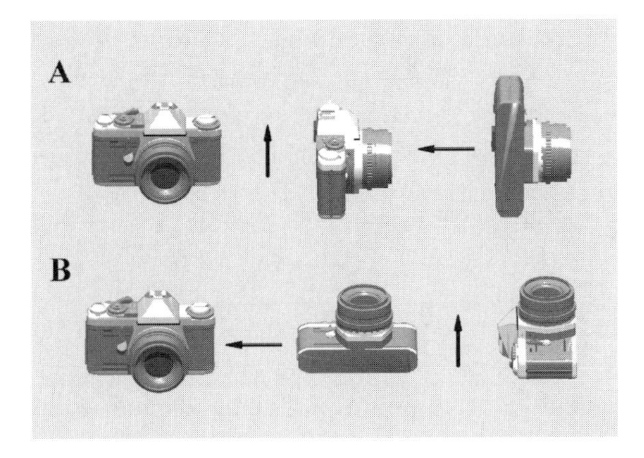

Fig. 6.1. The geometric definition of non-commutativity. The image on the right of each arrow is obtained by rotating the image on the left around an earth-fixed axis collinear with the arrow. The direction of rotation corresponds to the direction in which a right-hand screw advances. In panels A and B the camera, starting from the same initial orientation, undergoes two sequences of rotations. In panel A the camera first rotates 90° around a vertical axis and then 90° around a horizontal axis. In panel B the order of rotations is reversed. The final orientation of the camera is clearly different in the two cases.

function as pulleys. Why did such a complicated system of pulleys develop in the oculomotor plant? We will try to demonstrate here that this complexity in the orbit may simplify the brain's task in controlling 3-D rotations of the eye.

The control of ocular rotations in three dimensions is complicated because if a rigid object is rotated around a set of earth-fixed axes, say x, y, and z, the final orientation of the object will depend upon the sequence in which the rotations occur. For example, a 90° rotation around the x-axis followed by a 90° rotation around the y-axis brings the object to a different orientation than does the reverse sequence of rotations (Figure 6.1). Consequently, rotations are called non-commutative, i.e., dependent upon the rotation sequence. Note that this complication does not occur when translations of objects are considered; one step forward and one step leftward brings you to the same place as one step leftward followed by one step forward. Thus, translations are called commutative, i.e., independent of movement sequence.

Early models of the dynamics of saccadic eye movements (e.g., Robinson, 1975) concentrated on rotations around a single axis (e.g., the vertical axis for horizontal movements), which, as opposed to rotations around arbitrary axes, are commutative. The innervation required to make a saccade consists of a pulse that compensates for the presence of viscous elements in the plant and a step that holds the eye still by opposing the elastic forces in the orbit. In Robinson's model, the step (which is proportional to eye position) is obtained by integrating the pulse (which is proportional to eye velocity) because for commutative systems the derivative of position (i.e., single-axis orientation) is velocity (i.e., single-axis angular velocity). The extension of one-dimensional models of eye rotation to three-dimensions is not trivial because, for non-commutative systems, the derivative of orientation is not angular velocity (Goldstein, 1980) .

To extend Robinson's model from control of the eye around one axis to control around all three of its axes, Tweed and Vilis (1987) developed a model of the oculomotor system that embeds non-commutative, rotational operators. In sharp contrast to this approach, Schnabolk and Raphan (1994a; b) proposed that, in fact, a non-commutative controller is not needed for eye movements, and that a simple three-dimensional integrator can be used to obtain the Step from the Pulse. However, Tweed and colleagues (Tweed *et al.*, 1994; Tweed, 1997) showed that movements simulated using the commutative controller of Schnabolk and Raphan can be affected by large post-saccadic drifts, which are not observed when the same movements are produced by human subjects. Nonetheless, Tweed and colleagues (1994) pointed out that, if eye orientation and its derivative, but not angular velocity, are encoded neurally, the first signal could be computed by simply integrating the second. Under such an assumption a commutative controller could be used and it would then be up to the plant to convert the derivative of eye orientation into the appropriate angular velocity; however, Tweed *et al.* (1994) did not suggest any mechanism that would allow the plant to perform such a conversion. Nevertheless, several recent studies have relied on the hypothesis that the derivative of eye orientation, as opposed to eye velocity, is neurally encoded (Crawford, 1994; 1997; Crawford & Guitton, 1997).

Recently, Raphan (1997) introduced a new model which incorporates the orbital pulleys discovered by Miller and colleagues (1993). As pointed out by Miller *et al.* (1993; 1997), the pulleys make the axes of action of the extraocular muscles depend on the orientation of the eye. Raphan (1997) showed that a movement which, when simulated with the original Schnabolk and Raphan model, had a large post-saccadic drift was essentially drift-free when the effects of the pulleys, as proposed by Miller and colleagues, were taken into account. The first purpose of this chapter is to show quantitatively that the good steady-state and dynamic behavior of the saccades produced by simulations of Miller-Raphan type pulley models holds for all movements. The second goal of this chapter is to explain, in an intuitive way, why this is possible. A brief description of these results appeared elsewhere (Quaia & Optican, 1997a).

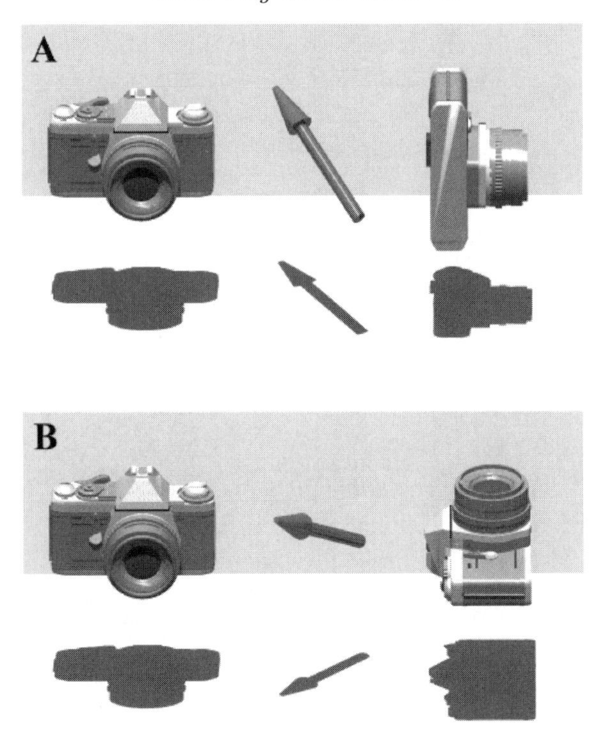

Fig. 6.2. The initial and final orientations of the cameras plotted in Figure 6.1 are repeated here. The arrows indicate the Euler axes that describe the orientation of the objects on the right. The Euler axis is the single axis around which the camera, starting from the orientation depicted on the left, has to be rotated in order to assume the orientation on the right by means of a single rotation. The final orientations in A and B are different, and the Euler axes reflect this difference.

6.2 Methods

Representation of orientation

The eyeball can be modeled as a sphere with its centre fixed in the head and capable of rotating around any axis. To describe the orientation of the eye, we make use of Euler's theorem: any orientation of a rigid body with one point fixed in space can be achieved, starting from a reference orientation, by a single rotation about an axis along a unit-length vector $\hat{n}$ through a positive angle Φ (Goldstein, 1980). For example, in Figure 6.2 we re-plot the leftmost and the rightmost panels of Figure 6.1; the vectors in the central column are collinear with the single *Euler axes* around which the cameras in the left column have to be rotated to assume the orientation represented in the right column. Thus, the vectors in the central column represent the Euler axes that describe the orientation of the cameras on the right; the top and the bottom vectors are different, as are the orientations of the rightmost cameras. The simulations reported in this chapter were performed using MATLAB (The Mathworks Inc., Mass.) programs running on a Challenge-L computer (Silicon Graphics Inc., California).

6.3 Results

The saccadic system in one dimension (rotations around a single axis)

To a first order approximation, the oculomotor plant in one dimension (one pair of muscles pulling around a single axis) can be described as a Voigt element (Robinson, 1964), i.e., the parallel connection of a viscous element and an elastic element. Newton's law requires that, if we indicate with $\Phi(t)$ eye orientation at time t and with $\omega(t)$ eye angular velocity at time t, the torque exerted by the muscles (which is equal to the product of the innervation signal $I(t)$ by the tension/innervation ratio S) is:

$$T(t) = S \cdot I(t) = B \cdot \omega(t) + K \cdot \Phi(t) \tag{6.1}$$

where B is the viscosity and K the stiffness of the plant (muscles and orbital tissues).

To rotate the eye from an initial orientation Φ_0 to a new orientation Φ_1, it would be sufficient to change the innervation signal in a stepwise manner. However, this would produce a slow movement, with the eye orientation changing from Φ_0 to Φ_1 with an exponential time course (Figure 6.3A), characterized by a time constant of approximately 200 ms (Robinson, 1975). Thus, for any eye movement, at least 600ms would be needed to foveate a target. This is clearly undesirable, because during this time retinal slip would strongly degrade vision.

However, recordings in the motoneurons show that the innervation signal does not change in a stepwise manner during saccades (Robinson, 1970; Luschei & Fuchs, 1972; Robinson & Keller, 1972; Fuchs *et al.*, 1988). In fact, the motoneurons' activity is composed of tonic (Step) and phasic (Pulse) components. During periods of fixation only the Step is present, whereas during saccades both the Pulse and the Step are present and the movement is considerably faster (Figure 6.3B). It is important to note that if the two signals (Pulse and Step) are not appropriately matched then when the Pulse is over the eye would have an orientation Φ_2, and then it would drift to Φ_1 with a time constant of 200 ms. In other words, a Pulse-Step mismatch would occur, and again the time during which the eye is not stable would be increased. To avoid such a Pulse-Step mismatch the Pulse and the Step must be precisely related to each other.

There is compelling physiological evidence that suggests that the Step is generated from the Pulse. For example, when a saccade is interrupted in mid-flight by stimulation of omnipause neurons which inhibit the pulse

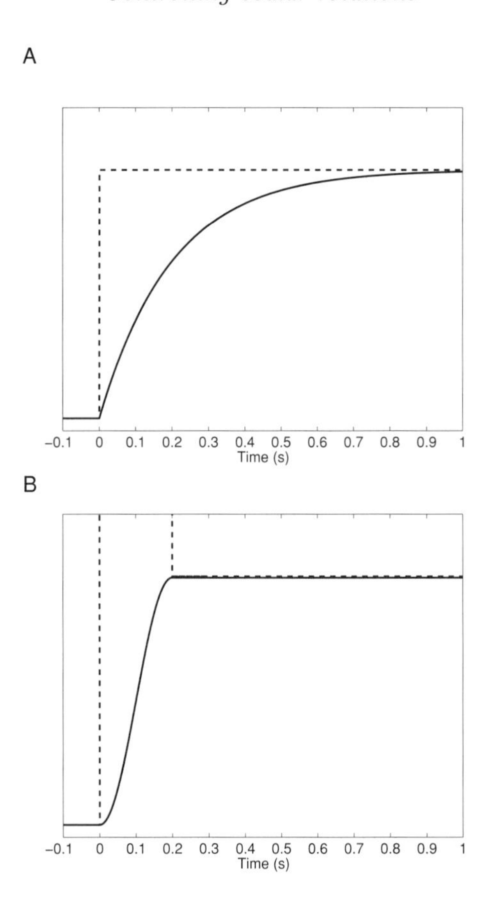

Fig. 6.3. **A**. When a step torque command is applied to the system described by Equation 6.1, a slow exponential movement is produced, and the final orientation is not reached before 600 ms (three times the time constant of the system). **B**. In contrast, if a Pulse-Step torque signal is applied (where the Step is obtained by integrating the Pulse), the attainment of the final orientation is much quicker.

generators, the eye does not drift toward the goal or the initial orientation but stays still (Keller, 1974). In other words throughout the movement the Step is always appropriate to keep the eyes where they currently are, i.e.,

$$Step = K_S \cdot \Phi(t) \tag{6.2}$$

where $K_S = K/S$. From Equations 6.1 and 6.2 it follows that:

$$Pulse = B_S \cdot \omega(t) \tag{6.3}$$

where $B_S = B/S$.

In the case of rotations around one axis, the angular velocity, $\omega(t)$, is the derivative of $\Phi(t)$. Thus the Step can be obtained simply by integrating the Pulse with an appropriate gain. This automatically guarantees a Pulse-Step match and thus the absence of post-saccadic drift.

Rotations around arbitrary axes

The Euler representation of orientation $(\Phi, \hat{n})$ has the characteristic of representing the shortest path that takes the eye from the primary orientation to a given orientation and vice-versa (Nakayama & Balliet, 1977). Thus, it is reasonable to say that the restoring torque due to the stiffness K of the plant will tend to realign the visual axis with the primary position acting along the unit-length axis $\hat{n}$, with an intensity proportional to Φ (Schnabolk and Raphan, 1994a). Accordingly, if the axis-angle form is used, converting Equation 6.1 into three dimensions is straightforward:

$$\vec{T}(t) = B \cdot \vec{\omega}(t) + K \cdot \Phi(t) \cdot \hat{n}(t) \tag{6.4}$$

where $\vec{T}(t)$ is the torque exerted by the muscles and $\vec{\omega}(t)$ the angular velocity.

The torque is applied to the eyeball by appropriately innervating three pairs of muscles. Using the simplification that each pair of muscles is collapsed into an equivalent ideal muscle, we will assume that three innervational signals are generated, forming a vector of innervation $\vec{I} = [I_1 \ I_2 \ I_3]^T$.

The torque exerted by the muscles can be evaluated by multiplying the vector of action of each pair of muscles, $\hat{m}_i$, (the unit-length vector along which the globe rotates under the action of a pair of muscles) by the corresponding innervation and by the tension/innervation ratio, S, which can be considered constant (Haustein, 1989) :

$$\vec{T_i}(t) = S \cdot I_i(t) \cdot \hat{m}_i \tag{6.5}$$

The global torque applied to the globe is then the sum of the three vectors obtained applying Equation 6.5 to each pair of muscles:

$$\vec{T}(t) = S \cdot (I_1(t) \cdot \hat{m}_1 + I_2(t) \cdot \hat{m}_2 + I_3(t) \cdot \hat{m}_3) \tag{6.6}$$

This is mathematically equivalent to multiplying the matrix that has as columns the vectors of action of the three muscles, by the vector of innervation:

$$\vec{T}(t) = S \cdot \begin{bmatrix} & & \\ \hat{m}_1 & \hat{m}_2 & \hat{m}_3 \\ & & \end{bmatrix} \cdot \begin{bmatrix} I_1(t) \\ I_2(t) \\ I_3(t) \end{bmatrix} = S \cdot \overline{M} \cdot \vec{I}(t) \tag{6.7}$$

We will suppose that the three pairs of muscles act in orthogonal planes and that the matrix $\overline{M}$ is the identity matrix.

Now, using the same line of reasoning used for the one dimensional case, it can be shown that, in order to avoid a Pulse-Step mismatch, from Equations 6.4 and 6.7 it follows that:

$$Step = K_S \cdot \Phi(t) \cdot \hat{n}(t) \tag{6.8}$$

$$Pulse = B_S \cdot \vec{\omega}(t) \tag{6.9}$$

where both the Step and Pulse are vectors of three components.

Thus, we return to the problem of generating the Step from the Pulse. Intuitively, the simplest way to extend the Robinson model from one dimension to three dimensions is to use three integrators, one for each pair of muscles. However, as pointed out in §6.1, rotations around arbitrary axes do not commute, and thus the orientation can not be obtained simply by integrating the angular velocity.

To account for this fact, Tweed and Vilis (1987) developed a model (the quaternion model) that uses non-commutative, rotational operators to compute the Step from the Pulse. Not much needs to be said about this quaternion model: it is a mathematically correct extension of the Robinson model, and, if the appropriate weights are selected, it produces perfect three dimensional movements, without any post-saccadic drift (the Pulse and the Step are perfectly matched).

In contrast, Schnabolk and Raphan (1994a) proposed that, because the innervation signals applied to the muscles produce torques, which commute, a commutative controller, in which the Step is obtained by integration of the Pulse, can be used. Now, it is true that in steady-state (i.e., after the dynamic transients of the movement have died out), the orientation is determined by Equation 6.8. If the order in which the three components of the Step change is modified, the final orientation remains the same, because Equation 6.8 does not depend on the sequence of the Steps.

Thus, the commutativity of torques guarantees what happens in steady-state, when the eye orientation is stable. However, torques do not introduce any constraint on the Pulse-Step matching issue (which controls the dynamic transients of the movement). Nonetheless, using simulations Schnabolk and Raphan showed that their model also behaves well dynamically. This raises the paradox of how a non-commutative plant can be controlled by a commutative controller. To resolve the paradox, a very simple question has to be addressed: just how large is the post-saccadic drift when a commutative controller drives the oculomotor plant?

How non-commutative do ocular rotations appear to the controller?

We will now show that the only information needed to decide whether a commutative controller can be used for eye movements is the degree of non-commutativity of the oculomotor plant, i.e., the magnitude of the Pulse-Step mismatch associated with the use of a commutative neural controller. To do so we will first develop an easy way to quantify this mismatch.

If the system were commutative, the integral of vectorial velocity would be orientation; consequently, the Step, as defined in Equation 6.8, would be the integral of the Pulse (Equation 6.9). Accordingly, one way to es-

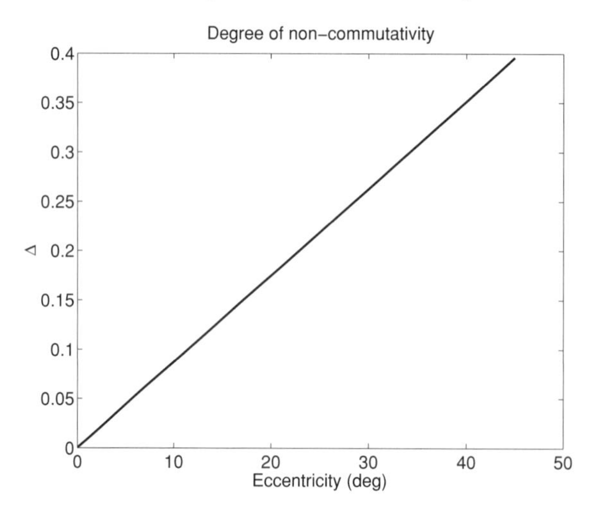

Fig. 6.4. The degree of non-commutativity, Δ, as a function of ocular eccentricity for a model with fixed axes of muscle action. Δ is a measure of the amount of Pulse-Step mismatch in the movement. Here the case when the angular velocity is orthogonal to the orientation (worst case scenario, see text) is considered. The amount of Pulse-Step mismatch increases almost linearly with the eccentricity.

timate the degree of non-commutativity of the system is to quantify the difference between the Pulse and the derivative of the Step. If this difference is small, the system is approximately commutative, and the Step can be computed simply by integrating the Pulse. Thus, we define the degree of non-commutativity, Δ, as:

$$\Delta = \frac{||Pulse - \frac{d}{dt}[Step]||}{||Pulse||} \tag{6.10}$$

where the operator $|| \; ||$ indicates the Euclidean norm of a vector. It can be shown (Quaia & Optican, in preparation) that Δ can be expressed as a function of the eccentricity Φ and of the angle α between the orientation vector $\hat{n}$ and the instantaneous velocity vector $\vec{\omega}$:

$$\Delta(\alpha, \Phi) = |\sin(\alpha)| \cdot \sqrt{(1 - \frac{\Phi}{2}\cot(\frac{\Phi}{2}))^2 + (\frac{\Phi}{2})^2} \cong |\sin(\alpha)| \cdot \frac{\Phi}{2} \tag{6.11}$$

where Φ is expressed in radians.

In Figure 6.4 we plot Δ as a function of the eccentricity Φ when α is equal to 90° (i.e., the worst case scenario for a given eccentricity), e.g., at the beginning of an upward saccade made from a rightward orientation. For small eccentricities (below 15°), Δ is less than 15%, so the system can be regarded as almost commutative (i.e., the difference between velocity and derivative of orientation is small). And this is why the movements simulated by Schnabolk and Raphan (1994a), which were executed at small eccentricities, were characterized by fairly good dynamic behavior, with a

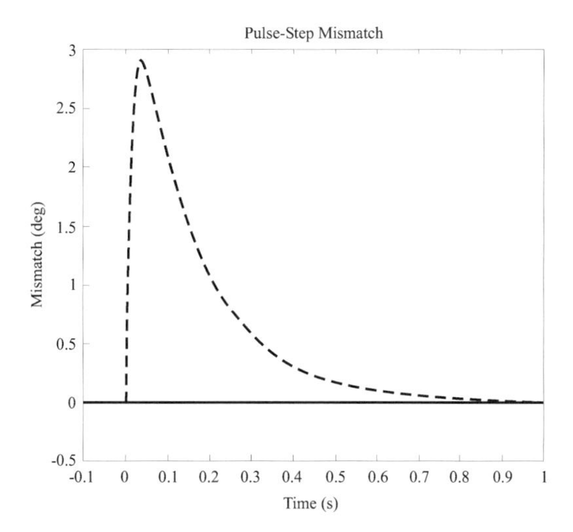

Fig. 6.5. Two simulations obtained using a commutative controller (to drive the plant described in Equation 6.4) are reported. Both movements are 20° in amplitude, one starting from straight ahead (solid line), and the other starting from 20° up (dashed line). The first movement is not affected by Pulse-Step mismatch, whereas the second shows some ocular drift (about 15% Pulse-Step mismatch).

limited, but not zero, Pulse-Step mismatch. In Figure 6.5 we show the mismatch for two movements of 20° to the right, one with the eye starting from straight ahead (solid line), and the other starting from 20° up (dashed line). When the movement starts from the primary position, α is equal to zero throughout the movement, and thus no mismatch occurs. However, when the movement starts from a secondary position α is not zero and a mismatch is present. When larger eccentricities are considered, Δ can become large, and large Pulse-Step mismatches are expected.

When Tweed *et al.* (1994) simulated the model proposed by Schnabolk and Raphan using large movements and eccentricities, they obtained movements with large post-saccadic drifts that were not observed experimentally. These observations led Tweed and coworkers to conclude that a commutative model cannot possibly be right and that the non-commutativity of rotations must be accounted for, either mechanically or neurally. The neural solution to the problem is to use the model developed by Tweed and Vilis (1987); as a mechanical solution, Tweed and colleagues (1994) proposed the so-called *linear plant model*, in which the Pulse encodes the derivative of eye orientation (and thus the Step can be computed by simply integrating the Pulse). However, Tweed and colleagues did not suggest any scheme for the physical implementation of this "linear plant".

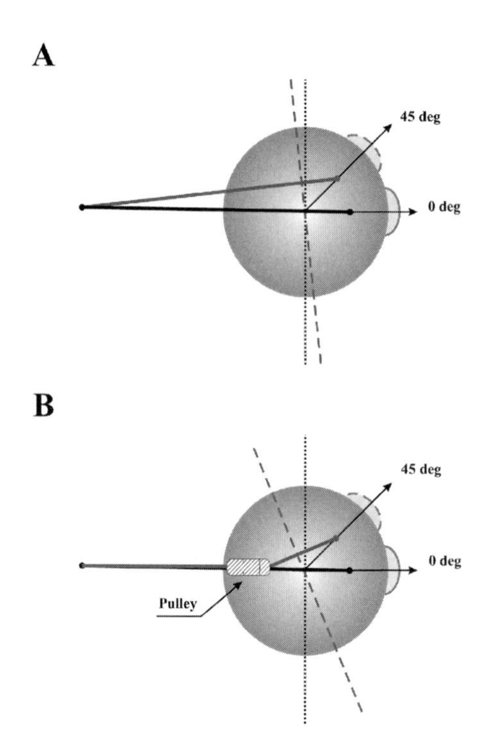

Fig. 6.6. Axes of action of the horizontal recti for two different models of orbital mechanics. The schematics are a scaled version of an actual human orbit (Miller & Robinson, 1984). **A**. If the muscles can move freely in the orbit, the muscular path does not change much whether the eye is in primary position (dark solid line) or elevated by 45° (gray solid line). Correspondingly, the axis of action (dotted and dashed lines) is approximately fixed in the orbit. **B**. If the path of the muscles through the orbit is constrained by pulleys, the axis of action of the muscles changes dramatically with orientation (dotted and dashed lines); the magnitude of this change is clearly a function of the position of the pulleys.

The Miller-Raphan Pulley Model

Recently, Raphan (1997) replying to the observation of Tweed *et al.* (1994), noted that the simulations by Schnabolk and Raphan were obtained under the assumption that the axes of action of the muscles do not change with the orientation of the eye. However, Miller *et al.* (1993) showed that this is not what happens; in fact, the muscles cannot slip freely within the ocular orbit but are constrained. Miller showed that the muscles do not change their path from their origin point to some point behind the insertion point; from there they go straight to the insertion point. This intermediate point corresponds to the position of a pulley and determines the relationship between the orientation of the eye and the axes of action of the muscles.

In Figure 6.6 the shift of the axis of muscle action for the horizontal recti is shown with a scale model of the human orbit (based on anatomic data from Miller and Robinson, 1984) . If no pulleys were present, the muscles would be able to move freely in the orbit, and an elevation of the eyes by 45° would cause a mild backward tilt of the axis of action of the horizontal recti (Figure 6.6A). If pulleys are introduced (Figure 6.6B), the geometry changes dramatically, and the axes of action of the muscles can change considerably with the orientation of the eyes. Thus, for the same innervation, the rotation produced by the muscles varies as a function of eye orientation. Including this idea, first proposed by Miller and colleagues (1993; 1997), in a new model, Raphan (1997) showed that the same movement simulated by Tweed *et al.* (1994) using the original Schnabolk and Raphan model, was now essentially drift-free. However, it is not clear how to generalize from this particular movement simulation to arbitrary movements. We will now show, using the same reasoning as above, that when the restricted slip of the muscles due to the presence of the pulleys is considered, the degree of non-commutativity of the plant is greatly reduced in general.

Rotations with pulley effects

In the case described above (rotations around arbitrary axes without pulleys), it was assumed that the axes of action of the three muscle pairs were orthogonal and the matrix $\overline{M}$ was the identity matrix. We will now assume that the planes of action of the muscles are not fixed in the orbit but are a function of the instantaneous orientation of the eyes. For the sake of simplicity suppose that the axes of action of the muscles rotate around the axis of rotation $\hat{n}$ by an angle that is a fraction, K_Φ, of the angle of rotation Φ (Raphan, 1997).

When such a partial muscular slip is introduced, the new matrix $\overline{M}$, having as columns the vectors of action of the three pairs of muscles, simply corresponds to the rotation matrix associated with a rotation of $\delta = K_\Phi \cdot \Phi$ degrees around the axis $\hat{n}$:

$$\overline{M}(t) = R\left[\delta(t), \hat{n}(t)\right] \tag{6.12}$$

Clearly, Equation 6.12 holds under the assumption that $\overline{M}$ is equal to the identity matrix when the eye is in primary position. If this condition does not hold, $\overline{M}$ is equal to the product of $R[\delta, \hat{n}]$ and $\overline{M}_0$, the muscle matrix in primary position.

So, Equation 6.7 for the torque applied to the globe can be rewritten as:

$$\vec{T}(t) = S \cdot \overline{M}(t) \cdot \vec{I}(t) = S \cdot \overline{M}(t) \cdot Step + S \cdot \overline{M}(t) \cdot Pulse \tag{6.13}$$

where both the Step and the Pulse components of the innervation are vectorial signals. Using Equations 6.13 and 6.4 we can now write modified versions of Equations 6.8 and 6.9 that take into account the partial slip of the muscles:

$$Step = K_S \cdot \overline{M}(t)^{-1} \cdot \Phi(t) \cdot \hat{n}(t) \tag{6.14}$$

$$Pulse = B_S \cdot \overline{M}(t)^{-1} \cdot \vec{\omega}(t) \tag{6.15}$$

Using the following properties of rotation matrices (α and $\hat{n}$ generic):

$$\begin{cases} R[\alpha, \hat{n}] \cdot \hat{n} = \hat{n} \\ R[\alpha, \hat{n}]^{-1} = R[\alpha, \hat{n}]^T = R[-\alpha, \hat{n}] \end{cases} \tag{6.16}$$

and taking into account Equation 6.12, we can now rewrite Equations 6.14 and 6.15 as:

$$Step = K_S \cdot \Phi(t) \cdot \hat{n}(t) \tag{6.17}$$

$$Pulse = B_S \cdot \overline{M}(t)^T \cdot \vec{\omega}(t) \tag{6.18}$$

From Equations 6.17 and 6.18 it is clear *that the effect of the pulleys on the Step and the Pulse are very different*. In particular, the Step encodes eye orientation even when the pulleys are included in the model. This very important finding will be discussed and illustrated later.

Defining Δ as above, it can be demonstrated that Δ can be expressed as a function of the eccentricity Φ, of the angle α and of $\delta = K_\Phi \cdot \Phi$

$$\Delta(\alpha, \Phi, \delta) = |\sin(\alpha)| \cdot \sqrt{(1 - \tfrac{\Phi}{2}\cot(\tfrac{\Phi}{2}))^2 + (\tfrac{\Phi}{2})^2 + \Phi\cot(\tfrac{\Phi}{2}) \cdot (1 - \cos\delta) - \Phi \cdot \sin\delta} \tag{6.19}$$

So, Equation 6.11 can now be derived as a particular case of Equation 6.19 where $\delta = 0$.

In Figure 6.7 we plot the value assumed by Δ when α is equal to 90^0, as was done in Figure 6.4. We plot different curves, showing the value assumed by Δ for five different values of K_Φ (0, 0.25, 0.5, 0.75 and 1). It appears clear from Figure 6.7 that the optimal value of K_Φ is 0.5. If the pulleys are located so as to produce a 50% muscular slip ($K_\Phi = 0.5$) the value of Δ always stays below 0.025, i.e. a Pulse-Step mismatch less than 2.5%.

An example of the mismatch produced using a commutative controller to drive the oculomotor plant with and without pulleys is shown in Figure 6.8. The solid line refers to a plant characterized by a K_Φ equal to 0.5; the dashed line represents the movement produced applying the same innervation signal to a plant without pulleys. This simulation shows for one movement what Equation 6.19 demonstrates mathematically for any movement, i.e., that a commutative controller is sufficient for eye rotations, provided that the pulleys are properly placed (i.e., so that $K_\Phi = 0.5$).

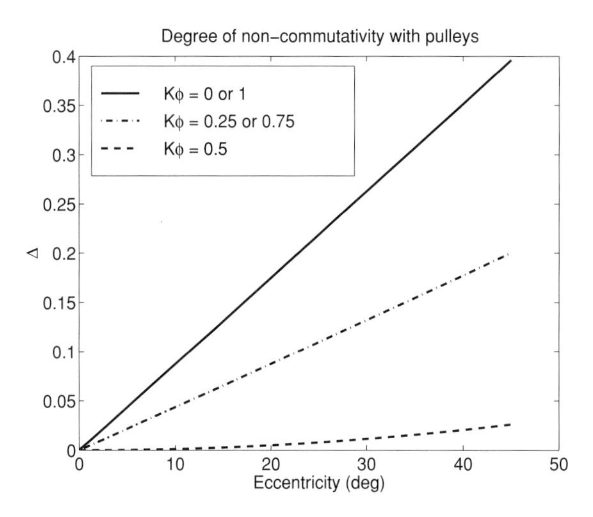

Fig. 6.7. The degree of non-commutativity in a pulley model with limited muscular slip. As in Figure 6.4, the worst case scenario is plotted (angular velocity orthogonal to orientation). Five values of K_Φ (0, 0.25, 0.5, 0.75 and 1) are considered, which correspond to different positions of the pulleys. The position of the pulleys has a dramatic effect on the degree of non-commutativity. The minimum effect is seen for $K_\Phi = 0.5$, when the system can be considered essentially commutative.

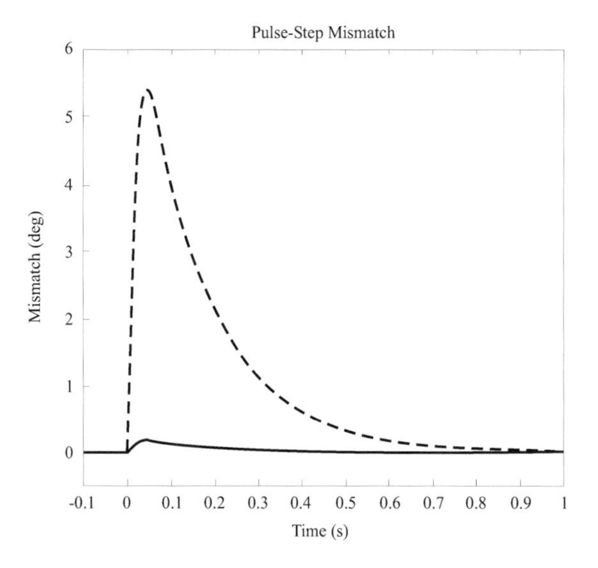

Fig. 6.8. Two simulations obtained using a commutative controller (to drive the plant described in Equation 6.13) are reported. Both movements are 40° to the right, and start from 20° up and 20° left. In one case K_Φ is set to 0.5 (solid line, optimal position of the pulleys), whereas in the other K_Φ is set to 0 (dashed line, no pulleys). The first movement is almost drift-free, whereas the second is affected by a quite large ocular drift.

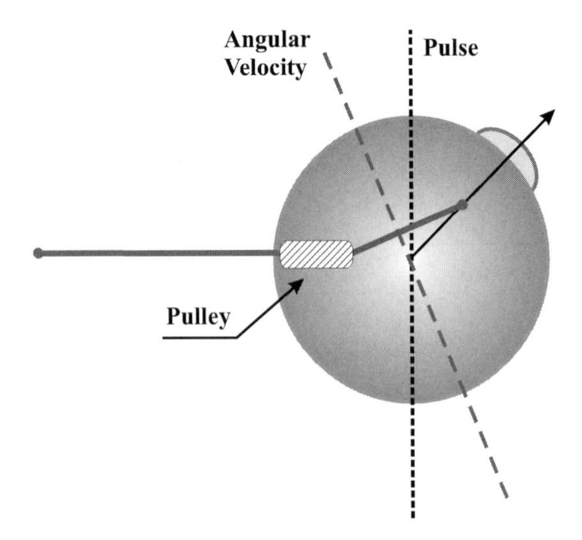

Fig. 6.9. When the eye is at zero elevation, and a Pulse is generated to innervate the horizontal recti, the Pulse and the angular velocity are collinear, because the eye rotates around the axis of action of the muscles. In contrast, when the eye is elevated, the same Pulse vector (black dotted line) will produce a rotation around a different axis (gray dashed line), which depends on the position of the pulleys.

6.4 Discussion

The major implication of the scheme proposed here is that to generate movements with appropriate dynamics, the brain does not need knowledge of the non-commutative mechanics of rotations. In fact, the plant can be treated as a translational (commutative) system. We will now explain in a more intuitive way how this is possible and how the pulleys affect the various signals implicated in eye movement control.

Functional role of the pulleys

We have shown here that, if the pulleys are in the right place, the integral of the Pulse is, to a very good approximation, equal to the Step. This statement is apparently at odds with the fact that rotations do not commute, and thus that the integral of angular velocity is not orientation. In fact, both statements are true. This is possible simply because *the Pulse does not represent angular velocity.* Instead, the Pulse vector is defined as the angular velocity vector $\hat{\omega}$ rotated by $-\delta$ degrees around the axis $\vec{n}$ (see Eq. 6.18); and we have demonstrated mathematically that this Pulse signal (but not angular velocity) is a very good approximation of the derivative of eye orientation. This concept is shown in Fig. 6.9, where the Pulse (dotted line) and the angular velocity (dashed line) are indicated. Suppose that the Pulse vector is collinear with the vertical axis (i.e., the Pulse is applied to the horizontal recti only). Now, because of the presence of the pulleys, the pulling direction of the horizontal recti changes as a function of the elevation of the eye. When the eye is in primary position the axis of rotation of the horizontal recti is vertical and thus Pulse and angular velocity coincide. In contrast, when the eye is elevated the axis of rotation of the muscles (dashed line) is tilted back, and the angular velocity vector (which is always collinear with the axis of rotation) does not coincide with the Pulse.

Up to this point we have described in great detail the effect of the pulleys on the Pulse; it is even more surprising to note the effect of the pulleys on the Step. Equation 6.17 implies that the pulleys have essentially no effect on the Step, i.e., the Step represents orientation with or without the pulleys (cf. Equations 6.8 & 6.17). The need for a different action of the pulleys on Pulse and Step was first pointed out by Crawford and Guitton (1997), but in that paper no solution was offered. We show here that the pulley model solves this problem; consequently, the pulley model can be viewed as a mechanical implementation of the purely conceptual linear plant model proposed by Tweed and colleagues (Tweed *et al.*, 1994; Tweed, 1997).

The behavior of the model can be intuitively explained by considering two different rotation sequences. Suppose that the eye is in primary position and that a Pulse is applied to a pair of muscles producing a rotation around the axis $\hat{y}$ (e.g., the vertical axis). Subsequently, a Pulse is applied to the muscles that, with the eye in primary position, act around the horizontal axis $\hat{x}$. However, because of the first rotation around $\hat{y}$, the eye is not in primary position anymore, and the second Pulse will thus result in a rotation around a new axis $\hat{x}'$. Suppose now that the order in which the pulses are applied is reversed; starting from the primary position with a rotation around $\hat{x}$; and then a rotation around a new axis $\hat{y}'$. Now, it can be shown (Quaia & Optican, in preparation) that, if the pulleys are in the right position, these two sequences of rotations result in the same final orientation. *Of course, different sequences of rotations around the same axes are non-commutative* (i.e., $\hat{x}$ followed by $\hat{y}$ as opposed to $\hat{y}$ followed by $\hat{x}$). However, we see here that when different sequences of rotations are made around *different* orientation-dependent axes ($\hat{x}$ followed by $\hat{y}'$ in one case, and $\hat{y}$ followed by $\hat{x}'$ in the other), they can reach the same final orientation. And this is why, even though rotations are non-commutative, the brain can use a simple, commutative, controller for eye rotations: the pulleys provide a mechanism that makes the plant (eye ball + extraocular muscles + pulleys) appear commutative to the controller.

Predictions of the model

We have shown that the pulley model predicts that the Pulse is closer to the derivative of orientation than to the angular velocity. To verify this prediction, recordings in the medium lead burst neurons (i.e., the neurons that carry the Pulse) during movements from secondary to tertiary positions are necessary. Because of the relatively large span of burst- neurons' on-direction, to have a good estimation of the signal that they carry it is important to average over a fairly large population of neurons (Quaia & Optican, 1997b) . Although the results of such an experiment have not been published yet, Van Opstal and colleagues (van Opstal, *et al.*, 1996) reported (Pg. 7294) that: "[...] recordings from both the riMLF and the oculomotor nucleus so far indicate that saccade-related burst activity is better correlated with the rate-of-change of 3-D eye position than with eye angular velocity (our unpublished observation)". These preliminary observations strongly support the model proposed here, especially because similar activity was recorded in both burst and motor neurons. In that case the only place left to perform the needed conversion of the Pulse into angular velocity would be the plant, as proposed here.

Conclusions

We have shown that properly located muscle pulleys simplify the neural control of eye movements by making the oculomotor plant appear commutative to the controller. This clearly holds if and only if the pulleys are properly placed, i.e., if K_Φ is close to 0.5. Unfortunately, detailed knowledge of the location and action of the muscle pulleys themselves is still not available but hopefully, thanks to the efforts of Miller, Demer and colleagues this gap will be filled in soon. Nonetheless, several predictions of the pulley model, only some of which have been presented in this chapter, have been confirmed, by either published results or by preliminary observations.

To conclude, we propose that the oculomotor plant evolved in this fairly complex way to allow the use of a simple neural controller. This would thus represent an example in which nature opted for a fairly complicated arrangement of the peripheral motor system in order to reduce the complexity of the central nervous system.

Acknowledgments

We thank Dr. Theodore Raphan for introducing us to his version of the pulley model, and providing us with a copy of his in-press 1997 book chapter before it was published.

References

Crawford, J. D. (1994). "The oculomotor neural integrator uses a behavior-related coordinate system", *J. Neurosci.*, 14: 6911-6923.

Crawford, J. D. (1997). "Geometric transformations in the visual-motor interface for saccades", In *Three-dimensional kinematics of eye, head, and limb movements*, M. Fetter, T. Haslwanter, H. Misslisch & D. Tweed (Eds.), 85-100, Harwood Academic Publishing, The Netherlands.

Crawford, J. D. & Guitton, D. (1997). "Visual-motor transformations required for accurate and kinematically correct saccades", *J. Neurophysiol*, 78: 1447-1467.

Fuchs, A. F., Scudder, C. A. & Kaneko, C. R. S. (1988). "Discharge patterns and recruitment order of identified motoneurons and internuclear neurons in the monkey abducens nucleus", *J. Neurophysiol.*, 60: 1874-1895.

Goldstein, H. (1980). *Classical Mechanics*, Addison-Wesley, Reading, MA.

Haustein, W. (1989.) "Considerations on Listing's law and the primary position by means of a matrix description of eye position control", *Biol. Cybern.*, 60: 411-420.

Keller, E. L. (1974). "Participation of medial pontine reticular formation in eye movement generation in monkey", *J. Neurophysiol.*, 37: 316-332.

Luschei, E. S. & Fuchs, A. F. (1972). "Activity of brain stem neurons during eye movements of alert monkeys", *J. Neurophysiol.*, 35: 445-461.

Miller, J. M. (1989). "Functional anatomy of normal human rectus muscles", *Vis. Res.*, 29: 223-240.

Miller, J. M. & Demer, J. L. (1997). "New orbital constraints on eye rotation", In *Three-dimensional kinematics of eye, head, and limb movements*, M. Fetter, T. Haslwanter, H. Misslisch & D. Tweed (Eds.), 349-357, Harwood Academic Publishing, The Netherlands.

Miller, J. M., Demer, J. L. & Rosenbaum, A. L. (1993). "Effect of transposition surgery on rectus muscle paths by magnetic resonance imaging", *Ophthalmology*, 100: 475-487.

Miller, J. M. & Robinson, D. A. (1984). "A model of the mechanics of binocular alignment", *Comp. and Biomed. Res.*, 17: 436-470.

Nakayama, K. & Balliet, R. (1977). "Listing's Law, eye position sense and perception of the vertical", *Vision Res.*, 17: 453-457.

Quaia, C. & Optican, L. M. (1997a). "A commutative controller is sufficient to rotate the eye", *Soc. Neurosci. Abstr.*, 23: 8.

Quaia, C. & Optican, L. M. (1997b). "A model with distributed vectorial premotor bursters accounts for the component stretching of oblique saccades", *J. Neurophysiol.*, 78: 1120-1134.

Raphan, T. (1997). "Modeling control of eye orientation in three dimensions", In *Three-dimensional kinematics of eye, head, and limb movements*, M. Fetter, T. Haslwanter, H. Misslisch & D. Tweed (Eds.), 359-374, Harwood Academic Publishing, The Netherlands.

Robinson, D. A. (1964). "The mechanics of human saccadic eye movement", *J. Physiol. (Lond.)*, 174: 245-264.

Robinson, D. A. (1970). "Oculomotor unit behavior in the monkey", *J. Neurophysiol.*, 33: 393-404.

Robinson, D. A. (1975). "Oculomotor control signals", In *Basic Mechanisms of Ocular Motility and Their Clinical Implications*, G. Lennerstrand & P. Bach-y-Rita (Eds.), 337-374. Pergamon Press, Oxford.

Robinson, D. A. & Keller, E. L. (1972). "The behavior of eye movement motoneurons in the alert monkey", *Bibl. Ophthalmol.*, 82: 7-16.

Schnabolk, C. & Raphan, T. (1994a). "Modeling three-dimensional velocity-to-position transformation in oculomotor control", *J. Neurophysiol.*, 71: 623-638.

Schnabolk, C. & Raphan, T. (1994b). "Modelling the neural integrator in three dimensions: an extension of the Robinson integrator in one dimension", In *Contemporary Ocular Motor and Vestibular Research: A Tribute to David A. Robin-*

son, A. F. Fuchs, T. Brandt, U. Büttner & D. Zee (Eds.), 320-328, Thieme, Stuttgart:.

Tweed, D. (1997). "Velocity-to-position transformation in the VOR and the saccadic system", In *Three-dimensional kinematics of eye, head, and limb movements*, M. Fetter, T. Haslwanter, H. Misslisch & D. Tweed (Eds.), 375-386, Harwood Academic Publishing, The Netherlands.

Tweed, D., Misslisch, H. and Fetter, M. (1994). "Testing models of the oculomotor velocity-to-position transformation", *J. Neurophysiol.*, 72: 1425-1429.

Tweed, D. & Vilis, T. (1987). "Implications of rotational kinematics for the oculomotor system in three dimensions", *J. Neurophysiol.*, 58: 823-849.

van Opstal, A. J., Hepp, K., Suzuki, Y. & Henn, V. (1996). "Role of the monkey Nucleus Reticularis Tegmenti Pontis in the stabilization of Listing's plane", *J. Neurosci.*, 16: 7284-7296.

7

Listing's Law: what's all the hubbub?

J. Douglas Crawford

Freely translating from the colloquial, the title of this chapter refers to the controversy and interest that has recently surrounded the study of three-dimensional (3-D) kinematics of the eye. Much of this has centred around the mechanisms and visual-motor implications of Listing's law, which describes how the eyes rotate in 3-D. One might well wonder why the obscure discovery of a nineteenth century physicist (Johannes Benedict Listing) is now driving so much experimental research and theoretical debate (reviewed in Crawford & Vilis, 1995; Hepp, 1994; Fetter *et al.*, 1997). To understand how this came about, this chapter will briefly review the history of this topic, and then summarize several reasons why an understanding of Listing's law is useful for people interested in vision and action. In the course of doing this I will review most of my own contributions to this field, but only in as much as they fit into the broader context of the ongoing history of Listing's law.

7.1 A Brief History of Listing's Law: 1848-1997

Frans Cornelis Donders (1848) is attributed with the observation (arising from measurements of visual afterimages at various eye positions) that when the head is upright and stable, only one unique 3-D eye orientation is employed for each direction of visual gaze. This is a non-trivial observation, because the eye is equipped with musculature that could potentially rotate it about the line of site, i.e. obtain any one gaze direction using a variety of 3-D orientations. In other words, the saccade generator has a "degrees of freedom problem" (Crawford & Vilis, 1995), and Donders' law says that this problem is solved in an orderly fashion. Moreover, if the eye always rotated about the most direct axis to go between any two directions of visual gaze, Donders' law would be violated (e.g. Tweed & Vilis, 1990 and see below). However, Donders' law only points out that one orientation is used for each eye position: itdoes not specify exactly *which* orientation is used.

The latter was provided at approximately the same time by Listing, who was best known in his time as a professor of physics at Göttingen (Henn,

1997). Listing's law states that if each eye position is characterized by an axis of rotation from some particular reference position, then these axes will form a plane (now called Listing's plane). This plane completely specifies 3-D eye orientation for each gaze direction. Interestingly, although Listing published several works on optics and psychophysics, he never published the conclusion for which he has now been immortalized. As a result, the means by which he came to this conclusion were lost even at the time of Helmholtz (see Henn 1997, for details). Moreover, 19th century investigators did not make the distinction between the different types of eye movement that have become so important today. Nevertheless, Helmholtz (1867) recognized the significance of Listing's theory, providing a detailed mathematical description and indirectly confirming it through the use of afterimages.

Owing to the lack of technology to record 3-D eye rotations, interest in Listing's law remained largely theoretical and obscure throughout the better part of the twentieth century. Westheimer (1957) introduced the idea of quaternion representations of 3-D eye position to the eye movement field, and also provided the simple description of Listing's law that we will employ below. Nakayama (1975, 1983) illustrated Listing's law as a minimum energy strategy, and explored several of its implications for vision. However, it was not until the mid 1980's that contact lenses equipped with dual search coils were used to verify Listing's law in humans directly. Ferman *et al.* (1987a, b) showed that Listing's law was obeyed well during fixations and almost as well during saccades. Unfortunately, because their data was expressed in Fick coordinates (a head-fixed vertical axis with eye-fixed horizontal axes), it was impossible to visualize Listing's plane, determine its precise orientation, or hence, definitively evaluate its precision (i.e. thickness).

The first direct visualization of Listing's plane occurred in 1986 when Douglas Tweed and Tutis Vilis developed the technology to convert scleral search coil signals into 3-D eye position vectors and consequently into angular velocity vectors that reveal the actual axis of eye rotation (Tweed *et al.*, 1990). These measurements confirmed that Listing's law is obeyed with remarkable precision during saccades and fixation when the head is fixed and the subject (either human or monkey) looks at far targets (Tweed & Vilis 1988, 1990a). Tweed and Vilis (1987, 1990b) also formulated the first hypothetical model for a neuromuscular implementation of Listing's law. Almost immediately after the original 3-D contributions of Tweed and Vilis, similar experimental and theoretical measures where taken up by Klaus Hepp, Volker Henn and their colleagues to aid in their studies of the brainstem oculomotor system (e.g. Straumann *et al.*, 1991; Van Opstal *et al.*, 1991), and this general approach is now employed by various laboratories in Canada, Europe, Australia, Japan, and the U.S.A.

To my mind, the initial theoretical and methodological contributions of Tweed and Vilis were crucial in establishing the current approach to this field, which was exemplified in a recent conference and associated book of proceedings (Fetter *et al.*, 1997). I was fortunate to join this lab in 1986, right at the origins of this movement, and was involved in the first direct recordings of Listing's plane from non-human animals (e.g. Crawford & Vilis 1988, 1989, 1991). Since then, I have continued to employ these techniques to address several questions regarding visual-motor control. It is from this insider's (and perhaps biased) perspective that I will address the current issues surrounding Listing's law.

7.2 Behavioural Description of Listing's Law

Although the primary function of saccades is to re-direct gaze direction, they do so using the constraints provided by Listing's law, which are not evident from observations of gaze direction alone. Figure 7.1 illustrates Listing's law according to the conventions developed by Tweed*et al.* (1990). Part A shows the two dimensional distributions of horizontal and vertical eye positions relative to the head during fixations between randomly directed saccades, in a subject that sat upright with the head fixed. Note that these data are not gaze directions, but rather the tips of 3-D position vectors emanating out from the origin. The origin is thus the zero vector which in this case is the *primary position* (defined below). The illustrated non-zero vectors describe eye position relative to primary position, as follows: their length is proportionate to the angle of rotation from primary position, and their direction describes the direction of rotation according to the right hand rule. Thus, points which appear upward on the plot actually designate leftward eye positions. From this perspective (A), we see a none-too surprising distribution of vertical-horizontal positions.

Figure 7.1B shows the same data from the "side" perspective, where Listing's plane becomes visually evident. In this case, we see the vertical axis components (for horizontal position) plotted against the torsional axis, where torsion is defined as a rotation about a head fixed axis aligned with gaze direction at the primary position. We can now define the primary position as that unique eye position where gaze is orthogonal to Listing's plane. We will continue to adhere to these definitions below, but note that this definition of torsion is different than that used to describe rotation of the eye about the line of sight. From this perspective (Figure 7.1B) it is clear that the entire 3-D distribution of eye positions is indeed planar, i.e. it forms Listing's plane. Furthermore, Listing's law can now be stated quite simply: for all gaze directions ocular torsion is maintained at zero. The actual variance of eye position about this ideal zero has been quantified several times in primates and humans, and is in the order of $\pm 1^o$ for positions moderately deviated from primary position (Tweed & Vilis, 1988; 1990a;

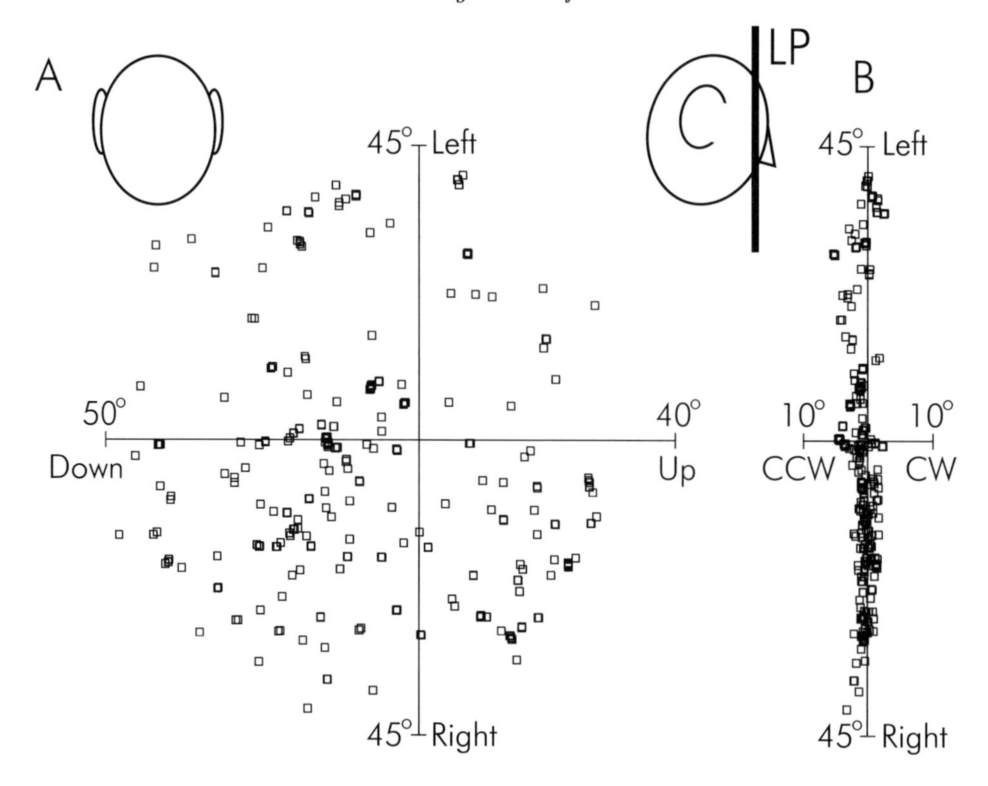

Fig. 7.1. Listing's plane in a human subject. Data represent the tips of 3-D eye position vectors emanating from the origin (primary position), as explained in the text. Head caricatures indicate the perspective used to view these vectors. **A**: Behind view of vertical and horizontal components. **B**: Side view of vertical and torsional components. The latter form a plane (Listing's Plane, LP), which may be tilted in the head as shown.

Crawford & Vilis 1991; Straumann *et al.*, 1991, 1995; Tweed *et al.*, 1994; De Souza *et al.*, 1997).

Although Listing's law is fully described as maintenance of eye position in Listing's plane, we need also to consider eye velocity to understand the full complexity of this phenomenon. Intuitively, one would expect angular eye velocities (which provide the instantaneous axis of rotation) to be confined to Listing's plane during saccades. However this is not the case. Figure 7.2 shows the angular velocity "loops" during 5 rightward saccades between two targets elevated $\sim 30^o$ above primary position (A), 5 rightward saccades between two targets located $\sim 10^o$ below primary position (B), and five more between targets at $\sim 50^o$ below primary position (which was somewhat elevated in the mechanical range of this subject). These vectors are plotted according to the right-hand rule in Listing's coordinates. Note that the axes of rotation for these saccades tilt out of Listing's plane (in contrast to the position vectors for the same saccades - not shown), such that they possess a considerable torsional component (e.g. $\sim 200^o$/s clockwise in A,

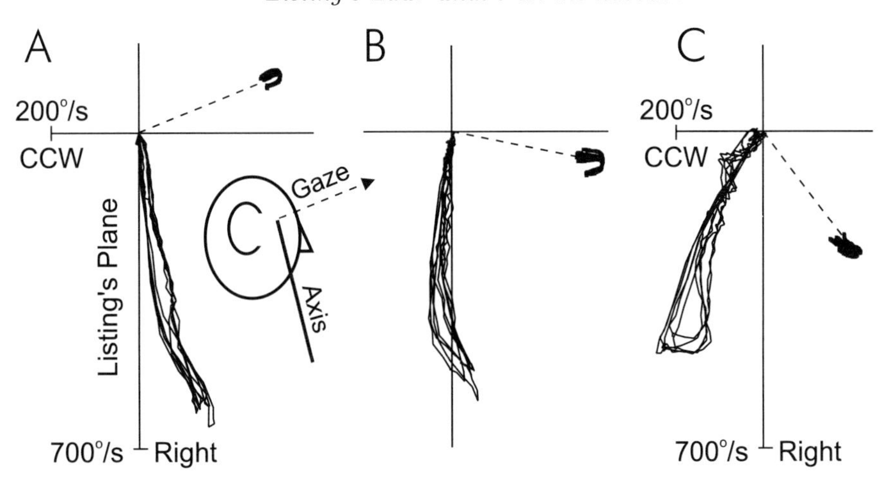

Fig. 7.2. "Half-angle rule" observed in Listing's law. Both angular velocity axes and gaze direction vectors (at ends of dashed lines) are viewed from the side in Listing's plane coordinates, during 60° rightward saccades at three vertical levels: $\sim 30^o$ above primary gaze direction (**A**), $\sim 10^o$ below primary position (**B**), and $\sim 50^o$ below primary position (**C**). The axis of rotation for saccades tilts with gaze direction but only by about half the angle.

and $\sim 250^o$/s counterclockwise in C) at peak velocity. Moreover, these tilts are systematic. According to the so- called "half-angle rule", the axis of eye rotation must tilt out of Listing's plane by half the orthogonal deviation of gaze from primary position in order to keep eye position vectors in Listing's plane (Tweed & Vilis, 1990a). As counter-intuitive as this may seem, it is a fundamental property of non-commutative rotational kinematics (Tweed & Vilis, 1987), which we will consider further below.

To summarize, Listing's law describes both the allowable range of eye positions during head-fixed saccades and the axes of rotation required to maintain this range. Although we tend to look at it from a negative point of view (i.e. that it does not allow torsion), it primarily describes which eye positions are used. For example, Donders' law could have been satisfied using any number of different types of position ranges, with different curvatures and orientations. This becomes particularly clear when one considers that the orientation of Listing's plane is not fixed in any trivial way. Surprisingly, the orientation of this position range varies considerably between subjects, and even across time within subjects (Helmholtz, 1867; Tweed & Vilis 1990a; Crawford, 1994). Therefore, the orientation of this position range must be taken into consideration in any basic science or clinical treatment of the control of eye position by neurons and muscles, or in considering the geometric relations between eye position and target direction in vision (Nakayama, 1983; Tweed, 1994; Crawford & Guitton, 1997).

Furthermore, it is misleading to think that Listing's law gets us off the hook in considering torsion. First, standard 2-D methods of measuring eye position do not give horizontal and vertical position in Listing's coordinates, but rather gaze direction. Just as torsion was always zero about our head-fixed definition of a torsional axis, conversely it is systematically non-zero about the visual-gaze definition of the torsional axis (so-called false torsion), in a pattern that is impossible to predict from 2-D measurements. Second, as we have seen, eye velocities do have torsional components (by our head-fixed definition) during saccades. Third, Listing's law only holds for artificially head-fixed eye movements while looking at far targets; for other more natural movements Listing's law is modified or violated in a systematic fashion.

To briefly expand on the latter point, Listing's law is obeyed during head-fixed saccades and pursuit movements (Haslwanter *et al.*, 1992; Tweed *et al.*, 1992), but (1) the Listing's plane for both eyes tilts outward when the eyes rotate inward during vergence (Mok *et al.*, 1992; Van Rijn & Van den Berg, 1993; Minkin & Van Gisbergen 1994), such that each eye now takes on different values of head-fixed torsion for any one gaze direction, (2) Listing's law is relaxed or abandoned during vestibuloocular (VOR) and optokinetic (OKN) slow phases (Crawford & Vilis, 1991; Fetter *et al.*, 1992; Misslisch *et al.*, 1994) so that the eye can rotate about an axis more collinear to that of the head, and (3) during head-free gaze shifts where such slow phase movements are routinely preceded by a saccade, the saccade normally drives the eye out of Listing's plane in an anticipatory fashion, such that it ends up coming back into Listing's plane (Crawford & Vilis, 1991; Guitton & Crawford, 1994; Tweed, 1997a). These observations indicate that the classic Listing's law masks a more general phenomenon of great subtlety and complexity that depends very much on a subject's behavioural state. In other words, the neural system goes to great lengths to actively choose Listing's plane as its preferred eye position range.

Figure 7.3 illustrates this point simply. These are data derived from an experiment in which we electrically stimulated the midbrain control centres for vertical and torsional eye rotations in the monkey (Crawford & Vilis, 1992). Stimulation of the right rostral interstitial nucleus of the medial longitudinal fasciulus (riMLF) nucleus produced smooth clockwise (subject's perspective) rotations (A), whereas stimulation of the left riMLF produced counterclockwise rotations (B). The important point here is that the final torsion at the end of the stimulation was held until a saccade occurred (indicated by * in Figure 7.3), which then invariably drove the eye torsionally back to Listing's plane (indicated by a dashed line). Clearly, the saccade generator has registered the presence of a torsional deviation from Listing's plane and actively corrected it. Since other motor systems (e.g. for the head, arm, and hand) show similar constraints when the third degree of freedom is unspecified by the task (Reviewed in Crawford & Vilis, 1995; Gielen *et al.*, 1997; Vilis, 1997), this seems to be a general, simplifying principle in

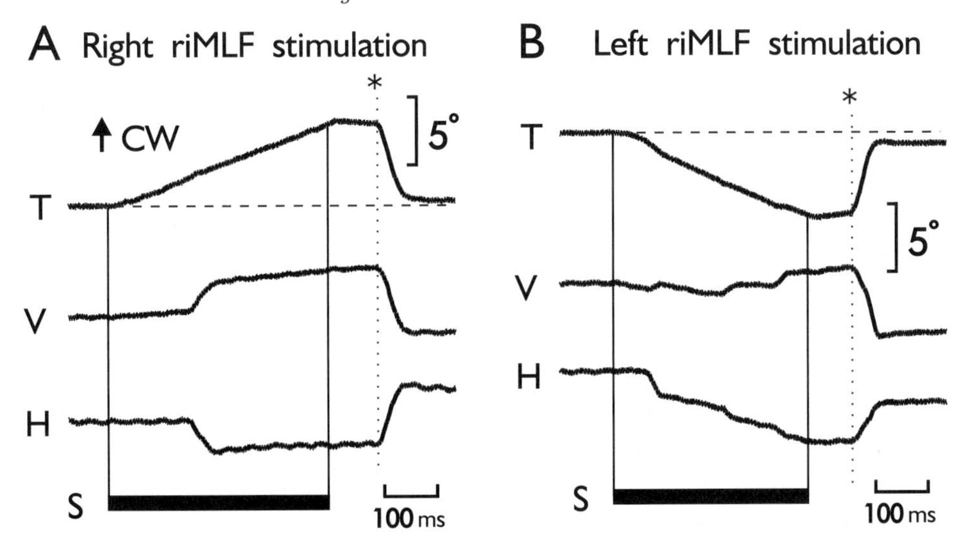

Fig. 7.3. Eye position during unilateral microstimulation of the (A) right and (B) left riMLF. Torsional (T), vertical (V) and horizontal (H) positions of the right eye are plotted as a function of time. The duration of the standard stimulus (S) (20μ A, 200 Hz) is indicated by the horizontal bar and solid vertical lines. Data is plotted in Listing's coordinates, so that torsion is relative to Listing's plane (—). Asterisks and vertical dotted lines indicate the first occurrence of a saccade after the stimulus-induced movement. CW: clockwise.

motor control (Bernstein, 1967), but what then is the physiological basis for such constraints?

This brings us to the first controversy associated with Listing's law: is it mechanical or neural in origin? Several earlier papers speculated that the eye muscles themselves might constrain the eye to Listing's plane (Ferman *et al.*, 1987a, 1987b). However, subsequent experiments, some of which are described above have shown that this is simply impossible. For example, eye muscles cannot generate a corrective torsional rapid eye movement like that shown in Figure 7.3 in the absence of a specific neural input. Nevertheless, some investigators have continued to insist that Listing's law is essentially mechanical in origin, (e.g. Schnabolk & Raphan, 1994; Raphan, 1997) for reasons that will be addressed in §7.6

In summary, the primary job of saccades and smooth pursuit is to redirect 2-D gaze direction according to the demands of behaviour and environment, but this can theoretically be accomplished in many ways. In order to know the range of 3-D eye orientations and axes of eye rotation used to accomplish this task, we must take Listing's law and its variations into account. Thus, Listing's law will influence every aspect of spatial vision, the visuomotor transformation for saccades (and pursuit), the motor control of eye movement and position, and the actions of the eye muscles themselves. I will now deal with each one of these stages in some detail.

7.3 Some Visual Consequences of Listing's Law

Most visual scientists take great care to establish the spatial relationships between stimuli, but then ignore the 3-D orientation of the eye, either as if it does not move at all, or as if the retina were a camera that translates in parallel to the stimulus screen. Listing's law potentially simplifies spatial vision by limiting the number of eye orientations used to view targets (Helmholtz, 1867), but even for eye positions in Listing's plane the spatial pattern of retinal stimulation is a non-trivial function of both objective target locations and 3-D eye orientation (Crawford & Guitton, 1997). Figure 7.4 demonstrates this point with the use of a very simple stimulus array commonly used in saccade studies. Part A represents five horizontal pairs of stimulus lights, at five vertical levels, plotted relative to the primary gaze direction. The subject fixates on the leftward member of each pair (▪), then the rightward member (×) flashes, and then the subject would saccade leftward toward it after it has been turned off, ie., with no visual feedback. The point on the retina stimulated by the flashed target defines the "retinal error" relative to the fovea. It might be assumed that this task stimulates "horizontal" retinal error at each of the five stimulus pairs, but this is not true.

Figure 7.4 B shows the associated retinal errors, now in eye coordinates. This has been simulated assuming that the primary gaze direction is centred at the origin in Figure 7.4A, and that the eye is in Listing's plane at the initial leftward fixation directions. The target direction relative to the eye is then computed by rotating target direction in head coordinates (A) by the inverse of 3-D eye orientation (Crawford & Guitton, 1997). This shows that retinal error depends not only on objective target displacement, but also on initial eye orientation, where the upper pair produced rightward-upward retinal error (in eye coordinates) and the bottom pair produced rightward-downward retinal error. A recent experimental quantification of this effect in humans has confirmed this effect, and shown that it can produce large ($> 20^o$) variations in retinal error as a function of eye position in the oculomotor range (Klier *et al.*, 1997).

This dependence of retinal error on eye orientation is particularly problematic for the assumptions about retinal coding made in all 2-D models of the saccade generator, but more generally it is a problem for any aspect of spatial vision: i.e., it is impossible to compute even relative target direction in space without taking 3-D eye orientation into account.

This becomes particularly relevant when one considers the geometric problems of binocular vision and stereopsis. On the one hand, Listing's law aids visual fusion because cyclotorsion (about the visual axis) changes the correspondence of retinal images between the two eyes. For example, the strong tilt observed in the vertical horopter (Nakayama, 1983) requires binocular

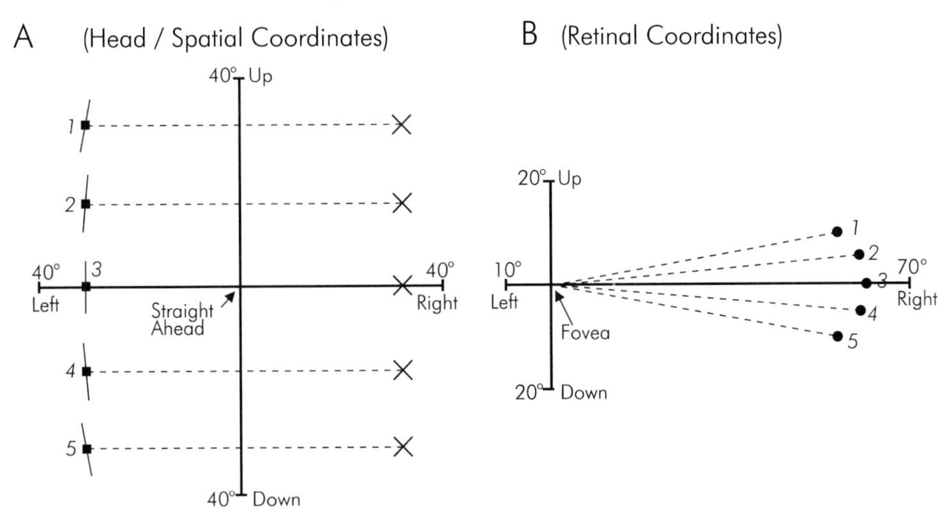

Fig. 7.4. Reference frame problem in comparing oculocentric retinal error with headcentric target displacements or gaze shifts. **A**: target directions in Listing's coordinates, i.e., horizontal and vertical components of unit direction vectors are plotted relative to primary gaze direction and projected onto a plane parallel to Listing's plane. Initial gaze direction (▪) was simulated at 30° to the left and at five different vertical levels $(1-5)$ from 30° up to 30° down through 15° intervals. Quasi-vertical lines indicate the approximate eye orientation, ie. as the lines of projection onto the vertical retinal meridian near the fovea. Five simulated targets $(\times)$ were presented to the right (in craniotopic coordinates) in the opposite quadrant. **B**: Retinal errors (•) computed from the five horizontal pairs of initial 3-D eye positions and target directions in A.

convergence between two non-parallel retinal lines. Geometry dictates that this horopter will degrade when the eyes rotate torsionally about their lines of sight, even if the movement is conjugate (Crawford & Vilis, 1991; Howard & Zacher, 1991). However, Listing's law is not the ideal solution for this problem, because it does produce so-called "false torsion" (torsion about the visual axis) at oblique eye positions. This is further complicated by the tilting effect of Listing's planes during vergence (Mok *et al.*, 1992, Van Rijn & Van den Berg, 1993; Minkin & Van Gisbergen, 1994). Based on these factors, Tweed (1994) formulated a binocular theory of Listing's law, which he suggests is the optimal compromise between ideal binocular alignment of images, and minimizing the angular displacement of the eyes from primary position. However, even in his model the visual system still needs to take 3-D orientation (now of both eyes) into account in order to fuse the two images correctly.

7.4 Implications for the Visuomotor Transformation for Saccades

The visual-motor transformation for saccades is an intensely studied area that is very well understood in some ways and very poorly understood in others. It is well understood in terms of the identification of the multiple cortical and subcortical areas and cell-types that contribute to the process, but poorly understood in terms of the signal transformations that occur between these areas. For example, a persistent debate has continued between the view that raw visual signals can be converted directly into displacement commands for saccades (Jürgens *et al.*, 1981; Hepp *et al.*, 1997; Raphan, 1997) and the view that this process requires intermediate representations of desired eye position and comparisons with current eye position (Zee *et al.*, 1976; Crawford & Guitton, 1997). This controversy touches on the mechanisms for space constancy when the eye rotates between the time that the subject *sees* a visual target and the time that he decides to act on it, e.g., with a saccade or arm movement (Duhamel *et al.*, 1992; Henriques *et al.*, 1998), but I will confine my discussion here to the more direct transformation from visual input about a target to a motor command for a saccade directly toward that target. Since Listing's law and the associated phenomena describe the input-output relationships for such movements, they place important constraints on the way that this process is performed.

In the original Tweed and Vilis (1990b) model of the 3-D saccade generator, an internal representation of desired target direction was fed into a "Listing's law operator" which computed the unique corresponding desired 3-D orientation of eye position within Listing's plane. Some now call this the Donders' law operator (e.g. Hepp), but for head-fixed saccades, Listing's law is the expression of Donders' law and so I will retain the original terminology used by Tweed and Vilis (1990b). This desired position was then compared multiplicatively with current 3-D eye position to compute a motor error command that explicitly encoded the desired *axis* of eye rotation, which would incorporate the half-angle rule for Listing's law (see Figure 7.2). Tweed and Vilis (1990b) and Crawford and Vilis (1991) predicted that the these motor error commands would be mapped within the deep layers of the superior colliculus. However, stimulation studies subsequently suggested that this map encodes 2-D gaze displacements, such that the position-dependent axis tilts seen in Listing's law must be implemented downstream* (Van Opstal *et al.*, 1991; Hepp *et al.*, 1993). From these data it was suggested, and widely accepted, that Listing's law is implemented downstream from the superior colliculus, and that the saccade generator

* This does not seem so surprising now that we have accepted that these axis tilts may be implemented by the muscles themselves (Demer *et al.*, 1995).

employs a direct "look-up table" mapping from retinal error vectors onto motor displacement commands for saccade vectors in Listing's plane (Hepp *et al.*, 1993).

Unfortunately, these conclusions were premature for two reasons. First, although the colliculus stimulation experiments do indicate that the half-angle rule for Listing's law is implemented downstream, this is most closely associated with the problem of velocity-position matching described in §7.6. Indeed, a plant equipped with pulleys for the half-angle rule will easily violate Listing's law for arbitrary neural inputs (Smith & Crawford, 1997; Tweed, 1997b). In contrast, the experiment says nothing about the actual Listing's law operator, which performs the 2-D to 3-D transformation, solves the degrees-of-freedom problem, and thus determines the width, shape and orientation of the allowable range of eye positions (i.e. Listing's plane). It would seem that this point was largely glossed over because these two different processes, the 2-D to 3-D transformation and the application of a half-angle rule, occurred in rapid succession in the Tweed-Vilis (1990b) model. Crawford and Guitton (1997) pointed out that these are most likely quite separate processes and have simulated experiments that could show the relative location of the Listing's law operator (see also Hepp, 1994; Crawford & Vilis, 1995). This 2-D to 3-D transformation would normally be quite invisible experimentally, but would show up more obviously in the act of correcting deviations from Listing's plane (e.g. Figure 7.3). See also Van Opstal *et al.* (1996).

The second problem with the vector look-up table idea is that retinal error and eye displacements are not the same in 3-D (Hepp *et al.*, 1993). Since the former are oculocentric and the latter are headcentric, they deviate from each other as a function of eye position in much the same way that eyecentric vectors deviated from headcentric vectors in external physical geometry (Figure 7.4. Crawford and Guitton (1997) offered a model (Figure 7.5) which solves this problem by rotating (here is where non-commutativity comes in again) retinal error by the inverse of current eye orientation before inputting it to the Listing's law operator (LL). The model simulated accurate saccades, whereas simulation of the vector look-up table model (Hepp *et al.*, 1993, 1997; Raphan, 1997) predicted position-dependent errors in saccade accuracy, primarily due to errors in saccade direction.

Both Crawford and Guitton (1997) and Hepp (1997) have pointed out that implementing the half-angle rule improves these errors compared to a direct mapping of retinal error onto head-fixed axes of rotation, but at the same time it prevents a complete oculocentric (or full-angle rule) solution to the problem. (For example, it prevents the saccade generator from trivially mapping the non-horizontal retinal error vectors in Figure 7.4 B onto the purely horizontal saccades (in terms of muscle activation) required to accurate foveate the targets in Figure 7.4 A). However, Hepp (1997) has

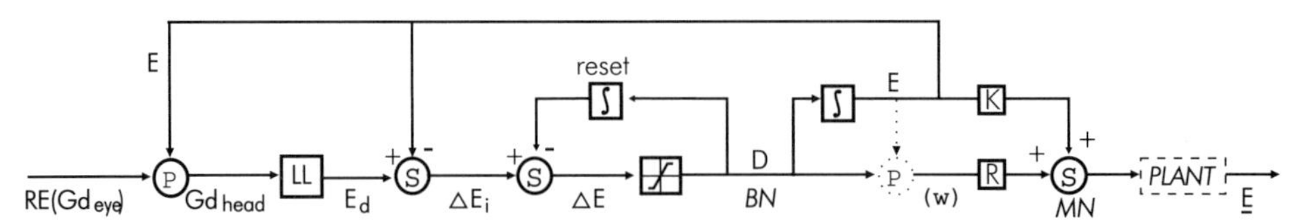

Fig. 7.5. A model for generation of saccades in 3-D. Retinal error (RE) is represented as the vertical and horizontal components of desired gaze direction relative to the eye (Gd_{eye}). Gd_{eye} is then rotated multiplicatively (P) by 3-D eye position (E) into the craniotopic reference frame (Gd_{head}). Here E is derived from the output of the downstream neural integrator, but it could also be derived from sensory organs in the eye muscles. Desired 3-D eye position (E_d) in Listing's plane, is then computed from Gd_{head}, using an operation (LL) first described by Tweed and Vilis (1990b). Subtracting E from E_d yields the initial desired change in eye position ΔE_i, which completed the visuomotor transformation. During the saccade, displacement feedback from a resettable integrator (S) is subtracted from ΔE_i to compute the instantaneous 3-D motor error (ΔE) that drives burst neurons (BN). The outputs of this feedback loop are the torsional, vertical and horizontal components of rate-of-position-change (D). Since D is the derivative of position with respect to time, its components are input directly into three integrators ($\int$) that generate the torsional, vertical, and horizontal components of E. D and E vectors are left multiplied by the plant viscosity (R) and elasticity (K) matrix constants respectively before summing componentwise at the motoneurons (MN). With the standard plant model (Tweed & Vilis, 1987; Crawford & Vilis, 1991), E is divided by a copy of E (dotted arrow) to produce a 3-D angular velocity command (w). With the linear plant model, the latter step was unnecessary. $\underline{E}$: actual eye position. After Crawford & Guitton, 1997.

suggested that this indeed is the purpose of Listing's law, i.e. to satisfy Donders' law while allowing a sensory-motor approximation that gives tolerable errors in the central oculomotor range. Although I agree with this basic principle, it cannot be the penultimate explanation of Listing's law because our recent experiments show that the brain does not make this approximation, i.e. the saccade generator makes accurate eye movements from all initial eye positions and therefore must perform the position-dependent reference frame transformation necessary to convert retinal error into motor commands correctly (Klier *et al.*, 1997). At this time however, the physiology experiments necessary to show where these processes occur have not yet been performed (but see Crawford & Guitton, 1997).

Thus, the input-output description given by Listing's law requires us to reject the vector look-up table hypothesis of the visuomotor transformation for saccades. Instead, this process must employ a position-dependent reference frame transformation and a 2-D to 3-D transformation that at least implicity encodes desired final eye positions (Figure 7.5; Crawford & Guitton, 1997). Although this latter model handles several of the new features described above and has a new configuration, it borrows heavily from the basic kinematic principles in Tweed and Vilis (1990b) and shows a remarkable, if superficial, resemblance to the original local feedback model of Robinson and colleagues (Zee *et al.*, 1976).

7.5 Implications for Neural Coordinate Systems

The term "neural coordinate system" may sound esoteric, but in this context I am simply referring to the way that the brainstem is compartmentalized for the differential control of various eye movement directions. As described above, Listing's plane provides a natural behavioural coordinate system for saccades and smooth pursuit. Moreover, even the behavioural coordinates of the VOR (although it does not obey Listing's law) seems to be influenced by Listing's law, in that the axis of minimal VOR gain aligns with the primary gaze direction (Crawford & Vilis, 1991). Do these behavioural coordinate systems have a neural correlate?

In fact they do. This has been treated at length elsewhere (Crawford & Vilis, 1992; Crawford, 1994; Crawford *et al.*, 1997) so I will only summarize the story here. It so happens that the premotor centres for saccade generation are compartmentalized within the brainstem into anatomically separate populations with directional control similar to the directions controlled by the individual extraocular muscles and sensed by the individual semicircular canals (Figure 7.6 B, C). The populations for torsional / vertical components are found mainly in the midbrain, divided into clockwise-up and clockwise-down populations on the right side and counterclockwise-up and counterclockwise-down on the left side (Henn *et al.*, 1989; Crawford *et al.*, 1991; Crawford & Vilis, 1992). In contrast, the centres for horizontal rotation are found in the pons (Luschei & Fuchs 1971) and medulla (Cannon & Robinson, 1986). By "centres" I mean both the "burst centres" that drive eye velocity during saccades and the separate "tonic centres" that are thought to mathematically integrate the former velocity signal to derive the signal that holds final eye position. It was previously debated whether the coordinates of these centres would align best with muscle or canal coordinates. However, there is good theoretical reason to believe that the neural coordinates might correlate best with Listing's plane, such that the vertical axis (for horizontal rotation) is aligned with Listing's plane and the horizontal / torsional axes are arranged symmetrically across Listing's plane (Crawford & Vilis, 1992; Crawford *et al.*, 1997).

The latter coordinate system is necessary for the vertical axis encoded by the horizontal populations to code rotations independently in Listing's plane, and for the clockwise and counterclockwise populations to cancel each other in order to encode zero head-fixed torsion at all points relative to Listing's plane (Crawford *et al.*, 1997). This particular sort of simplicity makes biological sense because Listing's plane is the default range of eye positions and should therefore take precedence. Since the orientation of

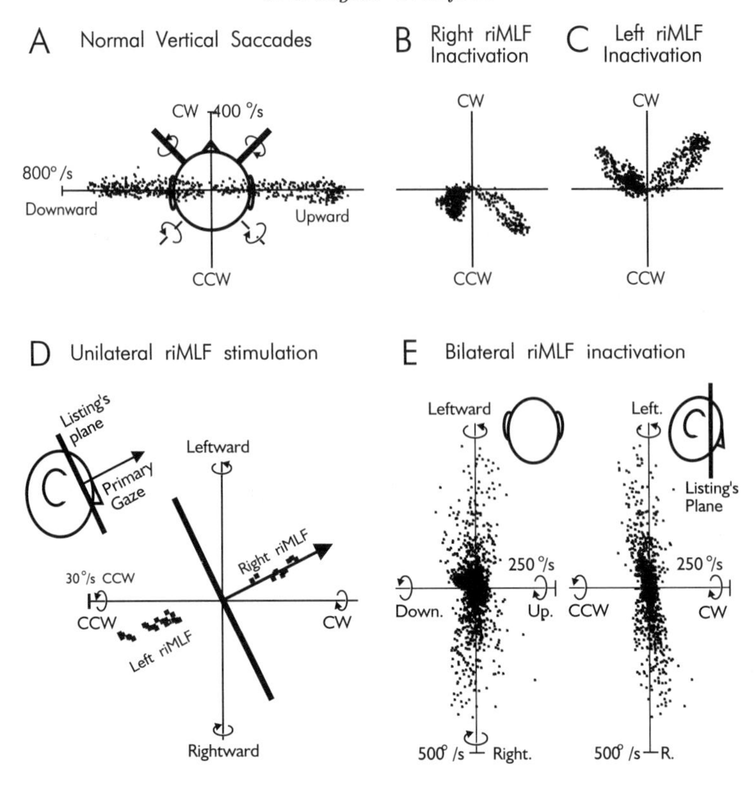

Fig. 7.6. Saccade-related burst neurons utilize a head-fixed coordinate system that resembles that of the vestibular canals or eye muscles, but aligns with Listing's plane. Top Row: The canal/eye muscle-like nature of burst coordinates. **A**. Instantaneous angular velocity "loops" during normal upward and downward saccades in a trained monkey, as viewed from above the head. Purely horizontal axes (for vertical rotation) are presumably generated by combining one of the clockwise-horizontal axes controlled by the right riMLF burst neuron populations (solid axes) with one of the counterclockwise-horizontal axes controlled by the left riMLF, such that the torsion cancels out. **B**, **C** The intact axes controlled by the opposite side are revealed when the neurons on the other side are pharmacologically inactivated and the animal attempts to make vertical saccades (Crawford & Vilis, 1992). These remaining axes are very similar to the axes of rotation sensed by the vertical semicircular canals, and those controlled by the vertical eye muscles. Lower Row: Alignment with Listing's plane. **D** During unilateral riMLF stimulation, axes of eye rotation (shown as points representing average angular velocities) align orthogonally to Listing's plane, independent of anatomic coordinates. **E** Similarly, following bilateral inactivation of the riMLF the remaining burst neurons (presumably those located in the PPRF) are essentially only able to generate a vertical axes of eye rotation aligned in Listing's plane (Crawford and Vilis 1992).

Listing's plane seems to be determined behaviourally and varies considerably compared to anatomic structures, this predicts that these neural coordinates should be measurably different from the anatomic coordinates.

We tested this prediction using electrical microstimulation and micro-injection of inhibitory drugs into the primate midbrain burst and tonic centres, and found that indeed, the movement deficits and/or induced movements did align with Listing's plane in the predicted fashion, and showed little quantitative correlations with anatomic coordinates (Crawford & Vilis, 1992; Crawford, 1994 and see Figure 7.6D, E). These observations are important because they demonstrate that non-trivial, physiologically-meaningful coordinate systems do exist at the level of neural populations, which was a matter of controversy in itself (Robinson, 1992), and show that such coordinate systems are probably optimized for behaviour. This leads to general hypotheses for the neural coordinates for various motor systems and behaviours (Crawford & Vilis, 1995; Flanders *et al.*, 1992; Crawford *et al.*, 1997), and confirms a basic assumption in the theoretical models discussed below.

7.6 Velocity-Position Matching and the Non-Commutativity of Rotations

Probably the most intense controversy surrounding Listing's law relates to the oculomotor mechanism for velocity-position matching (see §7.3 and also Optican and Quaia, this volume, for a more detailed and mathematically-rigorous description). The controversy began when Tweed and Vilis (1987) pointed out that the idea of neurally integrating the burst neuron velocity signal to obtain an accurate eye position signal is problematic in 3-D. Their central point was that 3-D rotations, being non-commutative (i.e. rotation about axis "A" followed by "B" is different from rotation about "B" then "A"), show a non-linear, position-dependent relationship between angular velocity and angular position. There can be no argument about these physical laws, but what do they portend for eye movement control?

To illustrate this, note that angular velocity vectors tilt out of Listing's plane (Figure 7.2) whereas eye position vectors stay in Listing's plane (Figure 7.1), even during head-fixed saccades. If the brain explicitly encoded 3-D angular velocity and integrated each of its components (i.e. there *is* a head-fixed torsional integrator; Crawford *et al.*, 1991), then an incorrect head-fixed torsional position signal would arise and cause eye position to drift out of Listing's plane. Tweed and Vilis (1987) solved this theoretical problem by positing that the eye velocity command is in effect multiplied by 3-D eye position feedback before it is input to the neural integrator(s). Although in outline this scheme was not radically different from the standard integrator hypothesis (i.e. there still was an integrator), it did represent a mathematically-important departure from the linear systems approach made famous by the seminal work of David A. Robinson. The Tweed-Vilis model made the correct behavioural predictions for both saccades (Tweed & Vilis, 1990a) and the VOR (Crawford & Vilis, 1991), but in retrospect, it would appear that this theory was received with a great deal of caution within the mainstream oculomotor community.

In the mean-time, a small group of investigators, including Doug Tweed, Klaus Hepp and myself, began to question some of the assumptions behind the Tweed-Vilis model. Based on the sketchy data available in the mid 1980's, Tweed and Vilis (1987) had assumed that saccade-related burst neurons do indeed encode angular velocity and that the eye muscle pulling directions are essentially independent of eye position. However, while still in the Vilis lab, Tweed was well aware that burst neurons might encode something else, namely the derivative of eye orientation (see the discussion in Tweed and Vilis 1990a) which *could* be input directly into a neural integrator without altering the basic Robinson scheme. This became more relevant when burst neuron and neural integrator data from my experiments and from those of Hepp, Henn, and associates (Crawford 1994; Hepp *et al.*, 1994; Reviewed in Crawford & Guitton, 1997) seemed to be more consistent with this derivative scheme.

Since these derivatives would ordinarily be parallel to Listing's plane, this would still require a non-commutative mechanism to supply the required position-dependent axis tilts (Figure 7.2), but this could conceivably be done by plant mechanics (Crawford & Vilis 1995). Indeed, by the early 1990's it was rumoured that fibro-muscular "pulleys" surrounding the muscle insertions could produce such a position-dependency (Demer *et al.*, 1995). With burst neuron and integrator coordinates aligned with Listing's plane (Crawford and Vilis, 1992; Crawford, 1994), this suggested a very elegant formulation of Listing's law wherein the neurons encoded zero-torsion position and movement vectors in Listing's coordinates and the muscles handled the position-dependencies (portrayed schematically in Figure 7.5). As a result, Tweed and I began tentatively to incorporate this scheme into our models (e.g. Crawford, 1994; Crawford & Guitton, 1994; Tweed *et al.*, 1994), employing variations of the simplistic "linear plant" model to emulate the complex plant. However, at that time this remained a matter of fairly esoteric and restricted interest.

This atmosphere changed when Schnabolk and Raphan (1994) published a paper which was intended as a direct criticism of the original model of Tweed and Vilis (1987). Schnabolk and Raphan felt that the Tweed-Vilis model was fundamentally flawed because they had failed to take into account the properties of the plant, in particular, that although rotations do not commute, torques do add commutatively. It has already been countered in print that the sequential non-commutativity of rotations and the simultaneous commutativity of torques are two separate issues with little to do with each other (Tweed *et al.*, 1994), but what of the claim that the internal representations of rotational kinematics are an unnecessary complication (Schnabolk & Raphan, 1994)? The model proposed by Schnabolk and Raphan is itself very complicated, but essentially boils down to a 3-D version of the commutative Robinson-style control system driving a non-commutative plant model with the same basic properties used in the original Tweed-Vilis model. When driven by a "velocity" vector parallel to Listing's plane, this model produced a 3-D "pulse - step mismatch" such that it drove the eye toward the correct final orientation, but using an incorrect trajectory, much like the predictions of the "straw-man" model simulated by Tweed and Vilis in 1987. This predicted (for either saccades or the VOR) that the eye would drift out of Listing's plane during the phasic part of the movement, but then drift back into Listing's plane at the end of the movement with the intrinsic time constant of the plant.

Three years before Schnabolk & Raphan's model was published, however, the primate VOR had been demonstrated not to behave in the way that the Schnabolk-Raphan model predicted. In 1991 Crawford & Vilis demonstrated after the VOR drives eye position out of Listing's plane, the eye does not drift back. And Tweed *et al.* (1994) demonstrated that the saccde generator also does not behave in the way that Schnabolk & Rapahan's model predicts: in fact only tiny deviations from Listing's plane are observed rather than the large drifts and subsequent corrections predicted by Schnabolk & Raphan. Nevertheless, the strong claims of Schnabolk & Rapahan's paper seemed to strike a deep chord within the oculomotor community, perhaps due to the long tradition of the linear systems approach to modelling.

Perhaps the most important contribution of the Schnabolk and Raphan (1994) paper was to raise the general consciousness of these problems amongst the mainstream oculomotor community, and particularly the importance of the oculomotor plant. (Tweed and Vilis had tried to do the same thing by pointing out that a simple plant would be non-commutative, but somehow the message did not stick). In response to this, Straumann *et al.* (1995) showed that altering the characteristics of the plant model can vastly improve the performance of the Schnabolk and Raphan (1994) saccade model. Furthermore, by incorporating "pulleys" into his plant model that induce approximately 50% muscular position dependency (Demer *et al.*, 1995), Raphan (1997) has now essentially eliminated the 3-D pulse - step mismatch of the saccade generator. Of course what this is actually doing (as rigorously documented by Quaia and Optican, 1997 and Optican and Quaia, this volume) is implementing the simpler "orientation derivative/linear plant" scheme proposed earlier by Tweed and colleagues (Tweed & Vilis, 1990; Crawford, 1994; Tweed *et al.*, 1994; Crawford & Guitton

1994, 1997). Thus, in fact, there is actually considerable agreement that this is a valid and reasonable scheme, although the details of mathematical representation differ and decisive proof that the plant behaves this way is lacking.

Supposing that eye muscles do implement the axis tilts shown in Figure 7.2 and thereby help to solve the problems of matching phasic and tonic commands in the saccade generator, does this mean that we can forget about non-commutativity? Absolutely not. First of all, eye rotations do not commute (that is a physical law); what we are talking about is a way to control eye rotations without taking their non-communtativity into account in the internal velocity-position matching mechanism for saccadic generation (Quaia & Optican, 1997). Second, because the vestibular canal afferents cannot encode eye orientation derivatives, the problems of velocity-to-position transformation demonstrated by Tweed and Vilis (1987) and Crawford and Vilis (1991) cannot be avoided in the VOR (Smith & Crawford, 1997; Tweed, 1997b). In other words, with the new plant model, the multiplicative position feedback proposed by Tweed and Vilis (1987) is not removed but rather re-configured (Smith & Crawford, 1997; Tweed, 1997b). This can be thought of as an internal "undoing" of the muscular half-angle rule, which is not good for an ideal VOR (Vilis 1997). The fact that actual VOR axes (in humans at least) show a "1/4 angle rule" in the dark (Misslisch *et al.*, 1994) may indicate that this compensation itself is less than ideal. Third, the new plant model cannot remove the need for a non-commutative reference frame transformation in the visual-motor interface for saccades that was discussed above (Crawford & Guitton, 1997). Finally, we have only recently pointed out (Henriques *et al.*, 1998) that a much more dramatic example of non-commutativity comes into play in the "remapping" of visual target directions during eye movements (e.g. Duhamel *et al.*, 1992). For example, the current "vector - subtraction" models for this process do not work in real 3-D space, where target representations must be multiplicatively rotated by the inverse of eye rotation in order to maintain the correct spatial registry, independent of plant mechanics (Henriques *et al.*, 1998).

7.7 Implications of Listing's Law for Orbital Mechanics

Each eye is equipped with a pair of muscles that rotate the eye horizontally and four muscles that participate in both vertical and torsional eye rotation. In our head-fixed coordinate system, these muscles have a relatively simple organization, rotating the eye clockwise-up, clockwise-down, counterclockwise-up and counterclockwise-down. With this arrangement, two muscles (i.e. one vertical rectus and one oblique muscle) must be co-activated to produce a purely vertical rotation, e.g., clockwise-up + counterclockwise-up to produce upward rotation (Crawford *et al.*, 1991;

Crawford & Vilis, 1992). Similarly, co-activation of muscles with similar torsional tuning provides for the torsional movements discussed above. Due to the apparent difference between muscle coordinates and Listing's plane, this scheme may only be approximately correct (e.g. equal coactivation of the superior rectus and inferior oblique may not give zero torsion in Listing's coordinates for all gaze directions), but would still hold in approximation. In contrast, the more common eye-fixed definition of torsion gives rise to a more confusing picture of vertical muscle actions: e.g. that the superior rectus produces vertical rotation when the eye is abducted and produces eye-fixed torsional rotation when the eye is adducted. This is true as far as that goes, but it gives rise to an error that arises again and again in basic ophthalmology and physiology textbooks. This is the idea that, for example, the superior rectus is preferentially used to generate upward eye rotations when the eye is abducted, whereas the inferior oblique is used when the eye is adducted. This may be true for the VOR, but during saccades this would cause the un-opposed headcentric torsional actions of these muscles to drive eye position out of Listing's plane. Since this does not occur, clearly the system uses the co-activation principle described above. Thus, a more accurate consideration of 3-D eye muscle actions during various oculomotor behaviours is warranted in the clinical literature. This surely has implications for considering the deficits associated with eye muscle damage, but only preliminary work has been done in this area.

A second consideration is the role of the eye muscles in implementing the saccade axis tilts illustrated in Figure 7.2. As described above it has been suggested that these are implemented by position-dependencies within the eye muscles themselves, in particular by the pulley-like actions of the tissues near the insertions of the muscles (Demer *et al.*, 1995). In addition, it would be useful to know if position-dependent viscosities contribute to these tilts (Crawford & Guitton, 1997). As described above, such mechanical position-dependencies would have important implications for understanding the neural processing for saccades, but so far much of our understanding of eye muscle physiology is currently derived from simulations based on static anatomic measurements (Miller *et al.*, 1997). Direct observations of eye muscle actions tilting dynamically with axes during saccades would be very valuable, but are not yet available.

Finally, in all 3-D models to date, it is assumed that the mechanical resting position for the eyes (e.g. Tweed *et al.*, 1987; Schnabolk & Raphan, 1994) aligns with the primary position of the eyes as defined by Listing's law. This is in line with Nakayama's (1983) interpretation of Listing's law as an energy saving mechanism, since it would minimize the eye's excursion from this point. It is conceivable that the neural system does indeed choose this position as the primary position for Listing's plane at far targets, but as far as I am aware, there is currently no proof or even serious theoretical consideration of this. Furthermore, it raises the question of why then would primary position (and the orientation of Listing's plane) vary over time in one subject. Clearly a lot of work has yet to be done on the relationship between eye muscle mechanics and Listing's law, but this is of primary importance for understanding oculomotor function.

7.8 Concluding Notes

There has probably been more work done (and certainly more written) on the three-dimensional kinematics of eye movements in the last ten years than in the preceding history of science. In this period we have gathered many new insights, and yet important questions remain. Probably the most embarrassing outstanding question concerns the precise purpose of Listing's law, i.e. why this and not some other form of Donders' law, such as the "Fick strategy" observed in head and arm movements (Gielen *et al.*, 1997; Vilis, 1997)? This chapter has touched upon several theories in passing, but a definitive answer remains elusive. Several investigators now believe that Listing's law is a compromise between various sensory and motor cost / benefit functions (Tweed, 1994; Hepp *et al.*, 1997), but this is a difficult story to tell and even more onerous to prove. And what of the neural mechanisms for Listing's law? More physiological work is needed to understand how the 2-D retinal code is mapped onto the 3-D burst neuron code, not only because this is intimately associated with the implementation of Listing's law, but more importantly because this is an integral part of all aspects (head-fixed and head-free) of controlling gaze direction (Crawford & Guitton, 1997). Finally, we will not be able to resolve the question of non-commutativity fully in the velocity-position matching mechanism for saccades until a complete description of functional position-dependencies in the plant is available.

Another important area for future research is the visual consequences of Listing's law and the related phenomena described above, taking both accurate descriptions of stimulus configurations and experimental measurements of 3-D eye orientation into account. This will be particularly important in patients with eye muscle damage or abnormalities. Since the neural system has only a limited capacity to compensate for orbital damage, this will be critical for understanding the resulting visual deficits (related to perceptual consequences of Listing's law described above and surgical treatment of eye muscle anomalies). However, such research will require a multidisciplinary approach to vision, kinematics, and eye muscle mechanics that has only been attempted in preliminary form at this time.

Finally, if the relatively well-understood saccade generator is to serve as a model for sensory-motor transformations in general, then it is of interest to note that most motor systems experience similar "degrees of freedom problems" and those that have been studied show solutions that are very much analogous to Listing's law (reviewed in Crawford & Vilis, 1995; Gielen *et al.*, 1997; Vilis, 1997). Furthermore, non-commutativity applies to every type of joint or segmental rotation. To me at least, it seems rather unlikely that the entire body musculature has been configured to compensate for this one computational problem, when it is really not much of a computational challenge for neural nets. It simply requires them (and us) to go beyond the standard math of the typical electrical engineering text, just as robotics engineers have been forced to do in order to deal correctly with the kinematics of artificial plants (e.g. Funda & Paul, 1990).

Acknowledgments

JDC thanks the following: M. Smith, E. Klier, D. Henriques, and M. Ceylan for helpful comments on this manuscript, and T. Vilis and D. Tweed for kinematic inspiration. This work is supported by grants from the Canadian Medical Research Council, the Canadian National Science and Engineering Research Council, and the Sloan Foundation. JDC is a Canadian MRC Scholar and an Alfred P. Sloan Fellow. This manuscript is dedicated to the memory of Volker Henn (1942-1997).

References

Bernstein, N. (1967). *The Coordination and Regulation of Movements*, Pergamon Press, Oxford, UK.

Cannon, S.C., Robinson, D. A. (1987). "Loss of the neural integrator of the oculomotor system from brainstem lesions in the monkey", *J Neurophysiol.*, 57: 1383-1409.

Crawford, J. D. (1994). "The oculomotor neural integrator uses a behavior-related coordinate system.", *J. Neurosci.*, 14: 6911-6923.

Crawford, J. D., Cadera, W. & Vilis, T. (1988). "The oculomotor velocity to position transformation involves the interstitial nucleus of Cajal", *Soc. Neurosci. Abstr.*, 14: 386-388.

Crawford, J. D., Cadera, W. & Vilis, T. (1991). "Generation of torsional and vertical eye position signals by the interstitial nucleus of Cajal", *Science*, 252: 1551-1553.

Crawford, J. D. & Guitton, D. (1994). "A Model of the sensorimotor transformations required for accurate 3-D saccades", *Soc. Neurosci. Abstr.*, 20: 234.

Crawford, J. D. & Guitton, D. (1997). "Visuomotor transformations required for accurate and kinematically correct saccades", *J. Neurophysiol.*, 78: 1447-1467.

Crawford, J. D., Vilis, T. & Cadera, W. (1989). "Quick phase planes anticipate violations of Listing's law produced by slow phases", *Soc. Neurosci. Abstr.*, 15: 211-215.

Crawford, J. D. & Vilis, T. (1991). "Axes of eye rotation and Listing's law during rotations of the head", *J. Neurophysiol.*, 65, 407-423.

Crawford, J. D. & Vilis, T. (1992). "Symmetry of oculomotor burst neuron coordinates about Listing's plane", *J. Neurophysiol.*, 68: 432-448.

Crawford, J. D. & Vilis, T. (1995). "How does the brain deal with the problems of rotational movement?", *J. Motor Behav.*, 27: 89-99.

Crawford, J. D, Vilis, T. & Guitton, D. (1997). "Neural coordinate systems for head-fixed and head-free gaze shifts", In M. Fetter, T. Haslwanter, H. Misslisch, & D. Tweed *Three Dimensional Kinematics of Eye, Head, and Limb Movements*, 43-56, Harwood Academic Publishers: Amsterdam.

Demer, J. L, Miller, J. M., Poukens, Y., Vinters, H. V. & Glasgow, B. J. (1995). " Evidence for fibromuscular pulleys of the recti extraocular muscles", *Invest. Ophthal. Vis. Sci.*, 36: 1125-1136.

DeSouza, J. F., Nicolle, D. A. & Vilis, T. (1997). "Task-dependent changes in the shape and thickness of Listing's plane", *Vis. Res.*, 37: 2271-2282.

Donders, F. C. (1848). "Bietrag zur Lehr von den Bewigungen des menschlichen Auges", *Holländeshen Beiträgen zu den Anatomischen und Physiologischen Wissenschaften*, 1: 105-145.

Duhamel, J.-R., Colby, C. L. & Goldberg, M. A., (1992). "The updating of the representation of visual space in parietal cortex by intended eye movements", *Science Wash.*, 255: 90-92.

Ferman, L., Collewijn, H., & van den Berg, A. V. (1987a) "A direct test of Listing's law. I. Human ocular torsion measured in static tertiary positions", *Vis. Res.*, 27: 929-938.

Ferman, L., Collewijn, H., & van den Berg, A. V. (1987b) " A direct test of Listing's law. II. Human ocular torsion measured under dynamic conditions", *Vis. Res.*, 27: 939-951.

Fetter, M. Haslwanter, T., Misslisch, H. & Tweed, D. (Eds.) (1997). *Three Dimensional Kinematics of Eye, Head, and Limb Movements*, Harwood Academic Publishers, Amsterdam.

Fetter, M., Tweed, D., Misslisch, H., Fischer, D.& Koenig, E. (1992). "Multidimensional descriptions of the optokinetic and vestibuloocular reflexes", *Annals of the New York Academy of Science*, 656: 841-842.

Flanders, M., Helms Tillery, S. I., & Soechting, J. F. (1992). "Early stages in a sensorimotor transformation", *Behav. Brain Sci.*, 15: 309-362.

Funda, J., & Paul, R. P. (1990). "A computational analysis of screw transformations in robotics", *IEEE Trans. Robot. and Autom.*, 6: 348-356.

Gielen, C. C. A. M., Vrijenhoek, E. J., & Flash, T. (1997). "Principles for the control of kinematically redundant limbs", In M. Fetter, Th. Haslwanter, H. Misslisch, & D. Tweed (Eds.), *Three Dimensional Kinematics of Eye, Head, and Limb Movements*, 283-296, Harwood Academic Publishers, Amsterdam.

Guitton, D., & Crawford, J. D. (1994). "Coordination of Monkey eye and head movements in three dimensions", *Soc. Neurosci. Abstr.*, 20: 1405.

Haslwanter, Th., Hepp, K., Straumann, D., Dursteller, M.R. & Hess, B.J.M. (1992)."Smooth pursuit eye movements obey Listing's law in the monkey", *Exp. Brain Res.*, 87: 470-472.

von Helmholtz, H. (1867). *Handbuch der Physiologischen Optik* (1st edn, Vol. 3). Voss., Hamburg, Germany *Treatise on Physiological Optics (English Translation)*, vol. 3, 44-51, translated by J. P. C. Southall (1925), Opt. Soc. Am., Rochester, NY

Henn, V. (1997). "History of Three-dimensional eye movement research", In M. Fetter, Th. Haslwanter, H. Misslisch, & D. Tweed (Eds.), *Three Dimensional Kinematics of Eye, Head, and Limb Movements*, 3-14, Harwood Academic Publishers, Amsterdam.

Henn, V., Hepp, K, & Vilis, T. (1989). "Rapid eye movement generation in the primate: Physiology, pathophysiology, and clinical implications", *Rev. Neurol. (Paris)*, 145: 540-545.

Henriques, D. Y. P., Klier, E. M., Smith, M. A., Lowy, D. & Crawford, J. D. (1998). "Gaze centered remapping of remembered visual space in an open-loop pointing task", *J. Neurosci.*, 23: 8.

Hepp, K. (1994). "Oculomotor control: Listing's law and all that", *Current Opinion in Neurobiology*, 4: 862-868.

Hepp, K., van Opstal, A. J., Straumann, D., Hess, B. J. M. & Henn, V. (1993). "Monkey superior colliculus represents rapid eye movements in a two-dimensional motor map", *J. Neurophysiol.*, 69: 965-979, 1993.

Hepp, K., van Opstal, A. J., Suzkui, J., Straumann, D., & Hess, B. J. M. (1997). "Listing's law: visual, motor, or visuomotor", In M. Fetter, Th. Haslwanter, H. Misslisch, & D. Tweed (Eds.), *Three Dimensional Kinematics of Eye, Head, and Limb Movements*, 33-42, Harwood Academic Publishers, Amsterdam.

Hepp, K, Suzuki, J., Straumann, D., & Hess, B. J. M. (1994). "On the 3-dimensional rapid eye movement generator in the monkey", In J. M. Delgado-Garcia, E. Godeaux & P.P.Vidal (Eds.), *Information Processing Underlying Gaze Control*, 65-74, Pergamon Press., Oxford, UK.

Howard, I. P., & Zacher, J. E. (1991). "Human Cyclovergence a function of stimulus frequency and amplitude", *Exp. Brain Res.*, 85: 445-450.

Jürgens, R., Becker, W., & Kornhuber, H. (1981). "Natural and Drug induced variations of velocity and duration of human saccadic eye movements: evidence for a control of the neural pulse generator by local feedback", *Biol. Cybern.*, 39: 87-96.

Klier, E. M., Lowy, D. & Crawford, J. D. (1997). "The brain accounts for 3-D eye position in an eye-to-head reference frame transformation for accurate saccades", *Soc. Neurosci. Abstr.*.

Luschei, E. S. & Fuchs, A. F. (1972). "Activity of brainstem neurons during eye movements of alert monkeys", *J. Neurophysiol.*, 35: 445-461.

Miller, J. M., Pavlovski, D. S., Shamaeva, I. (1997). "Orbit 1.8 Gaze Mechanics Simulation", Eidactics; Suite 404; 1450 Greenwich Street; San Francisco, CA 94109.

Minkin, A.W. H. & Van Gisbergen, J.A.M. (1994). "A three-dimensional analysis of vergence movements at various levels of elevation", *Exp. Brain Res.*, 101, 331-345.

Misslisch, H., Tweed, D., Fetter, M., Sievering, D., & Koenig, E. (1994). "Rotational kinematics of the human vestibuloocular reflex III. Listing's law", *J. Neurophysiol.*, 72: 2490-2501.

Mok, D., Ro, A., Crawford, J. D. & Vilis, T. (1992). "Rotation of Listing's plane during vergence, *Vis. Res.*, 32: 2055-2064.

Nakayama, K. (1975). "Coordination of extraocular muscles", In P. Bach-y-Rita & G. Lennerstrand (Eds.), *Basic Mechanisms of Ocular Motility and Their Clinical Implications*, 193-207, Pergamon Press, Oxford, UK.

Nakayama, K. (1983). "Kinematics of normal and strabismic eyes", In by C. M. Schor and K. J. Ciuffreda (Eds.), *Vergence Eye Movements: Basic and Clinical Aspects*, 543-564, Butterworths, Boston, MA.

Quaia, C. & Optican, L.M. (1997). "A commutative controler is sufficient to rotate the eye", *Soc. Neurosci. Abst.*, 23: 9.

Raphan, T. (1997). "Modeling control of eye orientation in three dimensions", In: M. Fetter, T. Haslwanter, H. Misslisch, & D. Tweed (Eds.), *Three Dimensional Kinematics of Eye, Head, and Limb Movements*, 359-374, Harwood Academic Publishers, Amsterdam.

Robinson, D. A. (1992). "Implications of neural networks for how we think about brain function", *Behav. Brain Sci.*, 15: 644-655.

Schnabolk, C. & Raphan, T. (1994). "Modelling three-dimensional velocity to position transformation in oculomotor control", *J. Neurophsyiol.*, 71: 623-638.

Smith, M. A. & Crawford, J. D. (1997). "Simulating neural processes in non-orthogonal coordinate system: a 3-D tensor model of the VOR", *Soc. Neurosci. Abstr.*, 23: 472.

Straumann, D., Haslwanter, Th., Hepp-Reymond, M. C., & Hepp, K. (1991). "Listing's law for eye head and arm movements and their synergistic control", *Exp. Brain Res.*, 86: 209-215.

Straumann, D., Zee, D. S., Solomon, D., Lasker, A. G. & Roberts, D. C. (19950. "Transient torsion and cyclovergence during and after saccades", *Vis. Res.*, 35: 3321-3335.

Tweed, D (1994). "Binocular coordination, stereo vision, and Listing's law", *Soc. Neurosci. Abst.*, 20: 1403.

Tweed, D. (1997a). "Three-dimensional Model of the human Eye-Head Saccadic System", *J. Neurophysiol.*, 77: 654-666.

Tweed, D. (1997b). "Velocity-to-position transformations in the VOR and saccadic system", in *Three Dimensional Kinematics of Eye, Head, and Limb Movements*, M. Fetter, T. Haslwanter, H. Misslisch, & D. Tweed (Eds.), Harwood Academic Publishers, Amsterdam, 375-386.

Tweed, D., Cadera, W., & Vilis, T. (1990). "Computing three dimensional eye position quaternions and eye velocity from search coil signals", *Vis. Res.*, 30: 97-110.

Tweed, D., Fetter, M., Andreadaki, S., Koenig, E. & Dichgans, J. (1992). "Three-dimensional properties of human pursuit eye movements", *Vis. Res.*, 32: 1225-1238.

Tweed, D., Misslisch, H., & Fetter, M. (1994). "Testing models of the oculomotor velocity to position transformation", *J. Neurophysiol.*, 72: 1425-1429.

Tweed, D. & Vilis, T. (1987). "Implications of rotational kinematics for the oculomotor system in three dimensions", *J. Neurophysiol.*, 58: 832-849.

Tweed, D. & Vilis, T. (1988). "Rotation axes of saccades", *Ann. N.Y. Acad. Sci.*, 545: 128-139.

Tweed, D. & Vilis, T. (1990a). "Geometric relations of eye position and velocity vectors during saccades", *Vis. Res.*, 30: 111-127.

Tweed, D. & Vilis, T. (1990b). "The superior colliculus and spatiotemporal translation in the saccadic system", *Neural Net.*, 3: 75-86.

Van Opstal, A. J. & Hepp, K. (1995). "A Novel Interpretation for the Collicular Role in Saccade Generation", *Biol. Cybern.*, 73, 431-445.

Van Opstal, A. J., Hepp, K., Hess, B. J. M., Straumann, D., & Henn, V. (1991). "Two- rather than three- dimensional representation of saccades in monkey superior colliculus", *Science*, 252: 1313-1315.

Van Opstal, A. J., Hepp, K., Suzuki, Y., & Henn, V. (1995). "Influence of eye position on activity in monkey superior colliculus", *J. Neurophysiol.*, 74: 1593-1610.

Van Opstal, A. J., Hepp, K., Suzuki, Y., & Henn, V., (1996). "Role of monkey nucleus reticularis tegmenti pontis in the stabilization of Listing's plane", *J. Neurosci.*, 16: 7284-7296.

van Rijn, L. J. & van den Berg, A. V. (1993). "Binocular eye orientation during fixations: Listing's law extended to induce eye vergence", *Vis. Res.*, 33: 691-708.

Vilis, T. (1997). "Physiology of three-dimensional eye movements: saccades and vergence", In M. Fetter, T. Haslwanter, H. Misslisch, & D. Tweed (Eds.), *Three Dimensional Kinematics of Eye, Head, and Limb Movements*, 59-72, Harwood Academic Publishers, Amsterdam.

Westheimer, G. (1957). "Kinematics of the eye", *J. Opt. Soc. Am.*, 47: 967-974.

Zee, D. S., Optican, L. M., Cook, J. D. & Robinson, D. A. (1976). "Slow saccades in spinocerebellar degeneration", *Arch. Neurol.*, 33: 243-251.

8

The visual control of steering

Michael F. Land

Abstract

Steering a car seems simple, but actually involves a double control system, first characterised from an engineering point of view in the 1970's. This has two separately identifiable components. A) An anticipatory feed-forward signal, derived from the angular discrepancy between the direction of motion and the edges of the road 10-20m ahead, is used to estimate upcoming road curvature, which is then translated into steering-wheel angle after an appropriate delay. B) A feedback signal from the lane edge much closer to the vehicle (<10m) provides fine-tuning to keep the vehicle in lane. Experimental evidence from eye-movement monitoring on real roads and simulators has made it possible to identify the regions of the road involved in these two sub-systems, the internal delays involved in their use, and the situations under which they are likely to fail. A key feature of the way vision is used in driving is the presence of an internal buffer through which the steering signal passes. This provides the 0.8-0.9s delay required to allow the vehicle to catch up with the point ahead at which feed-forward curvature estimates are made. It appears that when speed increases this delay remains more or less constant, and drivers allow for this by looking further ahead. The other function of the buffer is that it allows drivers to take 'time out' to look at road signs, other traffic and the scenery, without disturbing the steering signal itself.

8.1 Introduction

When driving along a winding road one has no feeling for the control processes that make steering possible; the driver's contribution seems so effortless and automatic that it barely intrudes into consciousness. However, its importance becomes immediately clear if one closes one's eyes even for a second. It does not take long to crash.

This paper examines a number of aspects of the steering process, particularly the rôle of vision. Accurate steering requires quite precise information about the future course of the road to be delivered more or less continuously on a short time scale. A general impression of the scene is not going to do the job, but equally a full 3-dimensional reconstruction of the world ahead is hardly necessary. Thus we need to ask questions that are specific to the steering process. What information does vision extract from the road

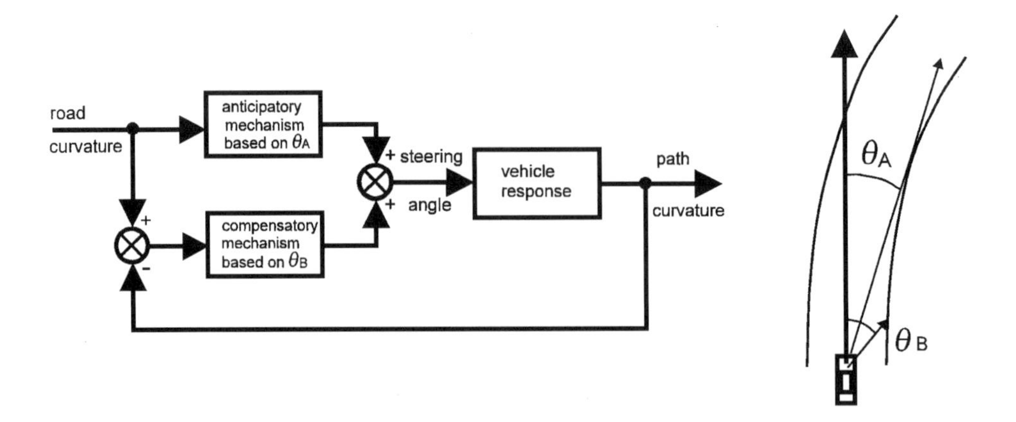

Fig. 8.1. Control diagram of the mechanism of steering, based on Donges (1978) and Godthelp (1986). Road curvature is obtained by determining the offset of some road feature from the current direction of travel at a suitable distance ahead (θ_A). This is used as an anticipatory feed-forward signal. Departures from the proper position in lane are measured by the angle of the road-edge closer to the vehicle (θ_B), and this is used as a feedback signal to correct residual errors left by the feed-forward sub-system.

ahead? Whereabouts on the road and its surroundings does this information come from? How is this information transformed into a command to turn the steering wheel? What are the delays in the transformation? What rôles do eye movements play in this process? Before tackling these questions it is useful to have an overview of the whole steering control system, and the next section deals with this.

8.2 Feed-forward and feedback in steering control

Early ideas about the control of steering concentrated on feedback mechanisms involving control by deviation correction (Biggs, 1966; Lee & Lishman, 1977). Features such as the observed distance to the lane edge markings, or the angle made by the lane edge to the estimated line of travel are obvious candidates for the input to such a feedback loop. Feedback of this kind is undoubtedly important in steering, and when learning to drive it may be the only source of control. However, it became clear by the mid-1970's that feedback on its own would not permit the kinds of performance that most drivers display. As we shall see, the half-second delay in the feedback loop makes control unstable if used in the upper range of ordinary driving speeds.

The most influential model of steering behaviour is still that of Donges (1978), who showed with the aid of a driving simulator that there were two components to the control system (Figure 8.1). There was visual feedback, as expected, but in addition there was anticipatory (feed-forward) control, based on the view of the road considerably further ahead in time and distance than the half-second involved in feedback. Donges envisioned that the anticipatory signal was the future road curvature, inferred from the curved appearance of the road ahead. This would translate directly into steering-wheel angle, since this itself converts directly to the curvature of the vehicle's path (see Godthelp, 1986). The anticipatory signal, however, refers to the curvature of the road well ahead of the vehicle, and so it is necessary to delay the steering response by an appropriate amount. In Donges' model the time delay involved was about 1 second. In the similar 'preview control model' of MacAdam (1988) the range of delays is wider, 0.8 to 3s. The evidence I present here suggests that real drivers employ delays of slightly less than a second.

The success of two-component driver models results from the way each component helps the other. If the anticipatory mechanism succeeds in measuring and matching curvature accurately, then in theory there will be nothing left for the feedback mechanism to do. However, having the right track curvature does not on its own insure that the car will keep its correct position in lane; indeed it is possible to have perfect curvature matching, but if the vehicle starts out of lane it will stay out of lane. Thus the rôle of feedback is to make the small adjustments required to keep the vehicle in lane. But because most of the steering work has already been done by the anticipatory mechanism, the feedback loop can operate at low gain, and so does not encounter the instability problems that occur when it operates unaided at high gain.

A crucial difference between feed-forward and feedback is that the former requires accurate calibration if it is to be effective. Drivers have to learn the relation between road appearance and how far to turn the steering wheel. Feedback, on the other hand, only involves taking action until some desirable condition is met, and then stopping that action. It is likely that a major feature of the early stages of learning to drive is the acquisition of appropriate calibrations for the feed-forward task.

8.3 Where do drivers look on the road?

A first step in trying to work out where drivers obtain the control signals needed for the estimation of road curvature and position in lane is to find out where they actually look. In a study with David Lee (Land & Lee, 1994), we asked three drivers to drive a car (a Jaguar with automatic transmission and power steering) along a tortuous road (Queen's Drive round Arthur's Seat in Edinburgh, Scotland), while wearing an eye-movement camera. This head-mounted device video-filmed both the image of the road ahead and

Fig. 8.2. Driver's view of Queen's Drive in Edinburgh, with a spot showing typical gaze directions on bends (top row), on a straight section (lower left), and fixating off-road on a passing jogger. Note that on the bends the fovea is directed to the tangent point on the inside the bend. Lower parts of each frame show the eye (inverted). The foveal direction is derived automatically from the position of the outline of the iris.

the subject's eye as it moved in its socket (Figure 8.2). After the drive, a computer model of the eye was fitted, frame by frame, to the image of the iris/sclera boundary, and from this fit the coordinates of eye direction relative to the head were recovered. These were used to provide the position of a spot that was added to the image of the road, thus providing a video record of the driver's direction of gaze. Head coordinates relative to the vehicle were also obtained by following the movements of markers on the car's windscreen, so that a continuous record of gaze direction relative to the vehicle (eye/head + head/vehicle) could be made. This was accurate to about 1^{O}. In addition, the car was instrumented, so that information on steering-wheel angle, speed and other parameters could be matched to the eye movement records.

The results were quite surprising. All three subjects spent much of the time fixating the edge of the road near a point known as the 'tangent' or 'reversal' point (Figures 8.2, 8.3a) where the inside edge of the bend changes direction (Riemersma, 1991). Their eyes usually sought out this point 1-3 seconds before entering the bend and returned to it from time

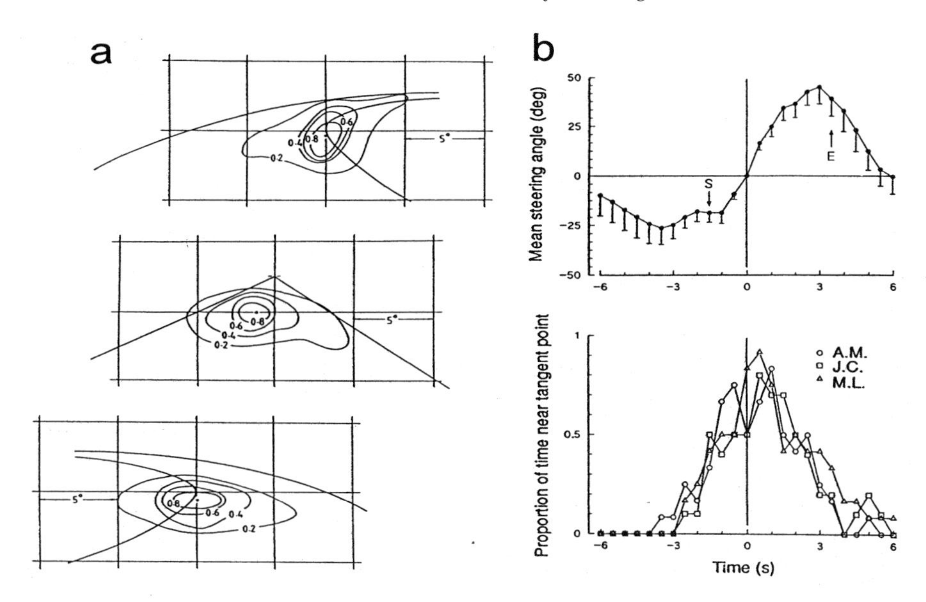

Fig. 8.3. a. Contour plots showing the location of fixations made on right and left-hand bends, and on straight road sections where no tangent point is visible. The contours give the density of fixations relative to the maximum (approximately 0.12 fixations.deg^{-2}.s^{-1}). Measurements on bends were made from the tangent point and on the straight road from the vanishing point. The 0.2 contour includes about 65% of all fixations. Data from three 1km drives by different drivers. Note that on bends the highest fixation densities are within 1^{O} of the tangent points. b. Time course of the use of tangent point. Top: steering wheel angle adopted around a 'typical' bend averaged from the ten largest bends on the drive, synchronised to the entry point of the bend (0 on abscissa) where the steering wheel angle crosses zero (see also Figure 8.8a). Bars indicate 1 s.e. Bottom: the portion of the total time spent by the gaze of each of three drivers within 3° of the tangent point. During the first second into the bend, all three drivers look at the tangent point almost all (>75%) of the time. Reprinted with permission from Land, M. F., & Lee, D. N. (1994). "Where we look when we steer", *Nature*, 369: 742–744.

to time throughout the bend. Perhaps the most telling statistic is that half a second after entering a bend all three drivers spent 80% of the time fixating within 3° of the tangent point (Figure 8.3b). It seems from this that there is something special about the tangent point that requires the drivers' attention at this time. This is confirmed by a study of Yilmaz & Nakayama (1995) who used reaction times to a vocal probe to show that attention was diverted to the road just before simulated bends, and that sharper curves demanded more attention than gentle ones. The likely explanation is that the position and/or motion of the tangent point in the field of view provides the signal needed to estimate the initial curvature of the bend, and so allows the driver to get the steering angle correct from the outset. Once in a bend of more or less constant curvature, steering can be adjusted by feedback, but the first part especially, requires an accurate anticipatory signal.

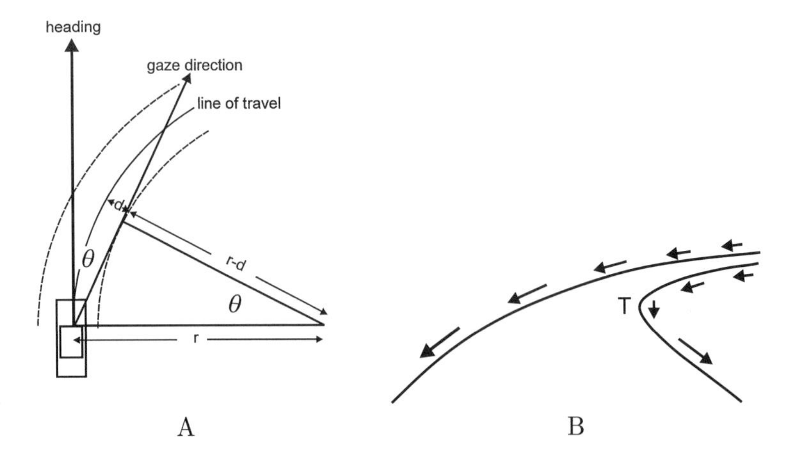

Fig. 8.4. A) Geometry of tangent-point steering. The curvature of the bend ($1/r$) can be obtained from the gaze angle θ, using the geometry of the right-angled triangle. Here, $\cos\theta = (r - d)/r$, where d is the distance of the driver from the lane-edge. However, the expansion of the cosine gives $\cos\theta \approx 1-\theta^2/2$. Substitution for $\cos\theta$ then gives: $1/r \approx \theta^2/2d$. B) Optic flow around the tangent point T on a road of constant curvature.

The relation of the tangent-point direction to lane curvature is shown in Figure 8.4. The angle θ between the current heading of the vehicle (and hence the driver's seat-belted body) and the tangent point, predicts the curvature (1/radius of curvature, $1/r$) of the future curvature of the car's track along the road by the simple formula:

$$\text{curvature} = 1/r = \theta^2/2d \tag{8.1}$$

where d is the distance to the lane edge. (Note that this is a somewhat simpler version of the relation given in Land & Lee, 1994, where r was taken as the inner radius of curvature of the road rather than the vehicle's track). To steer using this relation requires that the driver can square the angle q (although it is possible that a linear approximation would be adequate, see Murray *et al.*, 1996), and it requires a knowledge of the distance d to the lane edge. Even better would be the maintainance of a constant value for d, which then becomes a simple scaling factor. As this is the principal function of the feedback component of the Donges model (Figure 8.1), it seems likely that d may well be kept fairly constant - except when drivers deliberately cut corners! 'Tangent-point steering' has been used to guide robot vehicles (Raviv & Herman, 1993; Murray *et al.*, 1996), so it is certainly a strategy that works. The fact that drivers look at tangent points does not guarantee that they use them as beacons to guide steering, in the manner just suggested. Any other fairly distant point on the road could provide a curvature signal, although the tangent point has the special property that its distance from the driver does not have to be known. For other points on the vehicle's intended path the curvature is given by:

$$1/r = 2\sin(\theta/2)/D \quad (8.2)$$

where D is the distance of the point from the driver. Distance can be easily estimated, for example by the angular subtense of the lane at the appropriate point, or the declination of that point from the horizon, but arguably this extra measurement makes such points harder to use than the tangent point itself. There are additional reasons why drivers might favour the tangent point. Like the vanishing point on a straight road, the tangent point is a point in the visual field where optic flow is minimal. If a car is travelling around a bend of constant curvature the tangent point stays in the same place in the field of view, relative to the car's heading, and only moves if the curvature changes (Figure 8.4a, see also Raviv & Hermann, 1993). Of course, on the ground the tangent point is moving round the curve with the driver, and as seen by the driver objects in the vicinity of the road edge will 'flow' around the tanget point, as indicated in Figure 8.4b. At the tangent point, however, the flow is purely vertical in the field of view, and because the road edge itself appears vertical at this point, even this vertical motion will be almost invisible. Thus the tangent point is a place where gaze can rest without the eyes being carried elsewhere by flow-field motion, and eye movement records indeed show that the tangent point is often fixated and tracked smoothly for periods of several seconds. Finally, looking beyond the tangent point provides the driver with the longest uninterrupted view of the road, meaning that it is good place from which to detect future hazards.

Thus the status of the tangent point is somewhat uncertain. It may be the main source of the 'far road' feed-forward control signal, or it may just be a convenient location from which to view the road as a whole. We do know from simulations (see below) that drivers can steer adequately without the tangent point being visible, and it is not necessary for gaze to be directed at the tangent point, even when it is visible. Nevertheless, steering is less accurate when the segment of the road containing the tangent point is not visible, and it may well be that tangent point location is the preferred control signal, even if it is not the only one.

8.4 Where should drivers look? A simulator study

Although we have shown where drivers direct their gaze on bends, it is almost impossible to say where they are actually attending. We can, however, ask a related question: where should drivers look in order to get the best information by which to steer? If this coincides with where they actually look, then this gives us grounds for thinking that they are attending there as well.

The method involved a simple simulator (Land & Horwood, 1995), in which subjects drove round a skeletal version of the Queen's Drive road. Only the road edges were present, plus a horizon and a sketch of the car bonnet (hood), but no other scenery. The drive could be run at a variety of constant speeds, from 12.5 to 19.7ms^{-1} (28 to 44mph). Drivers found this similar to night driving, and had no difficulty negotiating the bends of the simulated road. The main measure of performance was the standard deviation of the subject's position in lane, taken over the whole 1km drive; under ideal conditions this was between 0.1 and 0.2m. After a few trials with the whole road outline visible, the view was restricted to either one or two 1^{0} high segments of the road edge (Figure 8.5a) which could be located at varying positions between 1^{0} and 10^{0} below the horizon, corresponding to distances between 63m and 6m from the vehicle. These segments behaved exactly as they would had the whole road been visible.

In the first set of experiments only one segment was visible (Figure 8.5). The principal result was that at each speed there was an optimum (vertical) position of the segment that gave the best steering performance. It was nearer to the vehicle at slow speeds and further at high speeds, but in terms of 'time ahead' it was close to 0.7s at all speeds. Except for the slowest speed, however, the performance was not as good as when the whole road edge was present, and at faster speeds the difference was big enough to make the vehicle stray from its lane. This implies that any one segment of the road is not capable of providing the whole of the required control signal. Another disconcerting feature of this study was that the part of the road that provided the optimum performance was somewhat closer to the vehicle than the region containing the tangent points, and closer than the region where drivers usually looked when the whole simulated road was present.

We also found that drivers behaved quite differently to near and far regions of the road. When only the far part of the simulated road was visible, drivers matched curvature well, but their lane keeping performance was poor; and when only the near part was visible lane keeping was better, but steering was unstable and jerky (Figure 8.5c). The drivers' control system had changed from smooth to 'bang-bang' (Land & Horwood, 1995). This suggested that far and near regions contribute to the overall control system in different ways.

A test of this came as a result of the comments of a perceptive reviewer (Bill Warren). He made the apparently obvious point that if we removed the 'optimum' regions from the road (Figure 8.5b), this should make steering performance worse than when only the optimum region was present. We did this, and far from worsening performance, it improved it! Taking out the 'middle distance' region of the road, but leaving the more distant and nearer

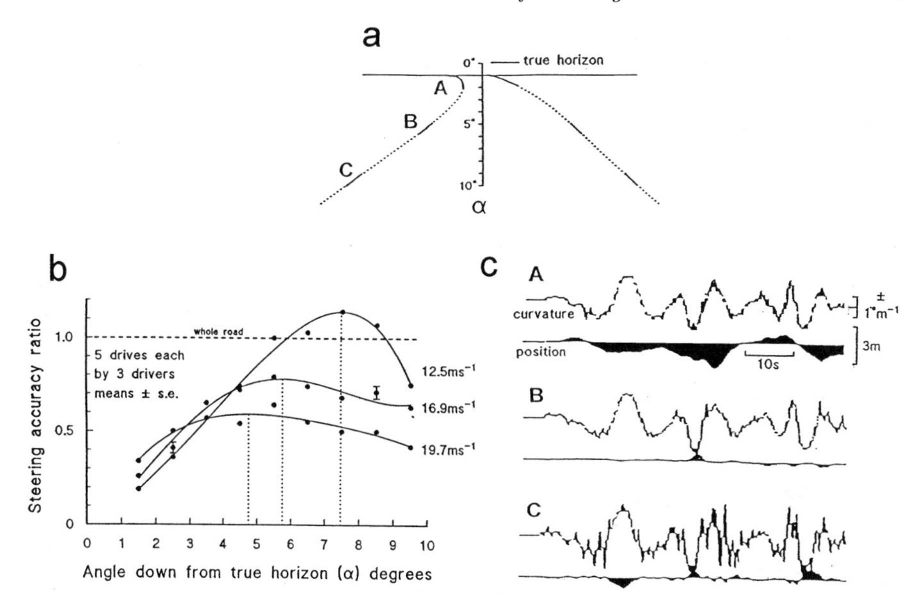

Fig. 8.5. Driving on a simulator with only parts of the road edge visible. a. Appearance of simulated road, showing the angular scale used in b and the three road segments used in c, each subtending 1° vertically. b. Accuracy of steering at three different speeds, as a function of the vertical position of the 1° visible section of the road edge. Accuracy is measured as the reciprocal of the standard deviation of position in lane, and accuracy ratio is the ratio of the accuracy with a single segment to that when the whole road edge is visible. The best position for the visible segment moves up the road (more distant) as speed increases. At the two higher speeds even the best segment gives a poorer performance than when the whole road edge is visible (dashed line), although at the slowest speed nearer segments give a better result. c. Differences in driver behaviour depending on the position of the visible segment. The upper trace in each pair gives the curvature of the road and the vehicle's track, with differences between them appearing as a solid black region. The lower trace shows position relative the road midline, with solid black indicating the extent of error. With only a distant segment visible (A) curvature matching is smooth and reasonably accurate, but position-in-lane accuracy is very poor. With only a near segment (C) curvature matching becomes jerky and unstable, although lane position is more accurate than in A. An intermediate distance (B) gives the best result. Redrawn from Land & Horwood (1995).

parts not only improved performance, but it made it indistinguishable from having the whole road present. So we repeated the experiment properly with two segments of the road, and the results are shown in Figure 8.6b. Here one visible segment was either at the most distant region (1-2° below the horizon) or the nearest (9-10° down), and the position of the other segment was varied. Broadly, the result shows that adding a second segment dramatically improves performance if and only if it is added to the opposite end of the road: the far segment if the near one is present, and vice versa. With the near-road segment present, the optimum location of the far-road segment was about 4° down from the horizon (at 16.9m.s^{-1}), and with the

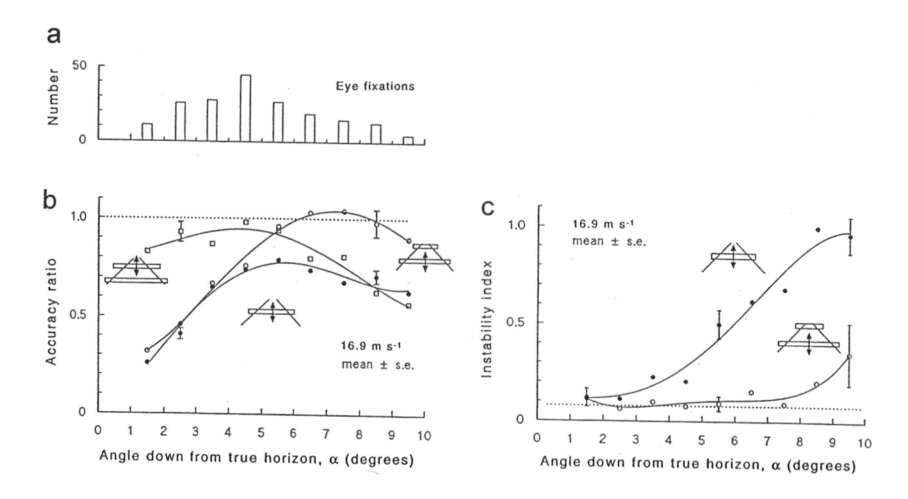

Fig. 8.6. Driver performance with two segments of road visible. a. Histograms of vertical gaze direction when viewing the whole road. Abscissa as in b and Figure 8.5a. The distribution peaks 4-5° down from the horizon (average of three drivers). b. With a single visible road segment (filled circles) the apparent optimum is 5.5^{O} down from the horizon, but it is not as accurate as when the whole road is present (dotted line). Adding a second segment to the near part of the road (squares) greatly improves performance, but only when the first segment is in the far part of the road; when a second segment is added to the far part of the road (open circles) performance is enhanced for segment positions in the near part of the road. With two appropriately spaced segments, performance is as good as with the whole road edge (average of five drives by three drivers). Axes as in Figure 8.5b. c. The presence of a distant segment reduces the instability of near-segment driving. The instability index is the product of the number and amplitude of the spike-like movements seen in Figure 8.5c(C), normalized to their maximum value. Reprinted with permission from Land, M. F., & Horwood, J. (1995). "Which parts of the road guide steering", *Nature*, 377: 339–340.

far-road segment, the best location of the near-road segment was about 7° down. I will refer to these regions as the 'far-road' and 'near-road optima'.

Of course, we should have expected this. The Donges (1978) model implies that feed-forward information comes from the distant road and feedback from the near road. There should be distinct regions of information uptake. Furthermore, as pointed out in §8.2, the two systems are synergistic, in particular the far-road feed-forward reduces the load on the near-road feedback system, and so stops it becoming unstable. This is indeed what happens (Figure 8.6c). Adding a far-road segment to a near-road segment, which on its own would give a very jerky performance, made the drivers' performance smooth and stable again. Thus we now see that the 'optima' in the single segment experiments (Figure 8.5b) are not real optima, but rather compromises in which both components of the double control system get some of the information they need.

Another comforting consequence of the two-segment experiment was that

speed m.s^{-1}	preview distance (m)		preview time (s)	
	far	near	far	near
12.5	9.8	7.8	0.78	0.62
16.9	14.2	8.6	0.84	0.53
19.7	18.3	8.5	0.93	0.43

Table 8.1. *Locations and times ahead of the far-road and near-road optima as a function of speed*

the location of the far-road optimum now coincided well with both the range of tangent-point locations, and with the region where drivers actually fixate in both real driving and simulations (Figure 8.6a). In general, drivers did not look down to the near-road optimum region, suggesting that they they take in information about position-in-lane from a somewhat peripheral region of the retina, about 5^{0} obliquely down from the fovea, whilst keeping their gaze on the far part of the road.

The distances and times ahead of the vehicle associated with the far-road and near-road optima are given in Table 8.1. The main result is that the preview time for the far-road mechanism varies very little with increase in speed. If drivers look a constant distance ahead, there would have to be a corresponding decrease in time ahead as speed increases, but if anything the opposite is true. Preview times increase slightly with speed, but preview distances increase very much more, almost doubling for a 58% increase in speed. When they drive faster it seems that drivers look further up the road, but don't change their internal delay. By contrast, the near-road optima (which are viewed peripherally) remain at a more or less constant distance of 8-9m, and the preview time does decrease, from 0.62 to 0.43s.

The far-road result is not quite consistent with the idea that drivers measure curvature from tangent point offset (see §8.3), since the distance of the tangent point is not affected by speed. Faster drivers could, however, pick up the tangent point at a greater distance and abandon it earlier, so that its use is restricted to a more distant range.

8.5 Two other methods of measuring preview time

If a vehicle's steering is suddenly disconnected, it will begin to crash immediately, crashing faster or slower depending on speed and the changes in road curvature. However, if vision of the road is suddenly cut off, the driver will be able to continue to steer for a short period until the visual information currently being processed is used up. The crash should be delayed by the time it takes to convert visual information to steering action (Figure 8.7).

These conditions were presented to four drivers using the road simulator. 'Crashes' were initiated at different points on ten of the bends of the simulated road. These points were the same in all trials, but the order of their occurrence was randomized to prevent the drivers guessing which would come next. There were three conditions.

(i) The steering was disconnected instantly from the program driving the simulator, and the image of the road on the screen also disappeared.
(ii) The image of the road disappeared, but the steering was not disconnected. The drivers were simply instructed to cease steering when the view of the road disappeared.
(iii) This was the same as (ii), but drivers were instructed to carry on steering for as long as possible (a 'shattered windscreen' situation).

The 'crashes' were monitored by measuring, via the program, the departure of the vehicle's position from the centre of the road at intervals of 0.14s, and the results were pooled to give standard deviations for each set of ten bends. Each driver made three of these drive sets (30 crashes in all per driver), for each of the three conditions.

Figure 8.7a shows the combined results from the four drivers. As control is lost during each crash, the resulting increase in standard deviation is approximately exponential. The time course of the crash that follows loss of vision (condition ii) is almost identical to that following the electronic disconnection of steering (condition i), except for a shift along the abscissa which represents a delay of 0.85s.

The fact that the curves almost superimpose implies that the delay is 'pure' rather than distributed, which means that steering information comes from a portion of the road ahead that is quite narrowly defined in time and space. If some information was obtained from the road closer than that implied by the delay of 0.85s, one would expect the delayed curve to begin its rise earlier, and conversely if information from further away were involved, the upper part of the curve would be delayed to a greater extent. Both of these effects would make the 'crash curve' following loss of vision (condition ii) shallower than that following steering disconnection, and that seems not to be the case. The implication is that steering information is not obtained from the whole road, but from a narrowly defined zone centred around 10-11m ahead, when the speed is 12.5 m.s^{-1}.

Interestingly, in condition (iii) (Figure 8.7a), where the instruction was to continue steering for as long as possible, the slope of the upper part of the curve was reduced, which means that some information from more distant parts of the road was available. This result was consistent for all four drivers. They all reported that a real cognitive effort was required to access this information - to try to recall whether the distant part of the bend

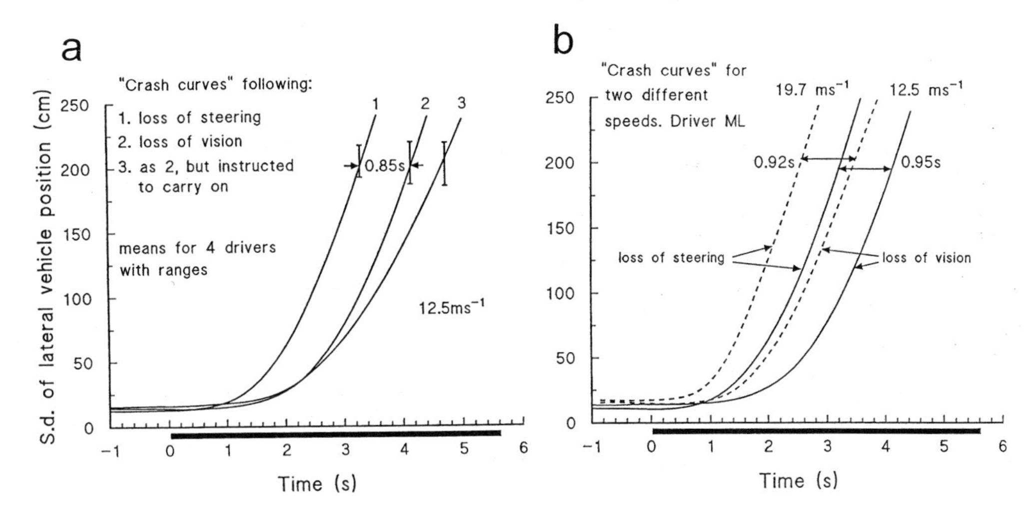

Fig. 8.7. Crashing without steering and without vision. a. The time course of crashes on the simulator, measured as the increase in standard deviation of distance of the vehicle from the road midline over a total of 120 crashes (30 per driver). In 1 the steering is electronically disconnected at time 0. In 2 steering remains intact but the screen goes blank. Drivers are asked to stop steering. 3 is the same as 3, but drivers are instructed to carry on for as long as possible. Curves 1 and 2 are identical but displaced by 0.85 seconds, which is presumably the time it takes visual information to pass through the sensory/motor buffer. Curve 3 is shallower than 2, indicating that some information from more distant parts of the road can be retrieved by the driver. b. Crashes at two different speeds. The crashes occur sooner at the faster speed, but the time gap between the loss of steering and loss of vision curves is almost the same. It seems that drivers do not change their internal delay to compensate for speed changes.

went left or right - and that this felt quite different from the automatic and effortless uptake of information involved in 'ordinary' driving.

What happens when speed is increased? As in the previous section, the driver has a choice of decreasing the preview time or increasing the preview distance. In simulations at faster speeds the crash occurs more rapidly, as expected, but the shift along the abscissa is almost the same for the faster and slower pairs of curves, implying that there is little difference in the delay involved (Figure 8.7b) . This means that at the faster speed information is taken from a more distant region of road - about 17m ahead at 19.7m.s^{-1} rather than 11m at 12.5m.s^{-1}. As noted above, it seems that drivers have a more or less constant internal delay of about 0.85s.

A third method of getting at the delay, which has more 'ecological validity' than the use of a simulator, is to compare gaze direction records from real drives with the steering records, and to see what delay there is between the two. The records obtained by Land & Lee (1994) showed that the two curves are very similar - as would be expected from equations 8.1 or 8.2. When cross-correlated over a range of delays (Figure 8.8), the best

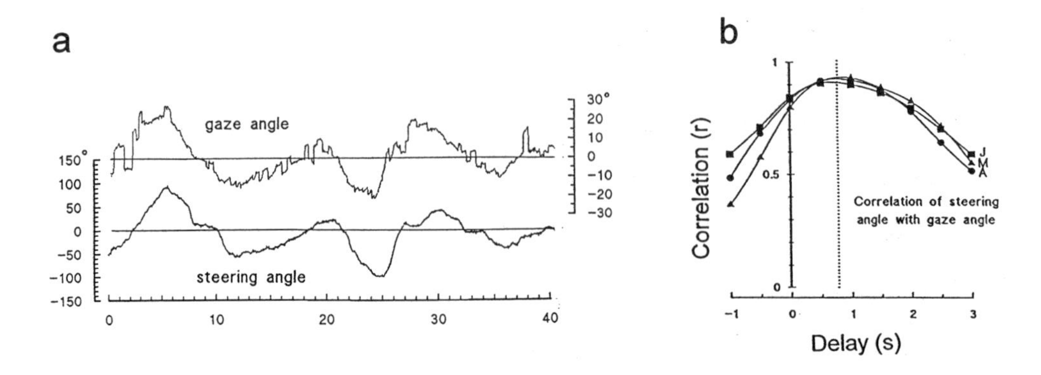

Fig. 8.8. Correlation between gaze direction and steering. a. Simultaneous record of gaze angle and steering-wheel angle for part of the drive depicted in Figure 8.2. Note the similarity of the curves, the slight delay of steering relative to gaze, and the occasional fixation off the road. Further records given in Land & Lee (1994). b. Correlations between gaze and steering wheel angle for different values of the delay between the two, for 1km drives by three drivers. Samples taken from the centres of all fixations within 3^{0} of the road edge (this restriction is to exclude fixations not concerned with steering, see Figure 8.9). The maximum correlation occurs after a delay between 0.5 and 1 second in all three cases.

correlations were obtained when gaze led steering by between 0.65 to 0.86s (three drivers, mean 0.73s). This is slightly shorter, by about 0.1s, than the estimate just given, and those obtained for the 'far-road optima'. A possible reason is that correlations were taken from the mid-point rather than the beginning of fixations, and this could reduce the apparent delay by about this amount, if information was actually taken in early in each fixation. In any event the discrepancy is not large, and confirms that drivers impose a continuous delay of about 0.8s between information uptake and steering action. Interestingly this figure of close to 1s is similar to that involved in other continuous activities such as reading aloud, reading music and touch typing (see Land & Furneaux, 1997).

The 0.8s delay corresponds to that required by the far-road anticipatory mechanism to allow the vehicle to catch up with the site of measurement, rather than the near-road compensatory mechanism. The latter would presumably operate best with as short a delay as possible - about 0.4s, given reaction times and mechanical delays. It unclear at present how drivers handle the different requirements of the two components of the overall system. They might just use one delay for both, or simultaneously employ different delays with each input, or they might alternate attention in some way between the two sub-systems. The second and third alternatives would imply a rather complex buffering arrangement in order to keep the contributions to steering synchronized.

8.6 Other aspects of steering: timesharing

The road used in this study, for both real driving and simulation, had many and frequent changes in curvature, forcing drivers to attend more or less continuously to the cues needed for steering. In more relaxed steering conditions, with wider roads and more predictable curves, drivers are able to take much more 'time out' to look at road signs, the scenery, other traffic and so on. The relationship between gaze direction and steering ceases to be a continuous one, and becomes sampled, with the eyes returning to the sources of steering cues only intermittently. This is the probable reason why the rôle of the tangent points has been little stressed in the past; they are visited relatively infrequently on wide roads and freeways. For example, Serafin (1993) in a comprehensive review of gaze direction on country roads concluded that there was only a vague relationship between where drivers looked and road curvature. A further complication is the fact that drivers can steer adequately when looking off the road, presumably by viewing the road itself with more peripheral vision. Using the simulator with a fixation target located away from the road, we found an almost linear decrease in steering performance (reciprocal of standard deviation of position in lane) with increase in angle. Accuracy fell to a half at an offset of about 15° from the centre of the far-road region, almost independent of the direction of displacement. This means that at modest offset angles of 5° or so drivers are still in useful visual contact with the road for steering purposes, but at large offset angles, such as those involved in adjusting car radios (typically $> 25^{\circ}$), the road is effectively out of sight. Under difficult steering conditions it is clearly important that drivers' vision stays close, in angular terms, to the far-road optimum region.

An obvious but intriguing point, relating to taking 'time out' from steering, is that the control processes implied in Figure 8.1 have to be suspended when drivers are not looking at the road itself. If this were not the case then drivers would steer towards whatever they were fixating. Figure 8.9 illustrates this with two short excerpts from one of the Edinburgh drives, one in which the driver is looking at distant hills, and the other in which he is checking on a cyclist while negotiating a bend. In both records it is clear that the steering record follows the gaze record when the eyes are viewing the road edge, but not the other features being viewed. This must mean that the steering control system is turned off - or better put on 'hold' - when the eyes are directed away from the road edge. In the case where the driver alternates between tangent point and cyclist, these switches occur every half second, so the engagement and disengagement process occurs on a fixation

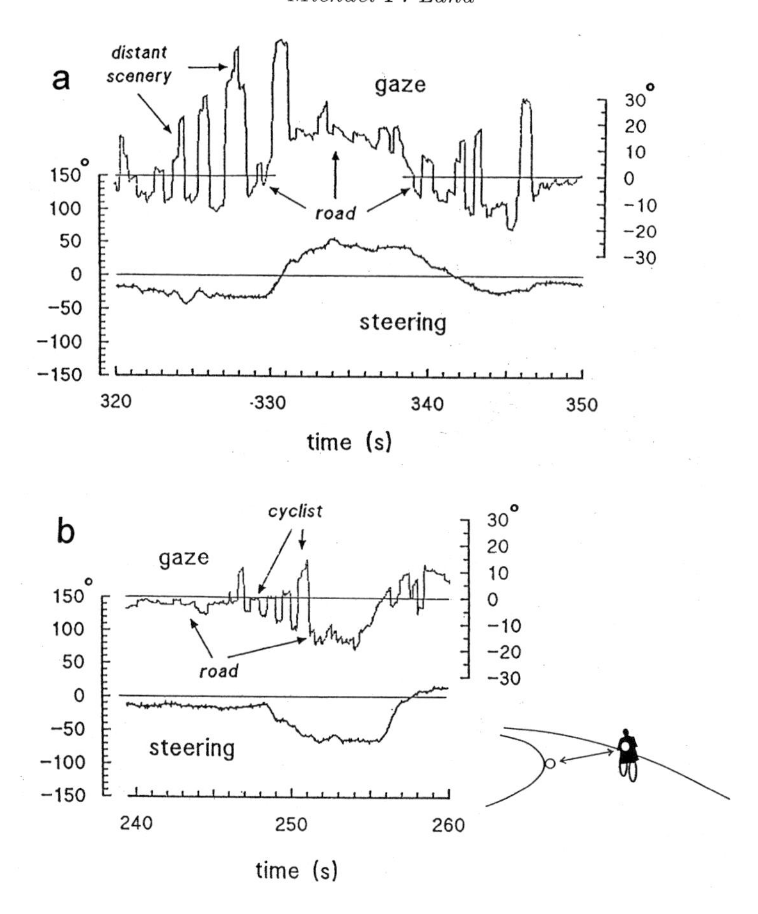

Fig. 8.9. Records of gaze angle and steering showing a driver taking 'time out' from steering to look at distant hills (a) and to check on a passing cyclist (b). Note that the steering only follows gaze angle when gaze is directed at the road. In the cyclist incident this means that steering is alternately disconnected from and reconnected to vision for periods of half a second each.

by fixation timescale. My dim recollection from learning to drive is that the knack of alternating is something one has to acquire; initially there is a tendency to steer towards whatever one is looking at.

8.7 Conclusions: beacons and flowfields

The locations of the edges of the road in the field of view appear to provide the principal visual cues for steering. They are necessary and sufficient, and the rest of the flow-field does not seem to be involved in any very direct or essential way. On poorly-marked roads, the differences in texture and motion at the road edge may substitute for discrete lines, but most drivers would agree that good white or yellow lines are a much better cue. One incidental consequence of the overall flow-field may well be that it forces drivers to fixate the tangent points rather than other road features - in preference to, say, the centre of the road - just because these are points of zero lateral flow when curvature is constant (see §8.3 and Figure 8.4b). The eye can take refuge there without being dragged about.

One of the two principal measurements drivers have to make involves the angle between the car's current heading and a distant region of the road. At first sight this would seem to involve actual knowledge of the vehicle's (or driver's) heading, as well as of the location of the appropriate road region, so that the difference can be derived. There has been much recent debate as to whether heading can be recovered from an optical flow-field involving mixed translation and rotation (e.g. Warren *et al.*, 1991). If this information were available it might contribute to the process involved in deriving the control signal. However, there is no indication from this study that it is necessary to know the vehicle's heading in absolute terms; the required difference signal can be obtained from a single direct measurement. Under most circumstances the angle between the driver's trunk and the appropriate part of the road is all that is required (see Figure 8.1) and this can be measured entirely from physiologically-derivable information. If the driver fixates the road ahead the trunk/head angle plus the eye/head angle gives the required measurement, and if the driver looks off the road the road/fovea angle, measured on the retina, must be added to this. Thus the rather difficult task of instantaneously determining heading from the flow-field is avoided. Other aspects of driving undoubtedly do involve aspects of the flow-field; braking for example requires a measure of expansion in certain identified parts of the field, notably the back of the vehicle ahead, but even there it is the local, not the global flow-field that is important.

Acknowledgments

I am grateful for a grant from the Joint Council Initiative on Cognitive Science and Human Computer Interaction, and from a grant from the BBSRC (UK) to the Sussex Centre for Neuroscience.

References

Biggs, N. L. (1966). "Directional guidance of motor vehicles – A preliminary survey and analysis", *Ergonomics*, 9: 193–202.

Donges, E. (1978). "A two-level model of driver steering behavior", *Human Factors*, 20: 691–707.

Godthelp, J. (1986). "Vehicle control during curve driving", *Human Factors*, 28: 211–221.

Land, M. F., & Lee, D. N. (1994). "Where we look when we steer", *Nature*, 369: 742–744.

Land, M. F., & Horwood, J. (1995). "Which parts of the road guide steering", *Nature*, 377: 339–340.

Land, M. F., & Furneaux, S. (1997). "The knowledge base of the oculomotor system", *Phil. Trans. R. Soc. (Lond.)*, 352: 1231-1239.

Lee, D. N., & Lishman, J. R. (1977). "Visual control of locomotion", *Scand. J. Psychol.*, 18: 224–230.

MacAdam, C. C. (1988). "Development of driver/vehicle steering interaction models for dynamic analysis", Univ. Michigan Transportation Res. Inst. Report UMTRI-88-53.

Murray, D. W., Reid, I. D., & Davison, A. J. (1996). "Steering and navigation behaviours using fixation", Proc. 7th Brit. Machine Vision Conf., Edinburgh, Vol 2., 635–644.

Raviv, D., & Herman, M. (1993). "Visual servoing from 2-D image cues", In Y. Aloimonos (Ed.) *Active Perception*, Erlbaum, Hillsdale, NJ. 191–226.

Riemersma, J. B. J. (1991). "Perception of curve characteristics", In A. G. Gale (Ed.) *Vision in Vehicles III*, North Holland/Elsevier, Amsterdam, 163–170.

Serafin, C. (1993). "Preliminary examination of driver eye fixations on rural roads: insight into safe driving behavior", Univ. Michigan Transport Res. Inst. Document UMTRI93-29: 1–63.

Warren, W. H. Jr., Mestre, D. R., Blackwell, A. W., & Marris, M. W. (1991). "Perception of circular heading from optical flow", *J. Exp. Psychol.: Human Perception & Performance*, 17: 28–43.

Yilmaz, E. R., & Nakayama, K. (1995). "Fluctuation of attention levels during driving" *Invest. Ophthal. & Vis. Sci.*, 36: S940.

9

Catching, hitting, and collision avoidance

D. Regan, R. Gray, C. V. Portfors, S. J. Hamstra,
A. Vincent, X. H. Hong, R. Kohly, and K. Beverley

To catch or hit a ball, and to avoid collision on the highway or in the air are actions that have in common the requirement that the sports player, driver, or pilot must, in effect, predict where an approaching object will be at some future instant. There is evidence that top sports players can estimate the time of arrival of an approaching ball to within ± 2 to 3 msec, and its instantaneous direction of motion to within 0.1 to 0.2 deg (Regan, Beverley & Cynader, 1979; Bootsma & van Wieringen, 1990). How this is done is a question that has attracted a considerable amount of experimental and theoretical research - and disagreement. Two lines of thought have emerged. When sufficient time is available for visual feedback, one possible method for achieving the desired goal-directed motor action is to exploit some relation between visual information and the progression of the required motor action.

A second possible method for achieving the desired goal is our topic here. This method is to anticipate the object's location at some well-defined future time on the basis of available visual information about the approaching object's time to collision and direction of motion. When sufficient time is available, an individual may use visual feedback to update the information on a moment-by-moment basis. In this way a games player might allow for changes in the speed and direction of motion that occur during the flight of the approaching ball. And in the special cases where anticipation is not possible and the time available to execute the required motor action is too short to allow visual feedback — the so-called "ballistic" action — the player must rely entirely on the information about time to collision and direction of motion that is available at the instant when he first sights the ball. (A sharp catch in the slips in cricket is one such situation. Even at international level there is a wide gulf between the great reflex catchers and everyone else).

In some situations, both of the two strategies discussed may be used. For example, when running to catch a high-hit ball in baseball or cricket it may be that the first method is used to bring the catcher close to the location where the ball will hit the ground (Todd, 1981; Saxberg, 1987a, b; Michaels

& Oudejans, 1992; McLeod & Dienes, 1993; McBeath, Shaffer & Kaiser, 1995), while the second method is important during the final 100's of msec before the catch is completed (Regan, 1998).

9.1 Monocular and binocular information about the time to collision with an approaching object

9.1.1 Monocular information about time to collision

9.1.1.1 Illusory perception of motion in depth produced by isotropic expansion of an object's retinal image

When a rigid sphere moves directly towards an observing eye, its retinal image expands. This much is self-evident. However, it is not self-evident that expanding an object's retinal image can produce the illusion that the object is moving in depth, even when the object is stationary (Wheatstone, 1852).

The rate of expansion of an object's retinal image is not a correlate of the time to collision; if a small sphere and a large sphere are at the same distance from the eye and approaching the eye at the same constant speed, the retinal image of the larger sphere will expand at a greater rate of deg/sec than the retinal image of the smaller sphere. However, the looming detector described below has a subtle property. It discriminates between isotropic and non-isotropic expansion of the retinal image. It achieves this by being sensitive to time to collision. In principle, therefore, a looming detector can signal time to collision.

In a series of papers, Regan and Beverley addressed the question whether human visual responses to isotropic expansion of the retinal image (i.e., looming) can be explained in terms of detectors of local unidirectional motion. Their experimental evidence showed that this is not the case.

In following the line of argument reviewed below it will be helpful to bear in mind the crucial role of the perception of motion in depth as distinct from the perception of changing size. By definition, looming detectors are sensitive to isotropic expansion of the retinal image. There is evidence that they signal isotropic expansion by creating a perception of motion in depth; when retinal image expansion is not isotropic, motion in depth perception can be weak or even absent (Regan & Beverley, 1978a, b; Beverley & Regan, 1979a).

There is a substantial literature on detectors of local unidirectional motion (LM filters in Figure 9.1) that are often modelled as "Reichardt Detectors" or as "Modified Reichardt Detectors" (Santen & Sperling, 1985). Reichardt detectors, however, respond to absolute rather than relative motion.

Evidence for a neural mechanism sensitive to unidirectional relative motion was obtained as follows (Beverley & Regan, 1979b). There were two

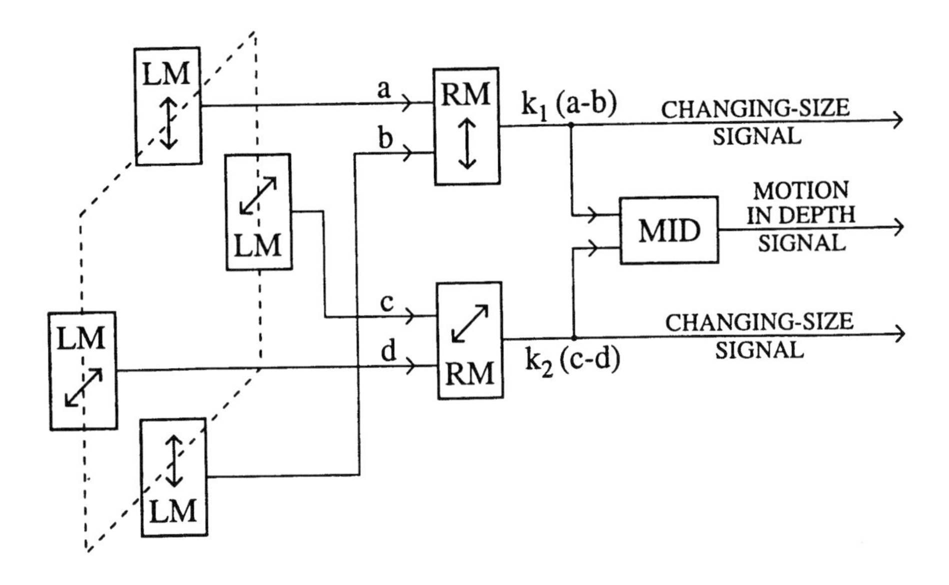

Fig. 9.1. Model of the processing of changing size and of the encoding of time to collision. The boundaries of a solid rectangular image are shown dotted. LM are filters sensitive to local motion along the directions arrowed. Their outputs (a, b, c and d) assume a magnitude that encodes local speed, and a sign that represents the direction of motion. RM are one-dimensional relative motion filters whose outputs signal the speed and sign (expansion versus contraction) of relative motion along some given retinal meridian. MID is a two-dimensional relative motion filter that is most efficiently excited by expansion of the retinal image, when the expansion is isotropic i.e. when $k_1(a-b) = k_2(c-d)$. We assume here that the amplitude of the output of the MID filter is equal to that of any one of its inputs from RM filters. If so, this output is inversely proportional to time to collision. From Regan, D. and Hamstra, S.J. (1993). Dissociation of discrimination thresholds for time to collision and for rate of angular expansion. *Vis. Res.*, 33, 447-462. Reprinted with permission.

adapting stimuli: a bright solid rectangle whose opposite vertical edges oscillated in anti-phase (so that the rectangle's location remained constant while its width oscillated), and the same rectangle but with opposite vertical edges oscillating in phase (so that its width remained constant while its location oscillated from side to side). After adapting to anti-phase oscillations, visual sensitivity to anti-phase oscillations was considerably reduced, while sensitivity to in-phase oscillations was little affected. A second finding was that adapting to in-phase oscillations had little effect on sensitivity to either anti-phase or in-phase test oscillations. The crucial point here is that the oscillations of either of the rectangle's edges were identical for the two adapting stimuli: the two adapting stimuli differed only in the relation between movements of the two vertical edges. Clearly, the results of this

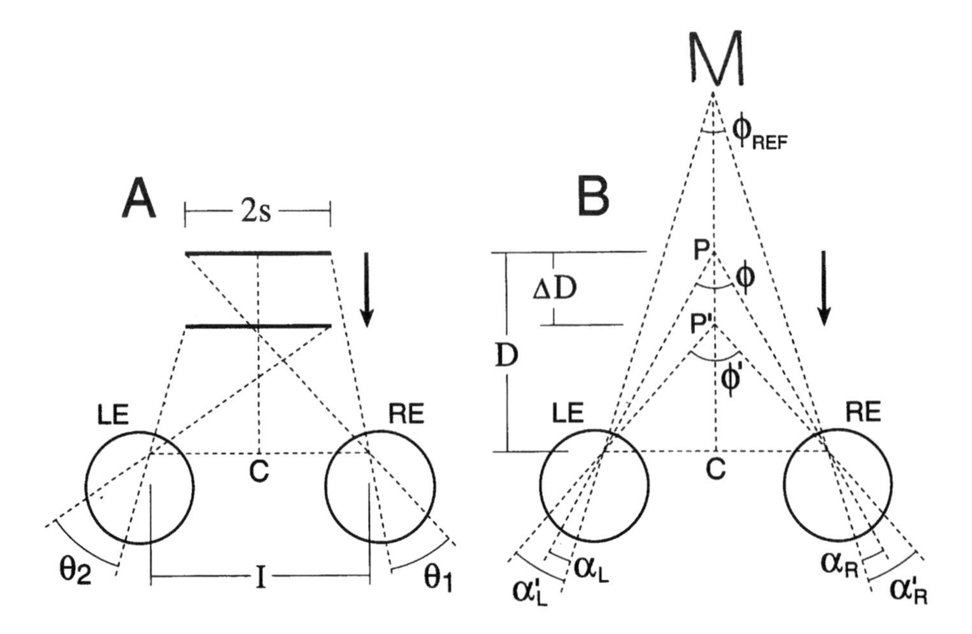

Fig. 9.2. A: An object of width 2s moves at an instantaneous velocity $\mathbf{V}_Z$ on a straight line through a point C midway between the eyes. The angular subtense of the object (θ) increases from θ_1 at time $t = 0$ to θ_2 at time $t = \Delta t$. B: A point object, located at P at a distance D from the eyes, moves at an instantaneous velocity $\mathbf{V}_Z$ on a straight line through a point C midway between the eyes. At time $t = \Delta t$, the object is located at P' an has traveled a distance ΔD. The disparity of the object (δ) relative to a stationary reference mark (M) changes from $\alpha R + \alpha L$ at time $t = 0$ to $\alpha'_R + \alpha'_L$ at time $t = \Delta t$. The change in disparity ($\Delta\delta$) is equivalent to $\phi' - \phi$. LE: left eye. RE: right eye. I: interpupillary distance.

experiment could not be explained entirely in terms of adaptation of local responses to unidirectional motion (LM filters in Figure 9.1). The conclusion was that the differential loss of sensitivity was caused by adapting to some *relationship* between the velocities of the rectangle's two vertical edges. In Figure 9.2 this relationship is computed by relative motion (RM) filters. Further evidence for a neural mechanism sensitive to relative unidirectional motion has subsequently been published (Freeman & Harris, 1992).

Now we turn to experiments on visual responses to isotropic looming. First, we describe a two-dimensional variant of the study just described (Regan & Beverley, 1978a). After adapting to isotropic oscillations in the size of a bright solid square, sensitivity to size oscillations was reduced five-fold, but sensitivity to oscillations of the square's location was unaffected. In addition, both thresholds were little affected by adapting to oscillations of the square's location. Again, the crucial point was that the oscillations of the square's edges (all four in this case) were identical for the two adapting stimuli: the two adapting stimuli differed only in the *relationship* between movements of opposite edges. Just as for the one-dimensional case reviewed above, these findings for the two-dimensional case could not be explained in terms of adaptation of local detectors of unidirectional motion. The most parsimonious explanation was that the reduction of sensitivity to isotropic oscillations of size was due to adaptation of a pair of relative motion filters (RM in Figure 9.1) one driven by the square's vertical edges and one by the square's horizontal edges.

The particular velocity relationship to which these relative motion filters were sensitive was indicated in a second study (Regan & Beverley, 1980). This showed that postadaptation reduction of sensitivity to looming was the same whether or not the square moved in the frontal plane at the same time as changing size. Another way of stating this finding is that the eye can abstract the line-of-sight component of an object's motion independently of the object's trajectory. This finding implies that the looming mechanism is sensitive to the algebraic difference between the velocities of the square's opposite edges*.

So far we have discussed how the LM and RM filters in Figure 9.1 can account for visual responses to change in the size of an object's retinal image. We have not yet taken into account the findings that visual responses to a rate of change of the size of the retinal image are strongly affected by any associated change in the shape of the retinal image (Beverley & Regan, 1979a , 1980). This finding implies that the visual system compares the rates of expansion along different meridia of the retinal image. We proposed that this comparison is carried out at a processing stage subsequent to the LM and RM filters (the MID stage in Figure 9.1) and it is the MID stage that, by generating an output z, causes a perception of motion in depth (Beverley & Regan, 1979a).

Next we discuss how the MID stage selects for isotropic expansion. The test for expansion with no associated change of shape (i.e. isotropic expansion) is that the value of $\theta/(d\theta/dt)$ is independent of meridian (θ is the instantaneous size and $d\theta/dt$ the instantaneous rate of expansion). A rough version of this test would be to compare the values of $\theta/(d\theta/dt)$ across two orthogonal meridia, say the horizontal and vertical meridia. There is direct experimental evidence for such a neural process. The neural process tests for the identity

$$\frac{\theta_V}{d\theta_V/dt} = \frac{\theta_H}{d\theta_H/dt} \tag{9.1}$$

where θ_V is the height and θ_H is the width of the retinal image (Beverley

* Subtraction of the velocities of opposite edges as proposed by Regan & Beverley (1980) is a linear process. Sensitivity to expansion or contraction is, therefore, based on a linear combination of velocity signals. However, as discussed below a motion-in-depth signal is influenced by a comparison of the times to collision signaled by different meridia in the object's retinal image and this requirement requires nonlinear processing.

& Regan, 1980). In the format of Figure 9.1 we can satisfy this equation by writing

$$k_1(a-b) = k_2(c-d) \tag{9.2}$$

and

$$k_1 = \frac{K}{\theta_V} \tag{9.3}$$

and

$$k_2 = \frac{K}{\theta_H} \tag{9.4}$$

where K is a constant.

As mentioned already, there is evidence that, provided that equation (9.2) is satisfied, the MID stage optimally produces a motion in depth signal z. If we assume that the amplitude of this signal is equal to the amplitude of any one of the several equal inputs, i.e.

$$z = K\theta_V/(d\theta_V/dt) = K\theta_H/(d\theta_H/dt), \text{etc.} \tag{9.5}$$

then

$$z = \frac{K}{T} \tag{9.6}$$

where T is the time to collision and K is a constant. In words, "the magnitude of the motion-in-depth output of the MID filter is inversely proportional to time to collision" (Regan & Hamstra, 1993). This hypothesis would make sense in terms of visually-guided motor action. For example, in highway driving the larger the output of the MID filter, the more urgent is the demand for evasive action. Again, when walking, driving or flying through space a containing several objects, locomotion could be guided by the strategy of continuously adjusting the direction of motion so as to prevent the output of any MID filter from rising above a safe level.

The hypothesis expressed in equation (9.6) leads to the prediction that altering the perceived speed of motion in depth produced by a given looming stimulus should change the estimated time to collision. This prediction is fulfilled when the effectiveness of a looming stimulus is changed by mismatching texture dynamics (Vincent & Regan, 1995; Gray & Regan 1998a). A second prediction – that adaptation to looming should change the estimated time to collision – has also been confirmed (Gray & Regan, 1998b).

9.1.2 Practical use of monocular information about time to collision

In a science fiction novel the distinguished astronomer and mathematician Fred Hoyle (1957) described a black cloud, located some distance outside the solar system, whose angular subtense was increasing though its centre remained in precisely the same visual direction. He pointed out that one possible interpretation of this pattern of motion is that the cloud was moving on a collision course with the Earth*. Assuming that the black cloud's speed was constant and that it was not expanding, and given that its angular subtense was less than roughly 10 deg, Hoyle derived the equation

$$T = \theta/(d\theta/dt) \tag{9.7}$$

where T is the time to collision with the Earth and θ is the object's angular subtense (i.e. the angle subtended at the eye by the object, see Figure 9.2A). We should also note that the ratio $\theta/(d\theta/dt)$ is an unequivocal correlate of the approaching object's time to contact only if the object is rigid and only if – except in special cases – the object is nonrotating. The special cases include a rotating sphere.

Following Lee (1976) several authors have suggested that humans take advantage of equation (9.7) in highway driving, in sporting activities and in aviation (Lee & Lishman, 1977; Lee, Lishman & Thomson, 1982; Lee, Young, Reddish, Lough & Clayton, 1983; Todd, 1981; Warren, Young & Lee, 1986; DeLucia, 1991; Savelsbergh, Whiting & Bootsma, 1991; Schiff & Detwiler, 1979; Bootsma & van Wieringen, 1990; Cavallo & Laurent, 1988; Regan, 1991a, 1995; Kruk & Regan, 1983; Karnavas, Bahill & Regan, 1990).

If humans can indeed utilize equation (9.7) to judge the time to collision with an approaching object of unknown size, distance and speed of approach, a minimum requirement is that humans possess the ability to discriminate variations of the ratio $\theta/(d\theta/dt)$ independently of simultaneous and independent variations of θ and $d\theta/dt$. Figure 9.3 shows evidence that this requirement is fulfilled. (Regan & Hamstra, 1993; Regan & Vincent, 1995; Gray & Regan, 1998d). Rather than using a real object, we simulated an approaching rigid sphere by creating a monocular retinal image that expanded in exactly the same way as the retinal image of a sphere that was moving at constant speed directly towards the observing eye. The simulated sphere disappeared after a randomly-varied presentation duration that was never longer than two-thirds of the initial time to collision.

The stimulus set consisted of 64 different combinations of time to collision, initial rate of expansion, starting size and presentation duration. For explanatory purposes this stimulus set can be visualized as an 8 x 8 square array whose sides are parallel to the x- and y- axes. Time to collision varied along the x- axis, but rate of expansion was constant along that axis. Rate of expansion varied along the y- axis, but time to collision was constant along that axis. Starting size varied in the same way along both axis.

* As noted by Poincaré (1913) an alternative interpretation would have been that the black cloud was expanding while remaining in the same place.

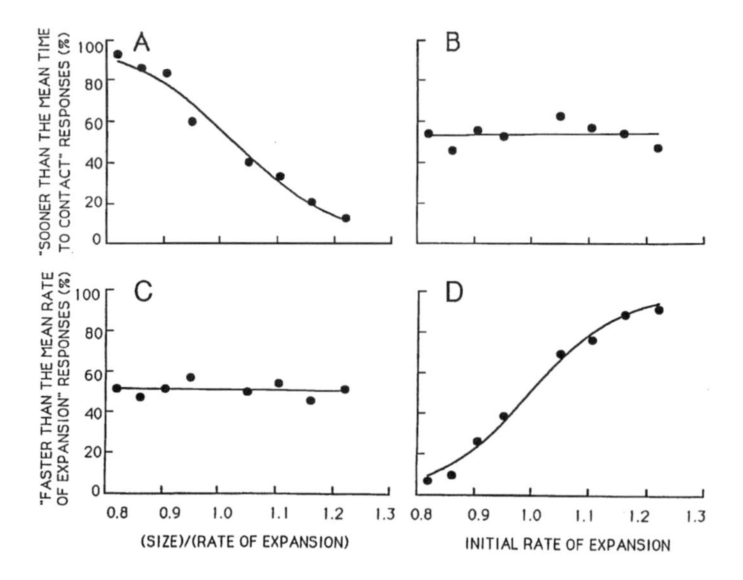

Fig. 9.3. Simultaneous discriminations of trial-to-trial variations of time to collision and rate of expansion. The percentage of "sooner than the mean time to collision" responses was plotted versus time to collision (A) and versus rate of expansion (B). The percentage of "greater than the mean rate of expansion" responses was plotted versus time to collision (C) and versus rate of expansion (D). From D. Regan & A. Vincent (1995). Visual processing of looming and time to contact throughout the visual field. *Vis. Res.*, 35, 1845-1857. Reprinted with permission.

Observers had three tasks. Following each presentation they were instructed to signal whether the time to collision was shorter than the mean of the stimulus set, whether the rate of expansion was faster than the mean of the stimulus set and whether the starting size was larger than the mean of the stimulus set. Responses were stored in three 8 x 8 arrays, each of which corresponded to the 8 x 8 stimulus array.

In Figure 9.3A & B the set of "shorter than the mean time to collision" responses were collapsed along the time to collision (i.e. $\theta/(d\theta/dt)$) axis (Figure 9.3A) and then the same responses were collapsed along the rate of expansion (i.e. $d\theta/dt$) axis. Figure 9.3A & B shows that the observer based his responses on trial-to-trial variations in the ratio $\theta/(d\theta/dt)$, while he ignored the co-varying rate of expansion. Further analysis showed that the observer also ignored trial-to-trial variations in starting size. In Figures 9.3C & D the set of "faster than the mean rate of expansion" responses were similarly collapsed along the $\theta/(d\theta/dt)$ axis (Figure 9.3C) and along the $d\theta/dt$ axis (Figure 9.3D). Figure 9.3C & D show that the observer's responses were based on trial-to-trial variations in the rate of expansion while he ignored the co-varying variable $\theta/(d\theta/dt)$. Further analysis showed that the observer also ignored trial-to-trial variations in starting size. We also found that the observer could discriminate trial-to-trial variations in starting size while ignoring simultaneous variations in time to collision and rate of expansion.

We concluded that observers can process time to collision, rate of expansion and size simultaneously, independently, and in parallel. (Regan & Hamstra, 1993; Regan & Vincent, 1995; Gray & Regan, 1998d). The physiological basis for this capability remains to be discovered. For example, it might be that the human visual system contains "hard-wired" filters sensitive to the three variables. On the other hand, it is possible that the simultaneous and independent processing of the three variables is possible only in the laboratory situation that an observer has several practice runs before data are collected, thereby allowing descending task-specific signals to modify neural connectivity in visual cortex so as to create temporary filters of the kinds required by the observer's designated task*.

So far we have discussed properties of foveal vision. The independence of responses to ratio $\theta/(d\theta/dt)$ and to rate of expansion $(d\theta/dt)$ shown in Figure 9.3A - D becomes progressively less as viewing eccentricity is increased up to at least 32 deg. In peripheral, but not in foveal vision, variations in rate of expansion (at constant time to collision) produced illusory variations in perceived time to collision (Regan & Vincent, 1995). Thus, when attempting to avoid collision with several approaching objects, it might be safer to glance rapidly form one to another rather than fixating on only one.

It has been suggested that, although visual responses to rate of expansion (looming) are present in early infancy, visual sensitivity to the ratio $\theta/(d\theta/dt)$ develops later and requires exposure to optic flow patterns created by self-locomotion (Regan & Vincent, 1995).

Now we turn from discriminating trial-to-trial variations of time to collision to estimating absolute time to collision. In a study reported by Gray & Regan (1998d) each trial consisted of one 0.7 sec presentation of the approaching target. At the designated time to collision, some time after the target had disappeared, there was a brief auditory tone whose timing could be set to an accuracy of 0.001 sec. The observer was instructed to press one of two buttons depending on whether the click occurred before the simulated approaching sphere would have hit the head. The value of $\theta/(d\theta/dt)$ was varied by the computer that controlled the experiment according to the observer's previous button presses. We used the staircase method described by Levitt (1971). Three designated times to collision (1.69, 2.09 & 2.72 sec)

* Task-dependent modification of neural characteristics has been observed in alert behaving non-human primates (Maunsell, 1995; Haenny & Schiller, 1988; Haenny, Maunsell & Schiller, 1988). Anderson & van Essen (1987) have attempted to explain theoretically how such changes might be produced.

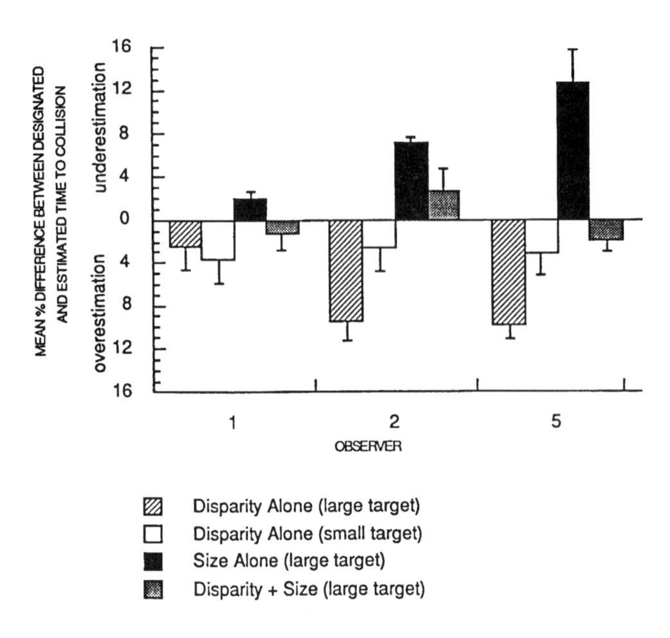

Fig. 9.4. The mean percentage difference (and standard errors) between designated and estimated time to collision for observers 1, 2 and 5. Hatched bars: estimates based on binocular information alone (large target). Open bars: estimates based on binocular information alone (small target). Solid bars: estimates based on monocular information alone (large target). Grey bars: estimates based on combined binocular and monocular information (large target). Reprinted from R. Gray & D. Regan (1998). Accuracy of estimating time to collision using binocular and monocular information. *Vis. Res.*, 38: 499-512. Reprinted with permission.

were combined with three starting sizes (0.41, 0.68 & 0.9 deg) to give a total of nine interleaved staircases.

The solid bars in Figure 9.4 show the mean percentage differences between the designated and estimated times to collision. Confirming previous reports (Schiff & Detwiler, 1979; Cavallo & Laurent, 1988), all observers consistently underestimated time to collision. Errors ranged from 2.0 to 12%. On the basis of these data plus the results of stepwise regression analysis we concluded that, provided the target is sufficiently large, observers can make accurate estimates of time to collision on the basis of the $\theta/(d\theta/dt)$ ratio alone, while ignoring simultaneous trial-to-trial variations in target size, rate of expansion and total change of size.

It is easy to see that judgements of time to collision based on the ratio $\theta/(d\theta/dt)$ will fail when the angular subtense of the approaching object is very small, because the rate of angular expansion would be undetectable. A less obvious point is expressed by the theoretical equation (9.8)

$$\frac{(d\theta/dt)}{(d\delta/dt)} \approx \frac{2S}{I} \tag{9.8}$$

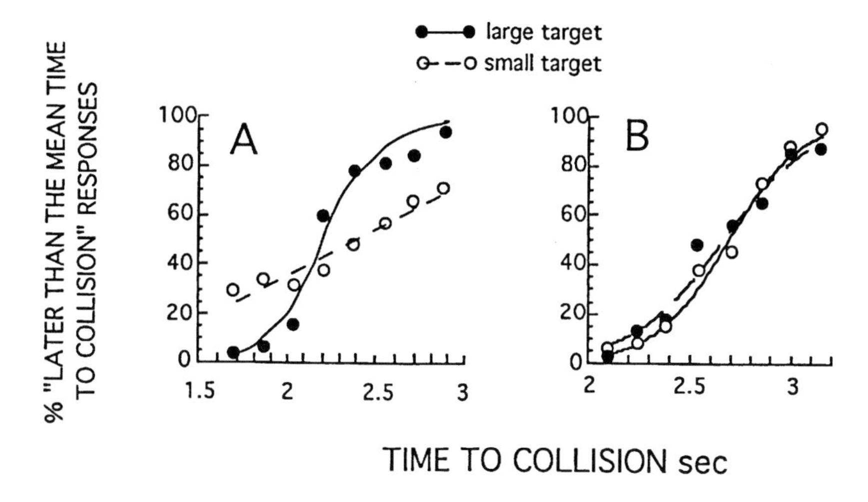

Fig. 9.5. Discriminations of trial-to-trial variations in time to collision in the case that estimates were based on monocular information only (A) and binocular information only (B). The percentage of "later than the mean time to collision" responses were plotted vs. time to collision. Filled circles are for the large target and open circles are for the small target. Reprinted from R. Gray & D. Regan (1998). Accuracy of estimating time to collision using binocular and monocular information. Vision Research, in press. Reprinted with permission.

where, as illustrated in Figures 9.2A & B, $\theta/(d\theta/dt)$ is the rate of increase of the angular subtense of the approaching object, $d\delta/dt$ is its rate of change of binocular disparity, $2S$ is the linear width of the object (expressed, for example, in cm), and I is the interpupillary separation of the observer's eyes (Regan & Beverly, 1979). In words: the ratio between the monocular $(d\theta/dt)$ and binocular $(d\delta/dt)$ correlates of motion in depth depend on the approaching object's linear width (as distinct from angular subtense) *independently of viewing distance.*

Gray & Regan (1998d) explored this point experimentally using a stimulus set that comprised 64 different combinations of time to collision [i.e. $\theta/(d\theta/dt)$], rate of expansion $(d\theta/dt)$, change of size during a presentation $(\Delta\theta)$, presentation duration and starting size. Observers were instructed to signal whether time to collision was shorter than the mean of the stimulus set. Filled and open circles in Figure 9.5A are for targets of mean starting size 0.7 deg and 0.03 deg respectively. Figure 9.5 A shows that discrimination threshold was considerably lower for the large target than for the smaller target. The difference between the two sets of data was, however, not merely quantitative. By subjecting the response data to stepwise regression analysis, it was found that, for the larger target size, observers based their responses on the task-relevant variables. In particular the ratio $\theta/(d\theta/dt)$

accounted for a high proportion of the variance (R^2 ranged from 0.72 to 0.80 for our four observers) and none of the other variables entered ($d\theta/dt$, $\Delta\theta$, duration, starting size) accounted for a significant amount of variance. On the other hand, for the smaller target the ratio $\theta/(d\theta/dt)$ accounted for only a small proportion of variance (R^2 ranged for 0.15 to 0.42 for the four observers), and for two observers *a task-irrelevant variable accounted for most variance*. We concluded that monocular information does not provide a reliable basis for either discriminating variations in time to collision nor estimating its absolute value when object size is small. This does not mean, however that it is impossible to estimate the time to collision with a small object. We will see next that binocular information can provide a basis for estimating time to collision with objects of any size.

9.1.2.1 Illusory Perception of an object's motion in depth produced by a rate of change of disparity of its retinal images in the left and right eyes

In some games the viewing distances involved can be considerably larger than the few metres over which the visual perception of relative depth is supported by binocular disparity. It is sometimes assumed that in such games binocular retinal image information is of negligible importance. The fallacy in this argument is that it does not take into account the distinction between the binocular visual processing of static depth and the binocular visual processing of motion in depth. The distinction has a geometrical aspect and a physiological aspect. First, while the relative disparity of an object depends on the viewing distance (equation 9.9), the rate of change of disparity depends on the object's velocity as well as the viewing distance (equation 9.10). Second, an individual's visual sensitivity to changing disparity cannot be directly predicted from that same individual's visual sensitivity to static disparity (Richards & Regan 1973; Regan, Erkelens & Collewijn, 1986b; Hong & Regan, 1989).

The geometrical aspect of the distinction can be understood as follows. Suppose we have one point object at a depth D, and a second point object at depth $(D - \Delta D)$. Provided that $D \gg I$, the difference between the disparities of the two point objects ($\Delta\delta$) is given by the well-known equation

$$\Delta\delta \approx \frac{I\Delta D}{D^2} \qquad (9.9)$$

where I is the observer's interpupilliary separation. Equation (9.9) indicates that a given difference in depth (ΔD) between the two points corresponds to less and less difference in disparity as the viewing distance (D) is progressively increased, so that the just-detectable difference in depth increases rapidly (with D^2) as viewing distance is increased. This is the basis for assertions that binocular depth perception is ineffective at viewing distances greater than 10m or so.

However, this is not necessarily the case for motion in depth. Figure 9.2B depicts an object moving at speed V_z along a straight line normal to the frontal plane that passes through a point midway between the eyes. The instantaneous rate of change of disparity ($d\delta/dt$) is given by Equation (9.10)

$$\frac{d\delta}{dt} = \frac{IV_Z}{D^2} \tag{9.10}$$

Equation (9.10) shows that, from geometrical considerations alone, the effective range for the detection of $d\delta/dt$ (i.e., for the perception of motion in depth) is set by a tradeoff between V_z and D^2.

As mentioned already (equation 9.8), binocular cues to an object's motion in depth grow relatively more important than monocular cues as the linear size of the object is decreased. For the limiting case of a point object, the $d\theta/dt$ monocular cue to the presence of motion vanishes, leaving only the $d\delta/dt$ binocular cue. Whether or not a given individual can actually use the binocular retinal image information about an object's motion in depth depends on the object's velocity as well as the viewing distance, and also on that individual's sensitivity to rate of change of binocular disparity.

Figure 9.6 indicates that different neural mechanisms support the perception of motion is depth and motion parallel to the frontal plane. Stereomotion visual fields are shown for two individuals, each of whom has an area of the binocular visual field that is specifically "blind" to stereomotion. The dotted and dashed lines show that thresholds for motion parallel to the frontal plane are unaffected by the presence of the stereo motion-blind area (Regan, Erkelens & Collewijn, 1986b; Hong & Regan, 1989).

Blindness to stereomotion can also exist in the presence of normal discrimination for static disparity (Richards & Regan, 1973; Regan, Erkelens & Collewijn, 1986b). This psychophysical finding is consistent with physiological evidence in animals that different neurons are sensitive to static disparity and to motion in depth (Cynader & Regan, 1978; Spileers, Orban, Gulyas & Maes, 1990)

Harris & Watamaniuk (1995) claimed that the human visual system does not contain a cyclopean speed-sensitive mechanism, and that speed discriminations of cyclopean motion in depth are based on trial-to-trial variations of disparity displacement ($\Delta\delta$) rather than on trial-to-trial variations in $d\delta/dt$. However, their evidence was based on the responses to a cyclopean target that passed through zero disparity and, therefore, disappeared and reappeared midway through its trajectory. Their conclusion conflicts with the evidence described next that acute discriminations of speed are possible when a cyclopean target remains visible throughout is trajectory (Portfors-Yeomans & Regan, 1996; Portfors & Regan, 1997). Figure 9.7A-D shows

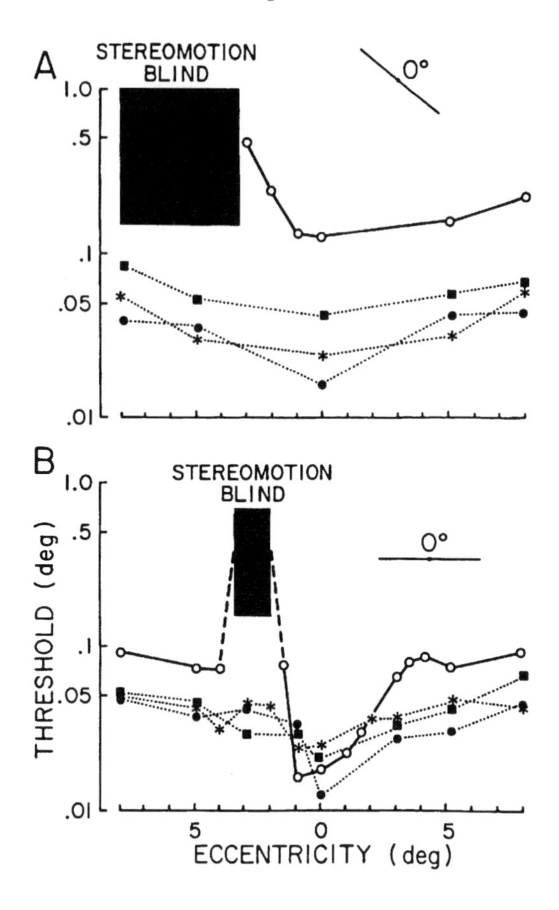

Fig. 9.6. Thresholds for detecting oscillatory motion in depth (open circles, continuous line) and oscillatory motion parallel to the frontal plane (filled symbols) plotted versus eccentricity along a line (see inserts) passing through both the fovea and a stereomotion scotoma. Filled circles signify left eye monocular stimulation, and stars signify binocular stimulation with motion parallel to the frontal plane. Zones of total stereomotion blindness are shown in black. A: author D.R. B: Caspar Erkelens. From D. Regan, C.J. Erkelens & H. Collewijn, (1986). "Visual field defects for vergence eye movements and for stereomotion perception", *Investigative Ophthalmology and Visual Science*, 27, 806-819. Reprinted with permission.

the results of an experiment in which the stimulus set comprised 64 different combinations of speed ($d\delta/dt$), displacement ($\Delta\delta$) and presentation duration. Following each presentation the observer was required to signal (a) whether the speed of the cyclopean target was faster than the mean speed of the stimulus set, and (b) whether the disparity displacement was larger than the mean of the stimulus set. Figure 9.7A & B show that the observer discriminated speed, while ignoring displacement. Figure 9.7C & D show that the observer discriminated displacement, while ignoring speed. We concluded that the speed and displacement of a cyclopean target's motion in depth are encoded independently. Weber fractions over six observers

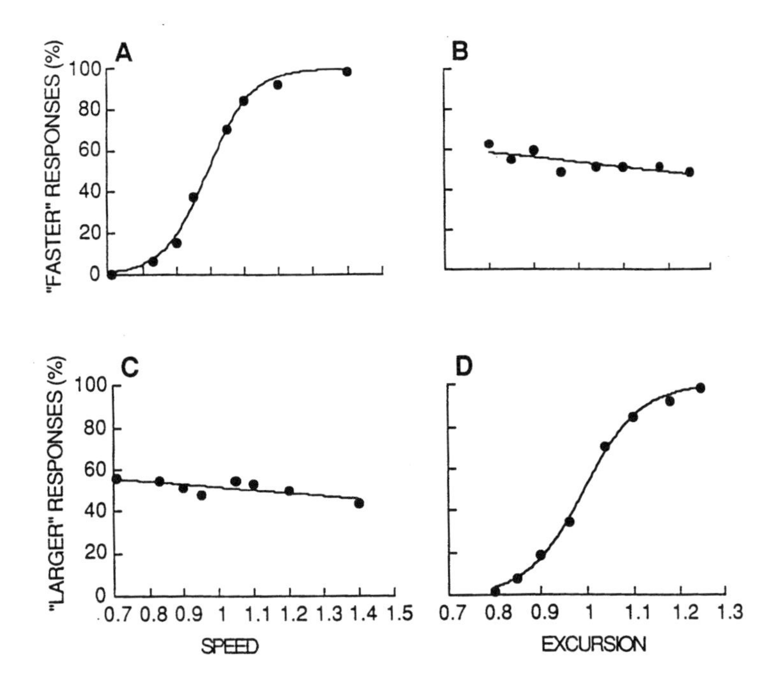

Fig. 9.7. Discrimination of simultaneous trial-to-trial variations of both the rate of change of disparity $d\delta/dt$ (A & B) and the displacement $\Delta\delta$ (i.e. excursion, C & D) of motion in depth for a cyclopean target. The target started at 5 min arc near disparity and moved towards the observer. From Portfors, C.V. & Regan, D. (1997). "Just-noticeable difference in the speed of cyclopean motion in depth and the speed of cyclopean motion within a frontoparallel plane". *Journal of Experimental Psychology: Human Perception and Performance*, 23: 1074-1086. Reprinted with permission.

ranged from 0.07 to 0.17 for speed and from 0.06 to 0.13 for displacement.

We proposed that the perceived speed of the motion in depth created by a rate of change of disparity is not directly related to the actual speed in m/sec (V_Z in equation 9.10). Rather, the perceived speed is inversely proportional to the time to collision (Gray & Regan, 1998d).

9.1.3 Binocular information about time to collision

Many previous studies on visual judgments of time to collision either eliminated binocular disparity information altogether (Todd, 1981; Schiff & Detwiler, 1979; McLeod & Ross, 1985; DeLucia, 1991; Regan & Hamstra, 1993; Regan & Vincent, 1995) or, when disparity information has been available, it was confounded with monocular information (Lee, Lishman & Thomson, 1982; Lee *et al.*, 1983; Warren, Young & Lee, 1986; Savelsbergh *et al.*, 1991; Bootsma & van Wieringen, 1990). Suggestive evidence that binocular retinal image information might aid judgment of time to collision has been scattered through the literature over a long period. For example, one hint was provided by Bannister & Blackburn (1931) who ranked 258 Cambridge undergraduates into "poor" and "good" categories according to their ability at ball games, and found that the group who were ranked "good" had a larger interpupillary distance than the group ranked as "poor". More recently, using high-speed photography it was found that, when catching a ball with one hand, the temporal organization of finger flexions was disrupted when the lights were switched off 275 msec before the ball arrived, that is when the ball was closer than 6 ft from the hand when binocular processing would be maximally effective (Alderson, Sully & Sully, 1974). These finger flexions are necessary if the ball is to be retained in the catcher's grip. Binocular vision seems to be important also at distances relevant to highway driving. Cavallo & Laurent (1988) compared the accuracy of time to collision judgments using binocular vs. monocular vision on a circuit under actual driving conditions. Accuracy was greater for binocular judgments providing that viewing distance was less than about 75m, but errors were still considerable (time to collision was consistently underestimated by at least 30%). However, at the considerably greater distances associated with landing a jet aircraft, occluding one eye during the landing approach had no detrimental effect on landing performance (Plaffman, 1948; Lewis & Kriers, 1969; Lewis *et al.*, 1973; Grosslight *et al.*, 1978).

It has been shown theoretically that a binocular correlate of time to collision (T) is available for small as well as for large objects (Regan, 1995; Gray & Regan, 1998d). In particular

$$T \approx \frac{I}{D(d\delta/dt)} \tag{9.11}$$

where D is the object's distance, I is the interpupillary separation and $(d\delta/dt)$ is the rate of change of relative disparity (see Figure 9.2B). However, although a substantial number of studies on stereomotion have been published [reviewed in Tyler (1991), Regan (1991b) & Collewijn & Erkelens, 1990)], as have a substantial number of studies on time to collision [reviewed in Tresilian (1995)], there have been very few reports of data on the use of binocular information in estimating time to collision. This might seem a curious omission given that the monocularly-available ratio $\theta/(d\theta/dt)$ is an ineffective indicator of time to collision for small objects (Regan & Beverley, 1979). Among the possible reasons for this omission are the following. (a) Viewing distance enters into equation (9.11), and the weight of evidence is that we are poor at judging the absolute distance of objects further than a few metres away from the head (Collewijn & Erkelens, 1990). (b) The sensation of motion in depth generated by a given rate of change of disparity is quite different in different visual spatial environments. In particular, the sensation of motion in depth is enhanced by the presence of stationary reference marks close to the moving object's retinal images (Tyler, 1975; Erkelens & Collewijn, 1985a, b; Regan, Erkelens & Collewijn, 1986a). (c) Many subjects have areas of the visual field that are selectively blind to stereomotion (Richards & Regan, 1973; Regan, Erkelens & Collewijn, 1986b; Hong & Regan, 1989).

We created retinal images in an observer's left and right eyes whose binocular disparity changed in exactly the same way as the retinal images of a spherical object moving directly towards the head at constant speed. Target size was constant. The stimulus set consisted of 64 different combinations of time to collision i.e. the ratio $I/D(d\delta/dt)$), and disparity displacement (i.e. $\Delta\delta$, the total change of disparity during a presentation), and presentation duration. Time to collision and disparity displacement varied orthogonally within the stimulus set. We chose to dissociate $d\delta/dt$ and $\Delta\delta$ because it has been claimed that discriminations of $d\delta/dt$ are based on $\Delta\delta$ rather than on $d\delta/dt$ (Harris & Watamaniuk, 1995) — though subsequent studies have shown this claim to lack general validity (Portfors-Yeomans & Regan, 1996; Portfors & Regan 1997). The observer's task was to signal whether time to collision was shorter than the mean of the stimulus set. Figure 9.5B shows that discriminations threshold for time to collision was the same for a large (0.7 deg) and small (0.03 deg) target — quite different from the situation when discrimination was based entirely on monocular information (Figure 9.5A).

When the same response data were plotted versus trial-to-trial variations in disparity displacement ($\Delta\delta$) the resulting psychometric function was flat, indicating that the observer ignored $\Delta\delta$. Stepwise regression analysis confirmed that discriminations were based entirely on time to collision, and that $\Delta\delta$ and presentation duration were ignored.

Although a low discrimination threshold is requisit for precise estimates of absolute time to collision, a low discrimination threshold does not necessarily imply *accurate* estimates: estimates might be consistently too long or too short.

We measured the absolute accuracy with which time to collision is estimated when only binocular information is available using a method similar to that already described for the monocular information case. Hatched and open bars in Figure 9.4 show errors for the large and small targets respectively.

From the data shown in Figure 9.4 plus the results of stepwise regression analysis we concluded that, on the basis of the ratio $I/D(d\delta/dt)$ alone, observers can make accurate estimates of time to collision with an approaching object, whether the object is small or large, and while ignoring simultaneous variations in disparity displacement ($\Delta\delta$), presentation duration and final disparity (Gray & Regan, 1998d). Errors were all overestimations.

The finding that the accuracy of judging time to collision can be higher for a small than for a large target might be due to the fact that the rate of change of disparity and the rate of change of size provide conflicting information, and that the conflict would be less for a small target, because the $d\theta/dt$ signal is much weaker.

We further concluded that in everyday situations when both monocular and binocular information are available simultaneously, accurate estimates of absolute time to collision, will be based almost entirely on binocular information when the approaching object is small and no more than a few meters away (Gray & Regan, 1998d).

We went on to measure the absolute accuracy of estimating time to collision in the situation that both binocular and monocular information was available exactly as in everyday conditions. We used the large (0.7 deg) target. The grey bars to the right of the black bars in Figure 9.4 shows that errors ranged from 1.3% to 2.7%. Subjecting the data to repeated-measures ANOVA and post-hoc Tukey test showed that, for the large target, errors in estimating time to collision were significantly lower when estimates were based on both binocular and monocular information than when they were based on either monocular or binocular information alone (Gray & Regan, 1998d).

If we assume that a cricket or table tennis player can use visual information up to about 300 msec before the instant of impact with the bat, a 1.3% error approaches the performance required to account for the 2.0 to 2.5 msec accuracy with which top sportsplayers judge the time to impact with an approaching ball (Regan, Beverley & Cynader, 1979; Bootsma & van Wieringen, 1990).

9.2 Monocular and binocular information about the direction of an approaching object's motion in depth

9.2.1 Monocular information about the direction of motion in depth

Two monocular cues to the direction of a sphere's motion in depth are the ratio $(d\alpha_1/dt)/(d\alpha_2/dt)$ and the ratio $(d\phi/dt)/(d\theta/dt)$, where $d\phi/dt$ is the angular velocity of the retinal image, $d\theta/dt$ is its rate of expansion and $d\alpha_1/dt$ and $d\alpha_2/dt$ are the angular velocities of opposite edges of the retinal image (Regan, 1986; Regan & Beverley, 1980; Regan & Kaushall, 1994). It is intuitively obvious that a trajectory for which either $(d\alpha_1/dt)$ or $(d\alpha_2/dt)$ is zero (i.e., $(d\phi/dt)/(d\theta/dt)$ equals +1.0 or -1.0) means that the centre of the sphere will pass exactly one sphere's radius (s) wide of the first nodal point of the eye (illustrated in Regan & Beverley, 1980, Figs. 4 & 5).

At a quantitative level, two monocularly-available correlates of the object's direction of motion in depth can be written as follows*.

$$ns \approx \frac{1 + (d\alpha_1/dt)(d\alpha_2/dt)}{1 - (d\alpha_1/dt)(d\alpha_2/dt)} \tag{9.12}$$

$$ns \approx \frac{d\phi/dt}{d\theta/dt} \tag{9.13}$$

Selective adaptation experiments provide no evidence for multiple channels sensitive to $(d\alpha_1/dt)/(d\alpha_2/dt)$ or to $(d\phi/dt)/(d\theta/dt)$ (Regan, 1986). In this respect, the processing of the monocular correlates of the direction of motion in depth seem to be quite different from the processing of the binocular correlates (Beverley & Regan, 1973, 1975).

Equations (9.12) & (9.13) express the direction of motion in depth in terms of the distance by which the centre of the ball will miss the eye. This distance is quantified as a number (n) times the ball's radius (s). Although equations (9.12) and (9.13) are mathematically equivalent, it does not necessarily follow that they are equivalent physiologically. We further note that, although the value of the ball's absolute size ($2s$) is required to estimate its direction of motion on the basis of monocular information alone, equation (9.8) above indicates that the value of $2s$ can, in principle, be obtained from binocular retinal image information.

In a laboratory study we measured monocular discrimination threshold for the direction of motion in depth (i.e. relative obliqueness of the trajectory) by generating on a monitor a target whose expansion and motion parallel to the frontal plane created retinal image dynamics that exactly mimicked the retinal image dynamics of a rigid object moving along a straight line at constant speed (Regan & Kaushal, 1994). It was important to ensure that subjects were forced to base their judgements of the direction of motion in depth entirely on the ratio (angular velocity of the retinal image)/(rate of change of angular size), i.e. on $(d\phi/dt)/(d\theta/dt)$ rather than some artifactual cue such as the direction or the speed of retinal image motion. Therefore, when measuring discrimination threshold for objects passing to the right of the eye we ensured that the reference and all the test stimuli passed to the right of the eye, and when measuring discrimination thresholds for objects passing to the left of the eye we ensured that the reference and all the test stimuli passed to the left of the eye, thus removing the direction of retinal image motion as a cue to the direction of movement in depth. We removed the speed of retinal image motion as a reliable cue to the direction of motion in depth by randomizing the magnitudes of $d\phi/dt$ and $d\theta/dt$ in a yoked manner so as to keep the ratio $(d\phi/dt)/(d\theta/dt)$ constant.

We found that subjects were able to unconfound variations in the ratio

* Bootsma and his colleagues have independently derived equation (9.13) and suggested that the human visual system represents an object's direction of motion in depth in terms of multiples of the object's radius (Bootsma, personal communication, 1991; Bootsma, 1991).

$(d\phi/dt)/(d\theta/dt)$, not only from variations in $d\phi/dt$, but also from variations in $d\theta/dt$, and could discriminate the simulated direction of motion in depth even though the only reliable cue to the task was the ratio $(d\phi/dt)/(d\theta/dt)$. We concluded that the human visual pathway contains a neural mechanism sensitive to the ratio $(d\phi/dt)/(d\theta/dt)$ independently of (at least small) variations in $(d\phi/dt)$ and $(d\theta/dt)$.

Discrimination threshold for the detection of motion in depth was remarkably acute. The best value of discrimination threshold was equivalent to about 0.06 deg. Threshold was approximately the same for the vertical, horizontal and oblique meridia out to at least $n = 4$ (Regan & Kaushal, 1994).

9.2.2 Binocular information about the direction of motion in depth

9.2.2.1 There is a cyclopean mechanism sensitive to the speed of motion within a frontoparallel plane

That the human visual system contains a cyclopean mechanism sensitive to the speed of motion parallel to the frontal plane is important for the following discussion. Since Harris & Watamaniuk (1996) have stated that the human visual system does not contain such a mechanism we must address this issue first.

We carried out an experiment for cyclopean motion parallel to the frontal plane that was analogous to the experiment whose results are shown in Figure 9.7. The psychometric functions looked much like those in Figure 9.7. We concluded that the observer discriminated speed, while ignoring displacement, and discriminated displacement, while ignoring speed. We further concluded that the speed and displacement of a cyclopean target's motion parallel to the frontal plane are encoded independently and in parallel. Weber fractions over six observer ranged from 0.09 to 0.20 for speed, and from 0.06 to 0.16 for displacement (Portfors & Regan, 1997).

These frontoparallel plane motion data were obtained using a short-edged square cyclopean target, whereas Harris & Watamaniuk (1996) used a stereo grating. When we repeated the experiment using a 3.5 x 2.8 deg cyclopean grating we found several observers who were able to discriminate speed while ignoring simultaneous variations in displacement, temporal frequency, spatial frequency and presentation duration (Kohly, Hajdur & Regan, 1997; Kohly & Regan, 1998). We suggest that the conflict between our findings and those of Harris and Watamaniuk (1996) might be because of their grating had very short (0.38 deg) bars.

9.2.3 Binocular discrimination of the direction of motion in depth

In Figure 9.8A the left and right eyes (LE, RE) fixate a nonious line (N) that forms part of a reference plane. At time $t = 0$ a target (T) is located on a line that is normal to the frontal plane and passes through point C midway between the eyes. Target T is located at distance D from point C, and is some distance in front of a reference plane of stationary marks. Target T is moving at a constant speed V along a straight line (bold arrow). At time $t = \Delta t$, target T will have moved through an absolute distance $V\Delta t$. Consequently, the angle between the retinal images of the target and any given mark in the stationary reference plane will change by $(\Delta\phi)_L$ in the left eye and $(\Delta\phi)_R$ in the right eye. If we let $\Delta t \to 0$ we can write the associated instantaneous rates of change as $(d\phi/dt)_L$ and $(d\phi/dt)_R$.

Figure 9.8B illustrates how the velocity ($\mathbf{V}$) of target T can be resolved into the following two orthogonal components: a component of magnitude V_z along direction TC (where $V_z = V \sin\beta$), and a component of magnitude V_x parallel to the frontal plane (where $V_x = V\cos\beta$). It will be convenient to discuss these two components separately.

Figure 9.8C illustrates that, at time $t = 0$, the disparity of target T relative to mark (N) in the reference plane is given by $\delta = \alpha_R - \alpha_L$. Figure 9.8D illustrates that, at time $t = \Delta t$, the V_z component of the target's motion has reduced the distance TC from D to $(D - \Delta D)$, and the instantaneous disparity of target T relative to mark N is now given by $\delta' = \alpha_R' - \alpha_L'$. The disparity displacement is $\Delta\delta$, where $\Delta\delta = (\delta' - \delta)$. (Note that this displacement is the same for any given mark in the reference plane). If we let $\Delta t \to 0$ we can write the associated instantaneous rate of change of disparity as $d\delta/dt$.

Figure 9.8E shows that at time $t = D\Delta Dt$, the V_x component of the target's motion has translated the target through distance Δx. This will alter the angular distance between the left eye's retinal image of the target and the left eye's retinal image of mark N by an amount $\Delta\phi$. Approximately the same change will occur in the right eye's retinal image. If we let $\Delta t \to 0$ we can write the associated instantaneous rate of change as $(d\phi/dt)$.

For motion contained within a plane that contains the eyes and is normal to the frontal plane, the magnitudes of the ratio $(d\phi/dt)/(d\delta/dt)$ and the ratio $(d\phi/dt)_R/(d\phi/dt)_L$ both vary with the direction of motion in depth. For example, both ratios vary as direction is changed from a through d in Figure 9.9A. For convenience we will term this case "motion within the horizontal meridian". The situation is different for motion confined to the vertical meridian in that the magnitude of $(d\phi/dt)/(d\delta/dt)$ varies as direction is changed from e through h in Figure 9.9B, but the ratio $(d\phi/dt)_R/(d\phi/dt)_L$ remains constant.

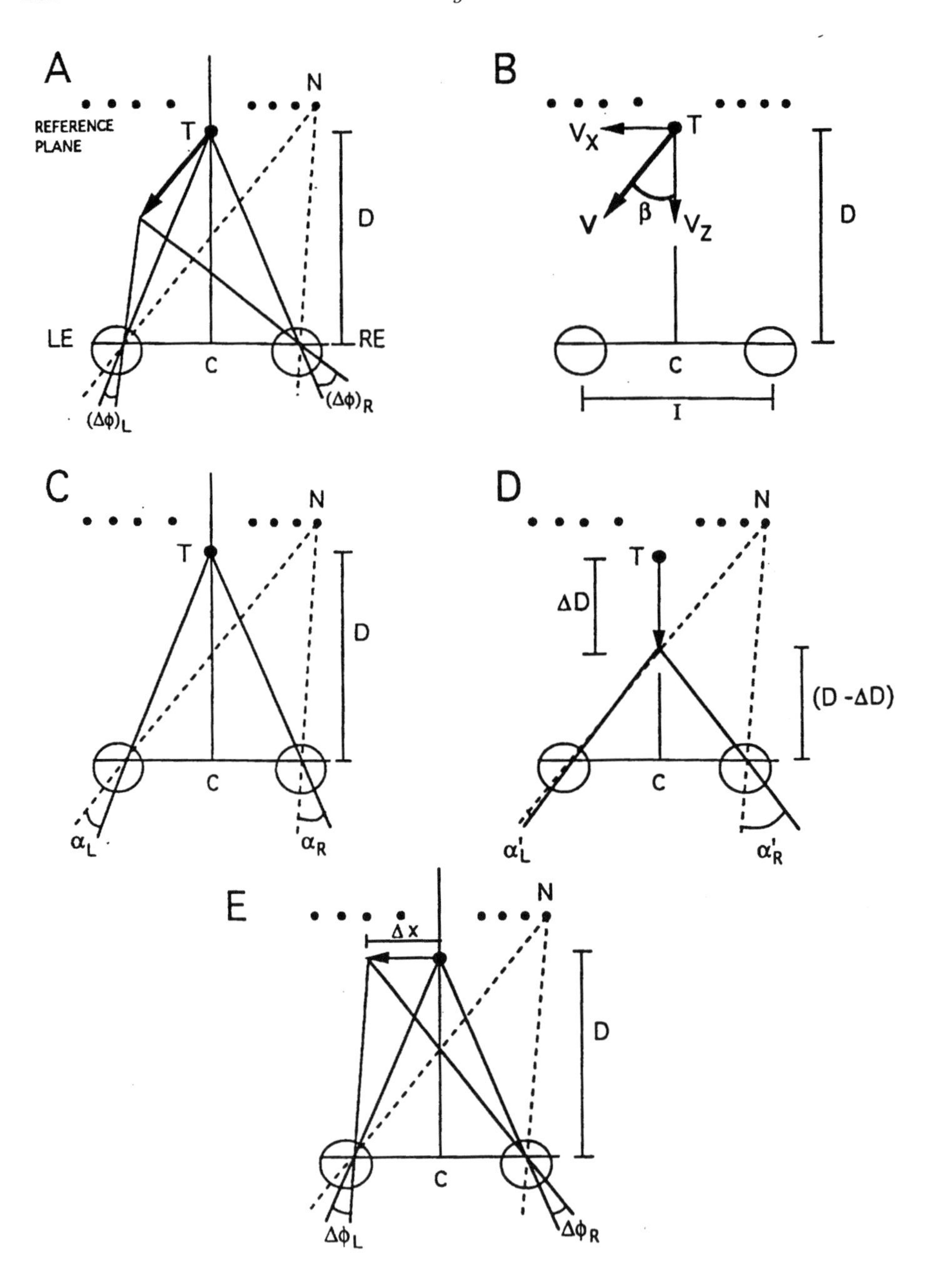

Fig. 9.8. Geometry of a target (T) moving in depth within a plane containing the left and right eyes. See text for details.

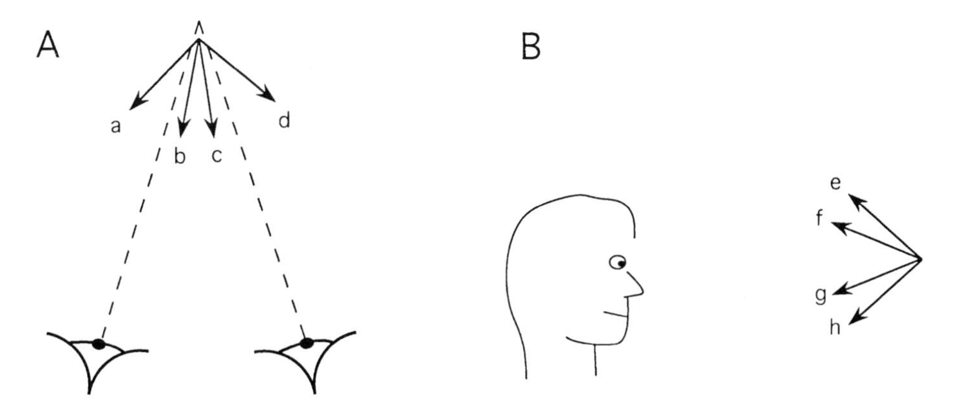

Fig. 9.9. (A) Different directions of motion within a plane that contains the left and right eyes and is normal to the frontal plane. For brevity, we will refer to this as motion within the horizontal meridian. (B) Different directions of motion within the vertical meridian. In all the experiments reported here the starting point was the same for all directions of motion.

In particular, there are two binocular correlates of the direction of motion in depth for monocularly-visible targets whose motion is confined to the horizontal meridian (Figure 9.9A). First, the direction of an object's motion in depth (β in Figure 9.9B) is given by the equation:

$$\beta \approx \tan^{-1}\left\{\frac{I\{[(d\phi/dt)_R/(d\phi/dt)_L]+1\}}{2D\{[(d\phi/dt)_R/(d\phi/dt)_L]-1\}}\right\} \tag{9.14}$$

(provided that $D \gg I$), where $(d\phi/dt)_R$ and $(d\phi/dt)_L$ are, respectively, the translational angular velocities of the object's retinal images in the right and left eyes, D is the object's distance and I is the observer's interpupillary separation (Beverley & Regan, 1973, 1975; Regan, 1986, 1993). [See Figure 9.9A.] However, even for monocularly-visible targets the $(d\phi/dt)_R/(d\phi/dt)_L$ cue is available only for motion contained within the horizontal meridian. A second binocular correlate of b is given by equation (9.15):

$$\beta \approx \tan^{-1}\left[\frac{I(d\phi/dt)}{D(d\delta/dt)}\right] \tag{9.15}$$

(again provided that $D \gg I$), where $(d\phi/dt)$ is the angular velocity of the binocularly-fused retinal image. (More exactly this is the angular velocity of the target's fused retinal image relative to the retinal image of some fixed reference mark. It is equal to $0.5[(d\phi/dt)_R+(d\phi/dt)_L]$, see Figure 9.9A). The quantity $(d\delta/dt)$ is the target's rate of change of relative disparity (Regan, 1993). This correlate is available for motion within any meridian*. It is the

* Equations (9.14) & (9.15) are valid only for an object in the straight-ahead position (i.e. an object located on a line that passes midway between the eyes and is perpendicular to the frontal

only correlate of the direction of motion in depth for cyclopean targets. It has been pointed out that, for an approaching object whose instantaneous location is straight ahead, this proposed representation of the direction of motion in depth directly indicates the point of arrival in the plane of the eyes (in particular whether the object will hit the observer's head), and that this property is independent of the direction of gaze and the angle of convergence (Beverley & Regan, 1973). This property follows from the fact that I/D radians is the angle subtended by the distance between the eyes from the viewpoint of the approaching object. In many everyday situations, and certainly in terms of evolutionary pressures, this information is considerably more important to the observer than is the ability to estimate the value of β. Turning back to our sporting context, equation (9.15) indicates where (in units of I) a catcher's hand should be placed to intercept a ball passing over, under, or to the side of the head. If we express in these terms the directional thresholds for monocularly-visible targets reported by Beverley & Regan (1973) and Portfors-Yeomans and Regan (1996), discrimination threshold for the point of impact of an approaching object is lowest for impact midway between the eyes and has a value of approximately 0.2 cm in the plane of the face, i.e. approximately $0.03I$ assuming $I = 6$cm. If, as seems to be the case (Portfors-Yeomans & Regan, 1996), visual processing of the terms in square brackets is independent of viewing distance, direction discrimination threshold expressed in terms of I will be independent of the approaching object's distance (providing that the speed of the approach is sufficiently high to raise $d\delta/dt$ well above threshold).

We previously reported that discrimination threshold for the direction of motion in depth for a monocularly-visible target moving in the horizontal meridian was 0.2 deg (Beverley & Regan, 1975). More recently we found that discrimination threshold was the same for motion within the vertical and horizontal meridian (Portfors & Regan, 1997). Given that the ratio of the left and right eyes' retinal image speeds provides no cue to the direction of motion in depth for motion within the vertical meridian, this finding is consistent with the idea that, for motion within both horizontal and vertical meridians, discrimination is based on the ratio $(d\phi/dt)/(d\delta/dt)$.

Our reason for measuring discrimination threshold for the direction of motion in depth of a cyclopean target (Portfors-Yeomans & Regan, 1996) was that the ratio between a cyclopean target's retinal images provides no cue to the direction of motion in depth, because the target is not visible monocularly. The set of 64 stimulus comprised different combinations of the direction of motion in depth [i.e. $I(d\phi/dt)/D(d\delta/dt)$], and $d\delta/dt$, $d\phi/dt$ and $\Delta\delta$. Following each presentation, observers were instructed to signal (a) whether the direction of motion in depth was directed wider of the head

plane). Different equations hold for an object that lies along a line that is considerably oblique to the straight-ahead directions.

than the mean direction of the stimulus set, and (b) whether the speed of motion in depth was faster than the mean of the stimulus set. We found that observers based their direction discriminations entirely on the task-relevant variable $I(d\phi/dt)/D(d\delta/dt)$, and ignored all task-irrelevant variables. They based their speed discriminations entirely on the task-relevant variable (i.e. speed), and ignored all task-irrelevant variables. Performance on both tasks was the same for motion within vertical and horizontal meridians. We repeated the experiment using a monocularly-visible target that was created by switching off all dots outside the cyclopean target. Thresholds were either the same or only slightly different for the cyclopean and monocularly-visible targets. We concluded that discrimination of the direction of motion in depth is determined by a mechanism sensitive to the ratio $(d\phi/dt)/(d\delta/dt)$ for both cyclopean and monocularly-visible targets. We also concluded that a single speed-sensitive mechanism determines speed discrimination threshold for both cyclopean and monocularly-visible targets.

Discrimination threshold for the direction of motion in depth of a monocularly-visible target can be as low as 0.14 to 0.2 deg (Beverley & Regan, 1995; Portfors & Regan, 1997). To explain this remarkably acute performance, Beverly and Regan (1975) proposed that the human visual pathway contains neural mechanisms tuned to different directions of motion in depth, and *that the relative activity of these mechanisms determines discrimination threshold*. The obtained evidence for such mechanisms by exploiting the finding that inspecting a monocularly-visible target moving along a line inclined in depth elevates threshold for detecting motion in depth, but only for a limited range of test directions (Beverley & Regan, 1973). These experiments, however, were restricted to motion in depth within the horizontal meridian so that it was not clear whether the proposed mechanisms were selectively tuned to the ratio $(d\phi/dt)_R/(d\phi/dt)_L$, to the ratio $(d\phi/dt)/(d\delta/dt)$, or to some combination of the two.

More recently we repeated and extended these experiments by using a cyclopean target, and using motion within both the vertical and horizontal meridian (Regan, Portfors, & Hong, 1997). Figure 9.10A shows baseline (preadaptation) thresholds for detecting oscillations of a cyclopean target along 9 test directions of motion in depth within the horizontal meridian. Figure 9.10G shows similar data for motion within the vertical meridian. Figure 9.10B - F shows threshold elevations caused by adapting to motion along 9 test directions within the horizontal meridian (arrowed on abscissae). Figure 9.10H - L show corresponding threshold elevations for motion within the vertical meridian.

The lines in Figure 9.10A - L join the data points. The lines in Figure 9.11A - L were derived theoretically from the sensitivity profiles shown in Figure 9.12A & B, using a simulation procedure described by Tyler, Bargh-

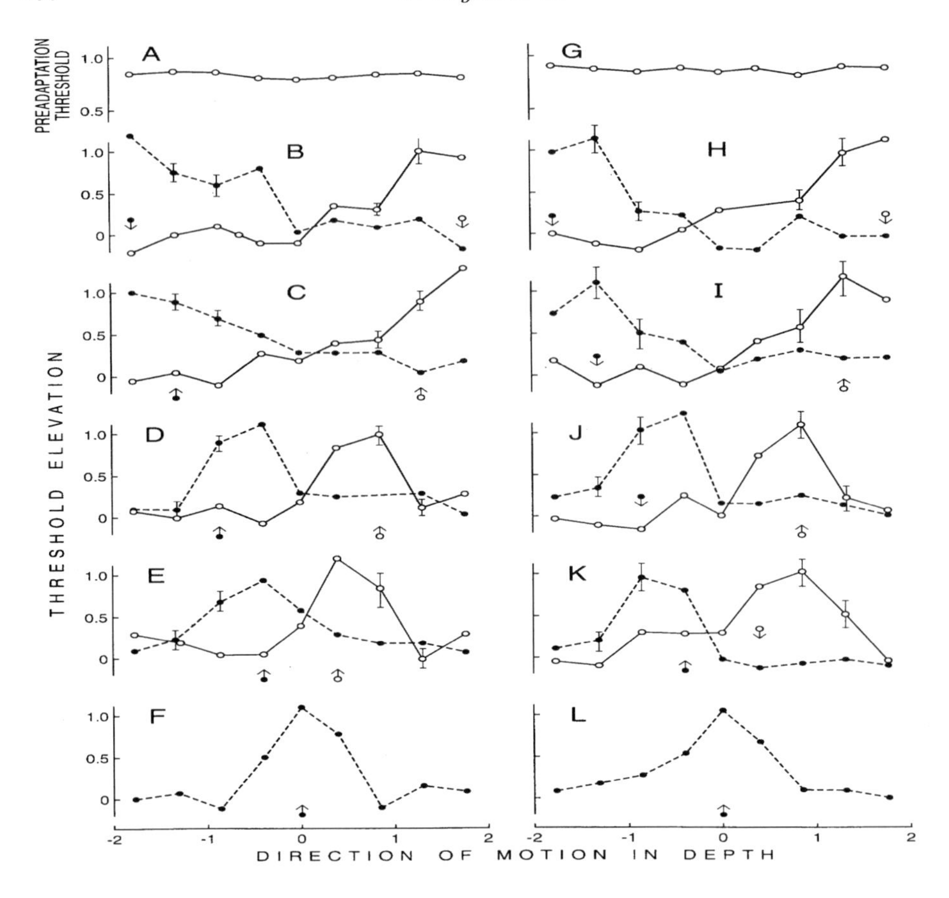

Fig. 9.10. Preadaptation and postadaptation thresholds for detecting the motion in depth of a cyclopean target moving with a horizontal meridian (A - F) and within a vertical meridian (G - L). A, G: Normalized preadaptation thresholds (ordinate) plotted versus the ratio $(d\phi/dt)/(d\delta/dt)$. B - F, H - L: Data points plot threshold elevations caused by adapting to different directions of motion in depth (arrowed). Positive values of $(d\phi/dt)/(d\delta/dt)$ mean rightward in B - F and upward in H - L. Vertical bars indicate ± 1 standard errors. The lines join point to point and are intended to guide the eye.

out & Kontsevich (1996). Except for panels 9.10F & L, the fits are tolerably close.

The profiles for the cyclopean mechanisms sensitive to motion within the horizontal meridian (Fig. 9.12(A)) are quite similar to the sensitivity profiles inferred by Beverley & Regan (1973) for mechanisms tuned to the direction of motion in depth within the horizontal meridian of a monocularly-visible target.

This finding supports our proposal that binocular information about the direction of motion in depth is processed by the same cyclopean mechanism whether the target is monocularly -visible or cyclopean.

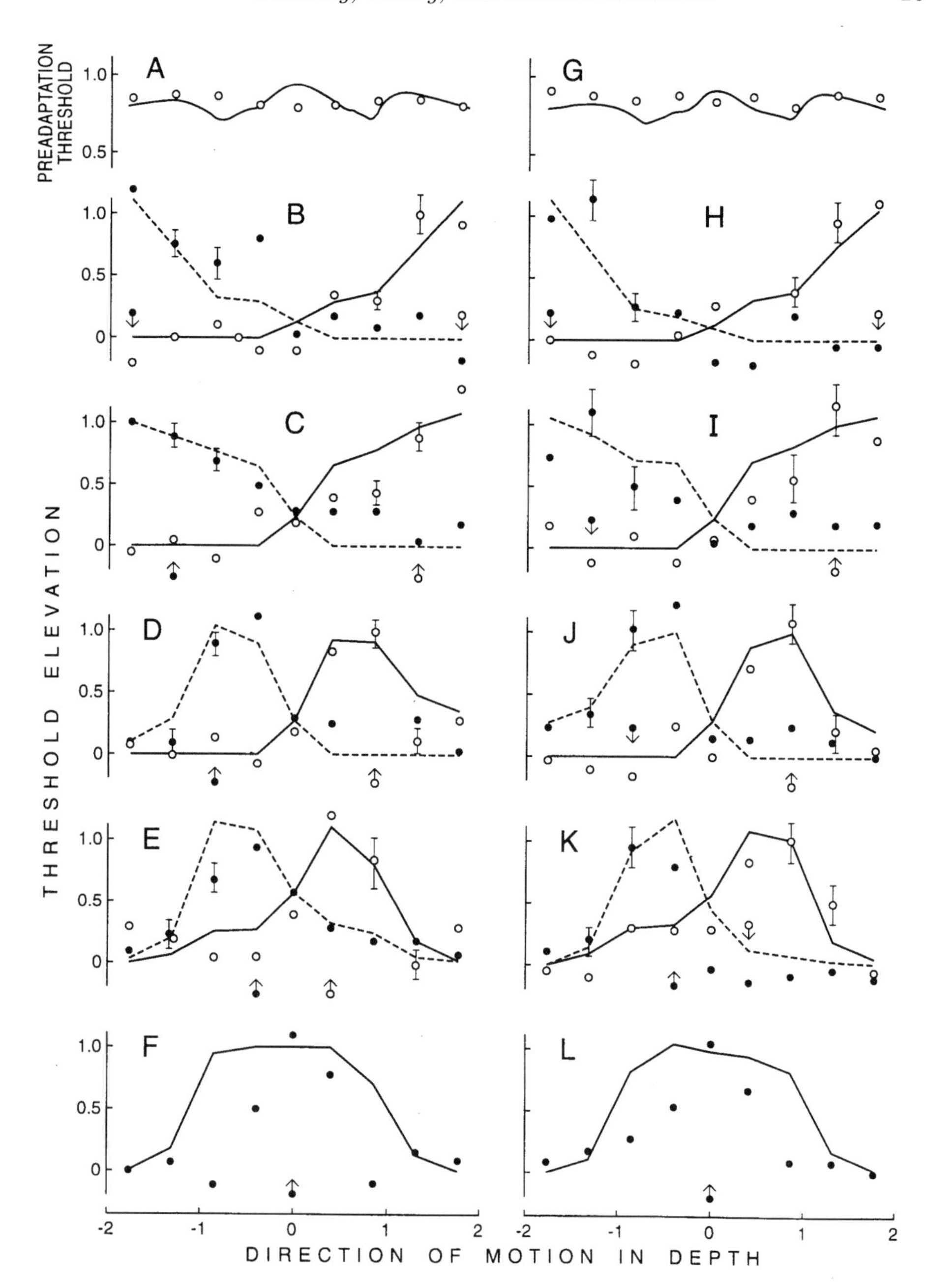

Fig. 9.11. Same data as in Figure 9.10. The lines are theoretical prediction based on the profiles in Figure 9.12.

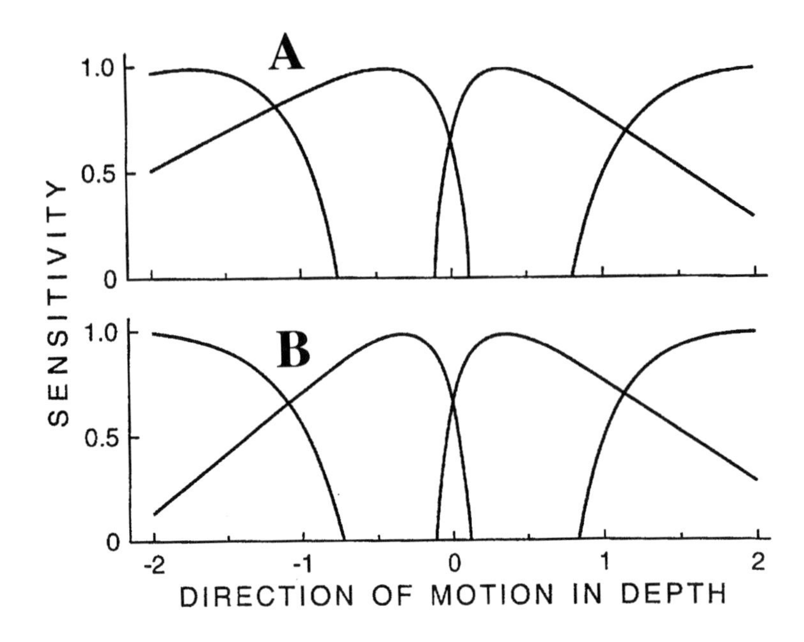

Fig. 9.12. Plots of relative sensitivity (ordinate) versus direction for the proposed mechanisms that best predict the data shown in Figure 9.11 A - L.

9.2.3.1 Calibration of cyclopean motion-in-depth mechanisms

Because the ratio $(d\phi/dt)/(d\delta/dt)$ indicates the direction of motion in depth for motion within any meridian, there is a clear advantage in encoding the direction of an object's motion in depth in terms of $(d\phi/dt)/(d\delta/dt)$ rather than in terms of the relative velocity of the object's left and right retinal images.

However, we are left to explain how the $(d\phi/dt)/(d\delta/dt)$ profiles align themselves to the particular directions of motion in depth that provide an optimal basis for judging whether an approaching object will pass to the left or right of the nose and whether it will hit or just miss the head (Beverley & Regan, 1975). For motion within the horizontal meridian, the magnitude of $(d\phi/dt)_{RH}$ passes through zero as the direction of the approaching object's motion passes through the right eye, and the magnitude of $(d\phi/dt)_{LH}$ passes through zero as the direction of the approaching object's motion passes through the left eye independently of the object's location, the direction of gaze and vergence angle. In contrast, the ratio $(d\phi/dt)/(d\delta/dt)$ offers no such convenient marker of trajectories that graze the right and left sides of the head (Figure 9.13). Given that, during early visual development, our experience of moving objects is of monocularly-visible objects, we propose that the conditions $(d\phi/dt)_{RH} = 0$ and $(d\phi/dt)_{LH} = 0$ are used to align the tuning of the horizontal-meridian $(d\phi/dt)/(d\delta/dt)$ mechanisms so that

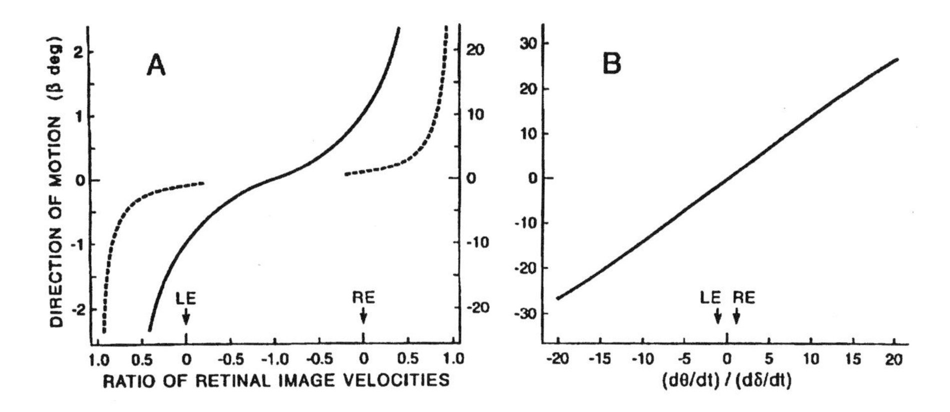

Fig. 9.13. Direction of an object's motion in depth (ordinate) plotted versus the ratio between the velocities of the retinal images in right and left eyes (panel A), and (panel B) versus the ratio $(d\phi/dt)/(d\delta/dt)$.

they prefer the appropriate directions of motion in depth, even though the mechanisms are cyclopean.

For motion within the vertical meridian we suggest that, in the absence of any other markers, the markers of absolute direction $(d\phi/dt)_{RH} = 0$ and $(d\phi/dt)_{LH} = 0$ are used to align the meridian $(d\phi/dt)/(d\delta/dt)$ channels during early visual development.

Taken together, the channels for motion within the vertical and horizontal meridians are adequate to provide a basis for judging whether an approaching object will hit the observer's head, whatever the object's trajectory.

9.2.3.2 Binocular neurons tuned to the direction of motion in depth

Neurons that, in response to binocular stimulation, fire most strongly for a particular direction of motion in depth have been found in the visual cortex of cat (Cynader & Regan, 1978, 1982; Regan, Beverley & Cynader, 1979; Regan & Cynader, 1982; Spileers et al., 1990; Ohzawa, DeAngelis & Freeman, 1996) and monkey (Poggio & Talbot, 1981). For motion within the horizontal meridian, a candidate physiological basis has been found for each of the four pairs of mechanisms proposed by Beverley & Regan (1973). In particular, all of the following kinds of neural response preferences have been reported: an approaching target moving along a line directed wide of the left eye, directed between the left eye and the nose, directed between the nose and the right eye and directed wide of the right eye; a receding target moving along one of the same four directions.

In the physiological reports just cited, the neural tuning was discussed in terms of the ratio $(d\phi/dt)_R/(d\phi/dt)_L$ and the tuning curves were plotted in the polar coordinate system introduced by Cynader & Regan (1978). According to this coordinate system, the ratio between the left and right retinal image velocities is plotted as a linear function of azimuth, and radial distance is a function of neural response. In the light of the findings discussed in the present paper, we raise the possibility that, in all the papers just cited, both the discussion and the analysis of the physiological data was incomplete. For example, it can be inferred from Figure 9.13 that neural tuning curves would look quite different when plotted as a function of $(d\phi/dt)/(d\delta/dt)$ than when plotted in terms of the ratios between left and right retinal image velocities within the horizontal meridian. As well, the mechanisms whose sensitivity profiles are depicted in Figure 9.12B are tuned to the direction of motion in depth for motion within the vertical meridian. A hint that such neurons might exist was reported by Cynader & Regan (1978) who found that neurons that preferred oblique orientations were tuned to the direction of motion in depth within the vertical meridian. Perhaps this question might be resolved by investigating neural sensitivity to the direction of motion in depth using cyclopean targets rather than the monocularly-visible targets that have been used in all physiological studies to date.

Acknowledgments

We thank Wei Hong for assistance in preparing this manuscript. This research was supported by the Medical Research Council of the U.K. and by the Natural Sciences and Engineering Research Council of Canada and sponsored by the Air Force Office of Scientific Research, Air Force Systems Command, USAF, under Grant Number F49620-97-1-0051. The U.S. Government is authorized to reproduce and distribute reprints for governmental purposes notwithstanding any copyright violation theron. D.R. holds the NSERC/CAE Industrial Research Chair in Vision in Aviation. X. H. is supported by the Institute for Space and Terrestrial Science.

References

Alderson, G. J. K., Sully, D. J. & Sully, H. G. (1974). "An operational analysis of a one-handed catching task using high-speed photography", *J. Motor Behav.*, 6, 217-226.

Anderson, C.H. & VanEssen, D.C. (1987). "Shifter circuits: A computational strategy for dynamic aspects of visual processing", Proceedings of the National Academy of Science, U.S.A. 84, 6297-6301.

Banister, H. & Blackburn, J. M. (1931). "An eye factor affecting proficiency at ball games", *British Journal of Psychology*, 21, 382-384.

Beverley, K. I., & Regan, D. (1973). "Selective adaptation for stereoscopic depth perception", *Journal of Physiology*, 232, 40-41P.

Beverley, K. I., & Regan, D. (1975). "The relation between discrimination and sensitivity in the perception of motion in depth", *Journal of Physiology*, 249, 387-398.

Beverley, K. I., & Regan, D. (1979a). "Separable aftereffects of changing-size and motion-in-depth: different neural mechanisms?", *Vis. Res.*, 19, 727-32.

Beverley, K. I., & Regan, D. (1979b). "Visual perception of changing-size: the effect of object size", *Vis. Res.*, 19, 1093-1104.

Beverley, K. I., & Regan, D. (1980). "Visual sensitivity to the shape and size of a moving object: implications for models of object perception", *Perception*, 9, 151-160.

Bootsma, R. J. (1991). "Predictive information and the control of action: what you see is what you get", *International Journal of Sport Psychology*, 22, 271-278.

Bootsma, R. J., & van Wieringen, P. C. W. (1990). "Timing an attacking forehand drive in table tennis", *Journal of Experimental Psychology: Human Perception and Performance*, 16, 21-29.

Cavallo, V., & Laurent, M. (1988). "Visual information and skill level in time-to-collision estimation", *Perception*, 17, 623-632.

Collewijn, H., & Erkelens, C. J. (1990). "Binocular eye movements and the perception of depth", In E. Kowler (Ed.) *Eye Movements and their Role in Visual and Cognitive Processes*, 213-261, Elsevier, Amsterdam.

Cynader, M., & Regan, D. (1978). "Neurons in cat prestriate cortex sensitive to the direction of motion in three - dimensional space", *J. Physiol.*, 274, 549-569.

Cynader, M., & Regan, D. (1982). "Neurons in cat visual cortex tuned to the direction of motion in depth: effect of positional disparity", *Vis. Res.*, 22, 967-982.

DeLucia, P. R. (1991). "Pictorial depth cues and motion-based information for depth perception", *Journal of Experimental Psychology: Human Perception and Performance*, 17, 738-748.

Erkelens, C. J., & Collewijn, H. (1985a). "Motion perception during dichoptic viewing of moving random-dot stereograms", *Vis. Res.*, 25, 583-588.

Erkelens, C. J., & Collewijn, H. (1985b). "Eye movements and stereopsis during dichoptic viewing of moving random-dot stereograms", *Vis. Res.*, 25, 1689-1700.

Freeman, T. C. A., & Harris, M. G. (1992). "Human sensitivity to expanding and rotating motion: Effects of complementary masking and directional structure", *Vis. Res.*, 32, 81-87.

Gray, R., & Regan, D. (1998a). "Motion in depth: adequate and inadequate simulation", *Perception and Psychophysics*, in press.

Gray, R., & Regan, D. (1998b). "Accuracy of estimating time to collision using binocular and monocular cues", *Vis. Res.*, 38: 499-512.

Gray, R., & Regan, D. (1998c). "Adapting to expansion increases perceived time to collision", *Vis. Res.*, in press.

Gray, R., & Regan, D. (1998d). "Estimating time to collision using binocular retinal image information alone, monocular retinal image information alone and a combination of the two", *Vis. Res.*, in press.

Grosslight, J. H., Fletcher, H. J., Masterton, R. B., & Hagen, R. (1978). "Monocular vision and landing performance in general aviation pilots: cyclops revisited", *Human Factors*, 20, 127-133.

Haenny, P. E., & Schiller, P. H. (1988). " State dependent activity in monkey visual cortex: I. Single cell activity in V1 and V4 on visual tasks", *Exp. Brain. Res.*, 69, 225-244.

Haenny, P. E., Maunsell, J. H. R. & Schiller, P. H. (1988). "State dependent activity in monkey visual cortex: II. Retinal and extraretinal factors in V4", *Exp. Brain Res.*, 69, 245-259.

Harris, J. M., & Watamanuik, S. N. J. (1995). "Speed discrimination of motion in depth using binocular cues", *Vis. Res.*, 35, 885-896.

Harris, J. M., & Watamanuik, S. N. J. (1996). "Poor speed discrimination suggests that there is no specialized mechanism for cyclopean motion", *Vis. Res.*, 35, 885-896.

Hong, X. H., & Regan, D. (1989). "Visual field defects for unidirectional and oscillatory motion in depth", *Vis. Res.* 29: 809-819.

Hoyle, F. (1957). *The Black Cloud*, Penguin Books,, London, pp. 26-27.

Karnavas, W. J., Bahill, A. T., and Regan, D. (1990). "Sensitivity analysis of a model for the rising fastball and breaking curveball", Proceedings IEEE Systems Man & Cybernetics, Los Angeles.

Kohly, R. P., & Regan, D. (1998). "Evidence for a mechanism sensitive to the speed of cyclopean form", *Vis. Res.*, in press.

Kohly, R. P., Hajdur, L. V., & Regan, D. (1997). "There is a cyclopean speed-sensitive mechanism", *Investigative Ophthalmology and Visual Science*, 38, S906.

Kruk, R., & Regan, D. (1983). "Visual test results compared with flying performance in telemetry-tracked aircraft", *Aviation, Space and Environmental Medicine*, 54, 906-911.

Lee, D. N. (1976). "A theory of visual control of braking based on information about time-to-collision", *Perception*, 5, 437-459.

Lee, D. N., & Lishman, J. R. (1977). "Visual control of locomotion", *Scandinavian Journal of Psychology*, 18, 224-230.

Lee, D. N., Lishman, J. R., & Thomson, J. A. (1982). "Visual regulation of gait in long jumping", *Journal of Experimental Psychology: Human Perception and Performance*, 8, 448-459.

Lee, D. N., Young, D. S., Reddish, D. E., Lough, S., & Clayton, T. M. H. (1983). "Visual timing in hitting an accelerating ball", *Quarterly Journal of Experimental Psychology*, 35A, 333-346.

Levitt, H. (1971). "Transformed up-down methods in psychoacoustics", *Journal of the Acoustical Society of America*, 49, 65-69.

Lewis, C. E. Jr., & Kriers, G. E. (1969). "Flight research program: XIV. Landing performance in jet aircraft after the loss of binocular vision", *Aerospace Medicine*, 44, 957-963.

Lewis, C. E. Jr., Blakeley, W. R., Swaroop, R., Masters, R. L., & McMurty, T. C. (1973). "Landing performance by low-time private pilots after the sudden loss of binocular vision — Cyclops II", *Aerospace Medicine*, 44, 1241-1245.

Maunsell, J. H. R. (1995). "The brain's visual world: Representations of visual targets in cerebral cortex", *Science*, 270: 764-769.

McBeath, M. K., Shaffer, D. M., & Kaiser, M. K. (1995). "How baseball outfielders determine where to run to catch fly balls" *Science*, 268, 569-573.

McLeod, P., & Dienes, Z. (1993). "Running to catch a ball", *Nature*, 362, 23.

McLeod, R. W., & Ross, H. E. (1985). "Optic-flow and cognitive factors in time-to-collision estimates", *Perception*, 12: 417-423.

Michaels, C. F., & Oudejans, R. R. F. (1992). "The optics and actions of catching fly balls: zeroing out optical acceleration", *Ecological Psychology* 4, 199-222.

Ohzawa, I., DeAngelis, G. C., & Freemen, R. D. (1996). "Encoding of binocular disparity by simple cells in the cat's visual cortex", *J. Neurophysiol.*, 75, 1779-1805.

Plaffmann, C. (1948). "Aircraft landings without binocular cues: a study based upon observations made in flight", *American Journal of Psychology*, 61, 323-335.

Poggio, G., & Talbot, W. H. (1981). "Neural mechanisms of static and dynamic stereopsis in foveal striate cortex of rhesus monkeys", *J. Physiol.*, 315, 469-492.

Poincaré, H. (1913). *The Value of Science*, Science Press, New York.

Portfors, C. V., & Regan, D. (1996). "Cyclopean discrimination thresholds for the direction and speed of motion in depth", *Vis. Res.*, 36, 3625-3279.

Portfors, C. V., & Regan, D. (1997). "Just-noticeable difference in the speed of cyclopean motion in depth and of cyclopean motion within a frontoparallel plane", *Journal of Experimental Psychology: Human Perception & Performance*, 23, 1074-1086.

Portfors-Yeomans, C. V., & Regan, D. (1997). "Discrimination of the direction and speed of motion in depth from binocular information alone", *Journal of Experimental Psychology: Human Perception and Performance*, 23, 227-243.

Regan, D. (1986). "Visual processing of four kinds of relative motion", *Vis. Res.*, 26, 127-145.

Regan, D. (1991a). "Specific tests and specific blindness: keys, locks and parallel processing", *Optometry & Vision Science*, 68, 489-512.

Regan, D. (1991b). "Depth from motion and motion in depth", In D. Regan (Ed.) *Binocular Vision*, 137-169, MacMillan, London.

Regan, D. (1993). "Binocular correlates of the direction of motion in depth", *Vis. Res.*, 33, 2359-2360.

Regan, D. (1995). "Spatial orientation in aviation: visual contributions", *Journal of Vestibular Research*, 5, 455-471.

Regan, D. (1997). "Visual factors in hitting and catching", *Journal of Sport Sciences*, 15: 533-558.

Regan, D., & Beverley, K. I. (1978a). "Looming detectors in the human visual pathway", *Vis. Res.*, 18, 415-21

Regan, D., & Beverley, K. I. (1978b). "Illusory motion in depth: aftereffect of adaptation to changing size", *Vis. Res.*, 18, 209-212.

Regan, D., & Beverley, K. I. (1979). "Binocular and monocular stimuli for motion in depth: changing-disparity and changing-size feed the same motion-in-depth stage", *Vis. Res.*, 19, 1331-1342.

Regan, D. & Beverley, K.I. (1980). "Visual responses to changing size and to sideways motion for different directions of motion in depth: linearization of visual responses", *J. Opt. Soc. Am .*, 11, 1289-96

Regan, D., & Beverley, K. I. (1980). "Visual responses to changing size and to sideways motion for different directions of motion in depth: linearization of visual responses", *J. Opt. Soc. Am.*, 11, 1289-1296.

Regan, D., & Cynader, M. (1982). "Neurons in cat visual cortex tuned to the direction of motion in depth: effect of stimulus speed", *Investigative Ophthalmology and Visual Science*, 22, 535-550.

Regan, D., & Hamstra, S. J. (1993). "Dissociation of discrimination thresholds for time to contact and rate of angular expansion", *Vis. Res.*, 33, 447-462.

Regan, D., & Kaushal, S. (1994). "Monocular judgement of the direction of motion in depth", *Vis. Res.*, 34, 163-177.

Regan, D., & Vincent, A. (1995). "Visual processing of looming and time to contact throughout the visual field", *Vis. Res.*, 35, 1845-1857.

Regan, D., Beverley, K. I., & Cynader, M. (1979). "The visual perception of motion in depth", *Sci. Am.*, 241, 136-151.

Regan, D., Erkelens, C. J., & Collewijn, H. (1986a). "Necessary conditions for the perception of motion in depth', *Investigative Ophthalmology and Visual Science*, 27m 584-597.

Regan, D., Erkelens, C. J., & Collewijn, H. (1986b). "Visual field defects for vergence eye movements and for stereomotion perception", *Investigative Ophthalmology and Visual Science*, 27, 806-819.

Regan, D., Portfors, C. V., & Hong, X. H. (1997). "Cyclopean mechanisms for the direction of motion in depth", *Investigative Ophthalmology and Visual Science*, 38, S1168.

Richards, W., & Regan, D. (1973). "A stereo field map with implications for disparity processing", *Investigative Ophthalmology and Visual Science*, 12, 904-909.

Santen, J. P. H., & van Sperling, G. (1985). "Elaborated Reichardt detectors", *J. Opt. Soc. Am. A*, 2, 300-321.

Savelsbergh, G. J. P., Whiting, H. T. A., & Bootsma, R. J. (1991). "Grasping Tau", *Journal of Experimental Psychology: Human Perception and Performance*, 17, 315-322.

Saxberg, B. V. H. (1987a). "Projected free fall trajectories: I. Theory and simulation", *Biological Cybernetics*, 56, 159-175.

Saxberg, B. V. H. (1987b). "Projected free fall trajectories: II. Human experiments", *Biological Cybernetics*, 56, 177-184.

Schiff, W., & Detweiler, M. L. (1979). "Information used in judging impending collision", *Perception*, 8, 647-658.

Spileers, W., Orban, G. A., Gulyas, B., & Maes, H. (1990). "Selectivity of cat area 18 neurons for direction and speed in depth", *J. Neurophysiol.*, 63, 936-954.

Todd, J. T. (1981). "Visual information about moving objects", *Journal of Experimental Psychology: Human Perception and Performance*, 7, 795-810.

Tresilian, J. R. (1995). "Perceptual and cognitive processes in time-to-contact estimation: analysis of prediction-motion and relative judgement tasks", *Perception & Psychophysics*, 57, 231-245.

Tyler, C. W. (1975). "Characteristics of stereomovement suppression", *Perception & Psychophysics*, 17, 225-230.

Tyler, C. W. (1991). "Cyclopean vision", In D. Regan (Ed.) *Binocular vision*, 38-74, Macmillan, London.

Tyler, C.W., Barghout, L. & Kontsevich, L. L. (1996). "Computational reconstruction of the mechanisms of human stereopsis", Proc. SPIE 2054: 52-68.

Verity, H. (1934). *Bowling Them Out*, Hutchinson, London.

Vincent, A., & Regan, D. (1995). "Parallel independent encoding of orientation, spatial frequency and contrast", *Perception*, 24, 491-499.

Vincent, A., & Regan, D. (1997). "Judging the time to collision with a simulated textured object: effect of mismatching rates of expansion of size and of texture elements", *Perception and Psychophysics*, 59, 32-36.

Warren, W. H., Young, D. S., & Lee, D. N. (1986). "Visual control of step length during running over irregular terrain", *Journal of Experimental Psychology: Human Perception and Performance*, 12, 259-266.

Wheatstone, C. (1852). "Contributions to the physiology of vision. II.", *Phil. Tran. Roy. Soc.*, 142, 1-18.

10

An analysis of heading towards a wall

Antje Grigo and Markus Lappe

Navigating towards a desired spatial location is an easy task, provided that you are capable of seeing your environment. Vision will guide your movement and allow you to reach your goal successfully. Vision in this situation encompasses various signals: the pattern of image motion on the retina (the retinal *flow field*), the variations in image motion of points in different distances in the environment (*motion parallax*) and the depth information itself relayed by cues such as texture gradients or binocular disparity. Sensory information is received from many other senses besides vision by the brain during locomotion. The vestibular organs transmit information about acceleration and orientation of the head relative to the ground, proprioception signals the posture of the limbs relative to the trunk, and the oculomotor reference signal yields information about the movement of the eyes.

It has been a long-lasting discussion whether visual information alone is sufficient to specify the direction of one's movement in space, or the *direction of heading*. Gibson (1950) was the first to investigate the visual motion that results from self movement through a structured environment. He introduced the term *optic flow* to refer to the temporal changes in the visual environment around a moving observation point (Gibson, 1950, 1966). This idealized observation point is thought to lie in front of an observer's eye. The optic flow therefore describes the visual movement that surrounds the moving eye. The movement pattern arising in the eye, i.e. on the observer's retina, is referred to as *retinal flow*. The retinal flow is the sum of the image motion resulting from self movement and image motion resulting from rotation of the observer's eye. Both optic and retinal flow patterns can be illustrated by vector fields in which each vector gives the optical velocity of a point in the environment relative to the head or eye respectively (see Figure 10.1). Based on geometric considerations of optic flow fields, Gibson (1950, 1966) suggested a way to determine the direction of heading. During linear translation, the heading direction becomes a singular point in the optic flow field. The movement of all other image points flows radially out from this point. It is therefore called the *focus of expansion*.

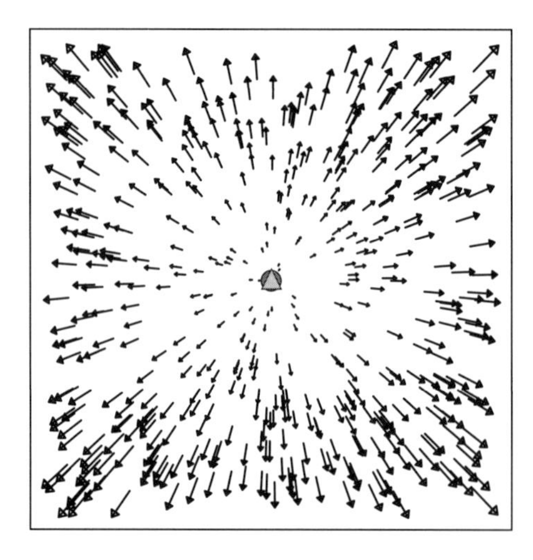

Fig. 10.1. When an observer approaches a vertical wall along the wall normal (i.e. on a path perpendicular to the wall) a radial optic flow pattern is created. In this vector plot of the optic flow pattern, each vector gives the optical velocity of a single point on the wall. The heading direction is marked by a triangle. It is located in the center of the radial motion pattern. It represents the focus of expansion of the optic flow field.

Figure 10.1 illustrates an optic flow field that arises when an observer moves perpendicularly towards a vertical wall. A vertical wall is a special case because all the points are in a plane. This means that the velocity and direction of all points can be accurately predicted from only a few parameters that describe the orientation of the wall. This is unlike the motion of points that make up a normal three-dimensional environment which do not have this redundancy. Around the heading direction, which is indicated by the triangle, a radial flow pattern appears with the focus of expansion marking the heading direction. With the discovery of the focus of expansion, a direct solution seemed to be found to explain the visual control of self–motion. The maintenance of a course would require locating the focus of expansion in the visual field, adjusting it to the desired aiming point, and then maintaining the coincidence between the focus of expansion and the desired goal during locomotion (Gibson 1950, 1966).

10.1 First approaches to the heading detection problem

During the decade after Gibson's initial work, these concepts were applied in theories that tried to explain the human ability to steer cars or to land air planes. In the seventies, the first experiments were performed to determine how accurately humans could locate the focus of expansion in an optic flow field. In these experiments, human subjects were presented with expanding random dot patterns with the focus of expansion at different locations in the display. Random dot patterns were used instead of natural scenes to avoid interference by object recognition processes and to restrict the information content of the stimulus to pure velocity fields. Because the real–time computer graphics used today were not available at that time, the experimental setups needed to be more inventive.

Llewellyn (1971) painted dots or glued circular discs on glass plates which were held in front of a point light source. The dots cast shadows on a translucent screen viewed by the subjects. To create expansion, the glass plate was moved towards the light source. After the movement had stopped the subjects had to adjust the light of a torch to the location on the screen where they had perceived the focus of expansion. Llewellyn (1971) found very poor accuracy with mean errors between 5° and 10°. A similarly disappointing result was found by Johnston, White, and Cumming (1973) who used an animated movie of a random dot pattern. A digital computer calculated the positions of dots seen from a moving vehicle approaching a vertically oriented, dotted surface. Single frames of that motion were then photographed from a visual display on to 16-mm film. This film was then projected by a fisheye lens onto a spherical dome 28 ft in diameter. The subject was seated at the center of this dome. When indicating the center of the projected expansional motion, subjects exhibited errors of 8° to 13°. Johnston, White, and Cumming (1973) as well as Llewellyn (1971) concluded that the precision in identifying the focus of expansion was too low to satisfy the locomotion strategy proposed by Gibson and hence called his proposal into question. They pointed out that in real life, locomotion takes place in surroundings with real objects of specific shapes, sizes and colours, the knowledge, recognition and observation of which during self–motion could help navigating (see also Cutting et al., 1992, and Vishton and Cutting, 1995).

Later, Warren, Morris, and Kalish (1988) examined the heading detection performance from flow fields, simulating translational movements towards a wall with computer–generated real–time simulations displayed on a video screen. In contrast to the earlier studies, they found mean detection errors of only about 1°. Besides the advanced experimental setup which allowed higher accuracy, Warren, Morris, and Kalish (1988) attributed the much better performance which they reported to the two-alternative forced-choice (2AFC) discrimination task they used instead of a pointing task. In a pointing task subjects have to judge absolute locations whereas in a 2AFC task they give their judgements relative to a target. However, it has to be considered that in such a discrimination task the average detection error depends on a threshold for percent correct, which has to be defined. Warren, Morris, and Kalish (1988) used a threshold of 75%. This means that 75% of the answers were better or equal to 1°. As proposed by Gibson (1950, 1966),

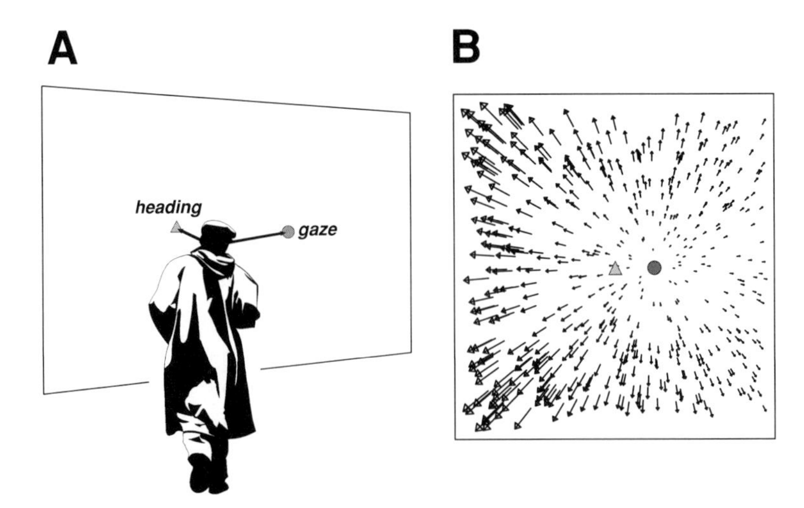

Fig. 10.2. A: While an observer moves towards a vertical wall his gaze may not be directed to the point on the wall that coincides with the heading direction. However, if he intends to keep gaze on another point on the wall during his forward motion, an appropriate tracking eye movement is required. B: The retinal flow pattern in this situation contains radial motion flowing away from the fixation point (circle) instead of from the direction of heading (triangle). The heading direction no longer corresponds to the focus of expansion. The gaze direction becomes a pseudo–focus of expansion.

they suggested that moving observers rely on the radial structure in the optic flow field to determine heading.

But is this radial structure really a good cue for the visual determination of heading during locomotion? When one considers retinal flow, i.e. the flow pattern that actually appears in the eye, the answer must be no. The best demonstration for this comes from considering the retinal flow arising from movement towards a fronto–parallel wall.

10.2 Retinal flow and the problem of eye movement

Regan and Beverley (1982) examined the retinal flow pattern that arises when an observer fixates and tracks a point on the wall that is offset from the direction in which he is heading. Such a movement is illustrated in Figure 10.2A. The observer walks towards the wall while directing his gaze to a point on the right of the heading direction. Regan and Beverley demonstrated that the radial expansion which indicates the heading direction in the optic flow field, is replaced in the retinal flow pattern by a quasi–expansional structure with its center at the fovea. This flow is illustrated in Figure 10.2B. The retinal flow field is shown that is created when the observer fixates a point to the right of the direction of heading. In this case the heading direction, indicated by the triangle, no longer corresponds to a focus of expansion.

Instead, a quasi–expansional structure appears around the gaze direction, which is indicated by the circle. We will refer to this point as the *pseudo–focus* of expansion. Near the center of the retinal expansion, the motion pattern differs only slightly from the pattern that would have been observed during real translation towards the direction of gaze (Figure 10.1). However, the deviation becomes stronger in the peripheral parts of the visual field.

This special case of movement towards a wall is a powerful demonstration that the radial structure in a retinal flow field does not simply indicate the direction of heading. The question then is: How is the heading direction deduced from the retinal flow when eye movements are involved?

Heading detection during eye movements: the need for depth or extraretinal information

To examine the influence of eye movements on heading detection, several authors (eg. Rieger & Toet, 1985; Warren & Hannon, 1990; Royden, Banks & Crowell, 1992, 1994; Banks *et al.*, 1996) used the paradigm of simulated eye rotations. They presented a flow field corresponding to translational movement of an observer, together with a rotation of the viewpoint and instructed subjects to fixate a stationary target. The retinal stimulus in this case should be the same as the retinal flow experienced during real translation and real eye movement. Rieger and Toet (1985) added various 3D rotations to flow fields that simulated an observer's approach to one or two vertical random–dot planes. However, the added rotations in this case did not actually simulate tracking eye movements but were varied in orientation and speed independent of the visual scene. Subjects had to judge the heading direction in a 4AFC task relative to the fixation point on the screen center. Rieger and Toet (1985) found that when the scene consisted of two planes separated in depth, subjects performed acceptably well. With only one plane corresponding to heading towards a wall heading detection was at chance.

The basic finding of all these studies was that depth differences in the environment help the visual system to interpret retinal flow in the presence of eye movements. It has been suggested that the observer relies on the relative optical motion of elements in different depths. The so called *motion parallax* contains information about the relative magnitudes of translational and rotational components in the movement. Whereas the observer's translational movement results in visual motion that depends on the depth layout of the scene, the velocity field resulting from an observer's rotation or eye–movement is independent of the element distances (Harris, 1994). Therefore, if the retinal flow results from motion within a rich, three-dimensional world, the translational and rotational components can, at least theoret-

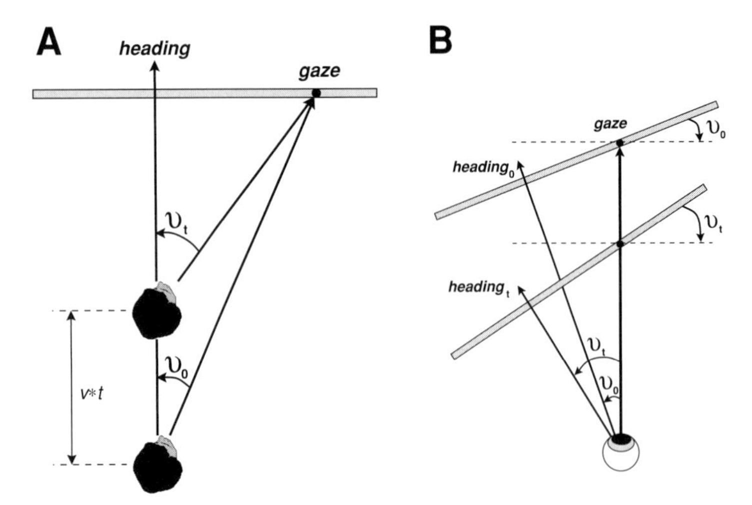

Fig. 10.3. A: Illustration of the movement simulated in the heading detection experiments. An observer approaches a vertical wall in direction of the wall normal while fixating a point to the right (or left) of the heading direction. He moves with a speed v for a time t. The angle between the direction of heading and the direction of gaze is ν_0 in the beginning and ν_t at the end of the movement. B: The same movement seen by the eye. During the approach towards the wall the eye holds gaze upon a specified point on the wall. The slant of the wall with respect to the line of sight changes because of the increasing retinal heading angle ν between the direction of gaze and the direction of heading.

ically, be separated (Rieger & Lawton, 1980; Koenderink & van Doorn, 1981; Longuet-Higgins & Prazdny, 1980; Harris, 1994). A wall is a very sparse three-dimensional world which makes the separation much more difficult.

Warren and Hannon (1990) used computer displays to present flow fields simulating an observer's linear forward motion combined with a tracking eye movement. They created different environmental scenes including a single vertical plane textured with dots, a 3D cloud of dots and a more natural scene of a dot covered ground plane to walk over. The geometry they employed in simulating movement towards a single plane is sketched in Figure 10.3A. The observer approaches the plane perpendicularly with a speed v. A specified point on the right or left from the heading direction is tracked during a movement of duration t. During the movement, the angle ν between the gaze and the heading direction changes from ν_0 to ν_t. Subjects viewed the simulated movement on a computer screen and subsequently had to judge the simulated heading direction. They were instructed to fixate a point which was either moved across the screen or was stationary. In the first case the subjects performed a real eye movement, whereas in the latter case such an eye movement was simulated within the flow field display.

Warren and Hannon (1990) found no differences in the subjects' ability to determine the simulated heading direction in both the simulated and real eye movement condition provided there were depth variations in the simulated environment. When movement towards a wall was combined with a tracking eye movement in the simulation, subjects responded at chance, similar to the findings of Rieger and Toet (1985). Warren and Hannon (1990) reported that in this case subjects tended to perceive themselves as heading towards the fixation point. As mentioned above and illustrated in Figure 10.2B, in the single plane layout, the fixation point becomes a pseudo–focus in the retinal flow field. Warren and Hannon suggested that subjects misperceived this pseudo–focus as the direction of heading. Thus, when heading towards a wall subjects seem to rely on expansion in the flow field as the indicator of their heading.*

Royden, Crowell, and Banks (1992) and Royden, Banks, and Crowell (1994) performed similar experiments to Warren and Hannon with varying environmental layouts and simulated as well as real eye movements. In their experiments, the eye rotation was not limited to the visual tracking of an object in the scene. Some conditions simulated the tracking of an independently moving object not attached to the scene. Consistent with Warren and Hannon (1990), Royden *et al.* (1992, 1994) found that during real eye movements, heading detection was possible and accurate. On the other hand, when the eye rotation was simulated, subjects always made large errors, except in 3D environments with very low rotation rates of below 1°/sec. In the simulation of heading towards a single wall with simulated eye movements, heading detection was always confused with gaze direction. Royden *et al.* (1992, 1994) concluded that to analyze retinal flow, extraretinal eye velocity information is necessary for the brain. When an oculomotor reference signal is available, the eye rotation can be compensated for and the flow field is analyzed correctly.

The view that emerges from these studies is the following: Heading detection from retinal flow fields simulating a linear approach to a vertical wall combined with a simulated tracking eye movement fails because of the lack of depth in the environment and the lack of an extraretinal signal. The retinal flow field contains a pseudo–focus of expansion at the fixation point. Because extraretinal input is absent no information is available to distinguish the pseudo–focus from a real focus of expansion. The approximately radial structure around the fixation point is taken as an indicator for the heading direction and leads to the systematic misjudgment.

In the following, we will describe experiments that require a revision of this view.

* As soon as depth is introduced in the environment the fixation point of course remains a singular point but no longer shows a clear expansional structure that could be mistaken for a focus of expansion. The fixation of a point on a floor as an observer walks over it for example, results in a spiral structure around the fixation point in the retinal flow field. When determining heading from such a flow field, subjects are forced to use cues other than radial structure.

10.3 Heading towards a wall: new considerations

When an observer approaches a vertical plane, minimal depth variations occur when the plane is perpendicular to the line of sight. If the plane is slanted at some angle with regard to the line of sight, parts of the plane will be nearer to the observer than others. If the observer performs an eye movement during his forward motion, the line of sight will change and in consequence the slant of the plane will change too. Sketched in Figure 10.3B is the orientation of the plane with respect to the eye of the observer of Figure 10.3A. The fixation of a point on the wall in a direction different from the wall normal, results in a slant ν of the wall with respect to the line of sight. During a movement of duration t, the eye approaches the wall. At the same time, the angle ν between the direction of heading and the direction of gaze becomes larger and so does the slant of the wall. Thus, if such a movement is presented over time, some subtle depth variations will inevitably occur. The question is: Is it possible to use these subtle changes to obtain a correct estimation of heading even during eye movement?

Given a homogeneously textured wall, depth information or information about the slant of the wall can be obtained from several visual cues. One is the visual density of points on the wall at different distances, i.e. the texture gradient. A second one is the speed gradient, i.e. the fact that points at different distances move at different retinal velocities. Third, if the scene is watched with both eyes, binocular disparity also yields depth information. Let us concentrate on the monocular cues first.

The importance of a large field of view

In Figure 10.4A a retinal flow field originating from the movement sketched in Figure 10.3 is shown. The wall is represented by a plane of dots which are projected on a visual field of $90° \times 90°$. The fixation point is projected on the fovea, that is on the center of the visual field, here marked by the circle. It remains fixed on the fovea while the images of the other dots move depending on the observer's movement. The figure shows the trajectories of the dot images during a typical movement. Each point indicates the position of a single dot image after a single frame. The thick black line marks the heading direction which also moves on the retina during the course of movement. The final angle between heading direction and gaze direction is always 6°. Around the fixation point, a clear expansional structure, the pseudo–focus, appears. Information about the changing slant of the wall relative to the eye is also available. A speed gradient can be observed since the velocities of the dots on the right side, which is the more distant side of the wall, are lower than of those on the left side. This can be recognized by the smaller spacing between points on each trajectory on the right side. Warren and Hannon (1990), as well as Royden *et al.* (1992, 1994), used a display consisting of homogeneously distributed dots which therefore, like our display, also

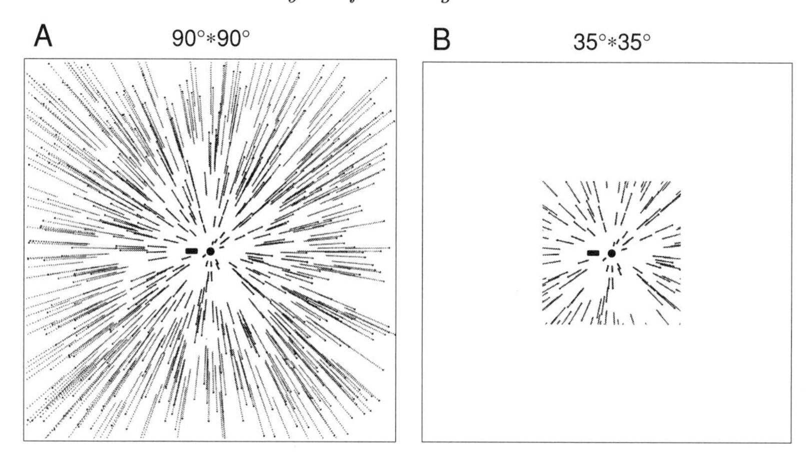

Fig. 10.4. Retinal motion resulting from a movement towards a random–dot plane in the direction of the plane normal in combination with a tracking eye movement. Displayed are the trajectories of the dot images during the movement. Each point indicates the position of a single dot image after a certain time interval. The circle marks the fixation point which is always projected on the fovea, i.e. on the center of the visual field. Also shown by the thick black line is the heading direction. It shifts from close to the direction of gaze at the beginning of the movement to 6° to one side in both cases. A shows a field of view of $90° \times 90°$, B shows a field of view of $35° \times 35°$.

contained texture and speed gradients. However, the strengths of both of these cues depend on two parameters. The first is the slant of the wall, which is directly related to the angle ν between heading and gaze direction. The second is the size of the field of view.

The displays in the described experiments presented a field of view of $40° \times 32°$ (Warren & Hannon, 1990) and $30° \times 30°$ (Royden, Banks & Crowell, 1992, 1994). The unrestricted visual field of a human observer covers roughly $180° \times 150°$. The flow field in Figure 10.4A extends $90° \times 90°$. In comparison, Figure 10.4B shows a field of view of $35° \times 35°$. It is difficult to observe texture or speed gradients in this small field of view. Therefore an extension of the field of view might be expected to influence the heading detection performance, as a larger field of view should be associated with a greater range of three-dimensional variation and therefore might make the task easier. This suggestion has been made previously on the basis of computer simulations of an ideal observer (Koenderink & van Doorn, 1987). However, surprisingly, an enlargement of the field of view up to $60° \times 40°$ in experiments of Banks *et al.* (1996) did not lead to significant improvements of the heading judgements.

We confirmed this negative result even with a field of view as large as $90° \times 90°$. We simulated a perpendicular approach towards a homogeneously dotted wall together with an eye movement that kept gaze on a point on the wall (Figure 10.3). The stimuli were generated on a Silicon Graphics workstation and were displayed by a video projector on a large tangent screen 60 cm in front of the subject. Subjects had to fixate in the middle of the screen. To re–examine the experiments of Warren and Hannon (1990) we chose stimulus parameters similar to theirs. The retinal heading angle ν, i.e. the angle between the direction of heading and the direction of gaze, was varied horizontally between $-6°$ and $6°$. The simulated speed of the observer was 2.0 m/sec. The stimulus presentation lasted for 3.2 seconds. However, we used a pointing task instead of the 2AFC task. After the movement had stopped, subjects were instructed to point at the location on the screen where they thought they would hit the wall if the movement had continued.

Despite of the large field of view, subjects were not able to judge the simulated heading direction correctly. Instead, their judgements were strongly biased towards the fixation point. Some subjects even pointed directly at the fixation point.

Warren and Hannon (1990) previously described that subjects often erroneously perceived their direction of gaze as their heading. In our case, however, usually only a bias towards the fixation point occurred. This difference is probably due to the larger size of our field of view Our display contained greater texture gradients appearing mainly in the periphery and helping to convey the changing slant of the wall relative to the line of sight.

We conclude that depth cues provided by an enlargement of the field of view alone are not sufficient to allow correct heading detection. We next asked whether binocular disparity might be exploited as a cue in this situation.

Does stereoscopic vision help?

Stereo vision has been shown to contribute to heading detection from optic flow by introducing more depth cues. Stereovision improves heading judgements from noisy optic flow fields in conditions where monocular observers perform only poorly (van den Berg & Brenner, 1994). A direct demonstration of an influence of disparity–based depth order can be observed in an "illusory optic flow" stimulus (Grigo & Lappe, 1998). In this illusion, subjects perceive a shift of the center of an expanding optic flow field when it is transparently superimposed on a unidirectional motion pattern (Duffy & Wurtz, 1993). This illusory shift can be explained by the visual system taking the presented flow pattern as a certain self–motion flow field and then determining the direction of heading as if the transparent translational motion had resulted from an eye rotation (Lappe & Rauschecker, 1995). The magnitude of the illusory shift depends strongly on the depth order of the two superimposed motion patterns. When the translating pattern is presented in front of the expanding pattern, a highly significant decrease of the illusory shift occurs, down to 25 percent of its magnitude at zero disparity. When the translation is presented behind the expansion, the illusory shift decreases only slightly (Grigo & Lappe, 1998).

We wanted to test whether binocular disparity can also help in perceiving one's movement towards a wall. We presented the same simulated observer movement as before, only now in a stereoscopic projection that contained realistic binocular disparities of the moving dots. We used active LCD shutter glasses for the stereoscopic perception. We found no improvements in the heading detection results compared to the experiments using monocular presentation. Presumably the disparities are too small and too evenly distributed compared to those of a rich three-dimensional stimulus.

Stimulus duration as a critical parameter

A further aspect of the experimental condition that could potentially influence heading judgements is the temporal duration of the simulated movement. As can be seen in Figure 10.4, the simulated heading direction shifts on the retina during an observer's movement towards a wall whilst tracking a point on the wall off to one side. The magnitude of the shift depends on the duration of the stimulus. Similarly, the magnitude of the change of the slant of the wall also depends on the stimulus duration. Therefore, we performed the experiment with a very short simulation time. Instead of 3.2 seconds as before, we used a stimulus duration of 0.4 seconds. The final retinal position of the direction of heading at the end of the trials was the same as in the previous experiments with the long presentation time. The initial retinal heading angles at the start of the trials were accordingly adjusted for the shorter stimulus presentations. Thus, the short trials effectively presented the last 0.4 seconds of the longer trials. Subjects had to indicate the final heading angle in the display.

Surprisingly, with these very short movement times, subjects were suddenly successful in detecting the direction of heading as being distinct from the direction of gaze and the pseudo-focus of expansion. However, this was only possible with the large field of view: with the short simulation time but only a small field of view, performance remained poor. Subjects verbally reported low confidence in their heading judgements in all conditions, but found the short stimulation condition easier to perform. Thus, there seem to be two main constraints on successfully detecting heading direction in a simulated movement towards a wall: the field of view has to be large and the presentation time has to be short.

Surprised by this finding, we measured the relationship between stimulus duration and heading direction performance. We presented optic flow patterns with durations between 0.4 and 3.2 seconds. The final angle between the direction of gaze and the direction of heading was varied in five steps

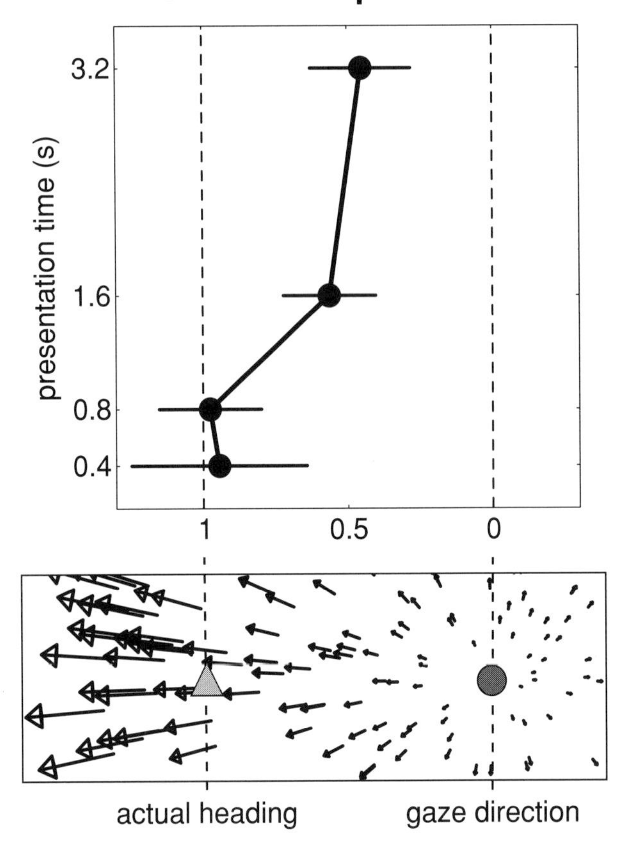

Fig. 10.5. Dependence of the accuracy of heading judgements on the duration of the stimulus. The final difference between heading direction and fixation direction was varied in five steps between −6° and 6°. The relative performance is the slope of the regression line of judged heading direction against actual heading direction. A value of 1 corresponds to correct heading detection. A value of 0 indicates pointing towards the direction of gaze. The responses for each heading angle were averaged over 9 subjects. Performance clearly drops for prolonged movement times as a bias towards the fixation point appears. Whereas with short presentations subjects performed correctly, with stimulus durations greater than one second, they pointed closer to the fixation point.

between −6° and 6°. Each combination of presentation time and heading angle was displayed five times. Nine subjects participated in the experiment. For the analysis, the responses for each heading angle were averaged over the subjects. Then the relative performance was calculated as the slope of the regression line over the five heading angles. A slope of unity indicates correct heading perception and zero corresponds to pointing towards the pseudo–focus at the fixation point.

The results are displayed in Figure 10.5. We found that with movement times over about one second the tendency to point closer to the fixation point increased dramatically. With stimulus durations less than one second subjects performed very well, but with longer simulated movements, subjects confused heading directions with the fixation point.

These results indicate distinct temporal dynamics in the process of heading detection from retinal flow. Previous studies had linked the erroneous percept of moving along the direction of gaze to the absence of extraretinal input, because heading detection was accurate when real eye movements were performed, i.e. extraretinal input was available (Warren & Hannon, 1990; Royden et al., 1994). However, in these experiments the stimulus duration always exceeded 1 sec. In contrast, with durations below 1 sec heading detection is possible without the help of extraretinal oculomotor signals. The visual system obviously relies on the visual input during brief stimulus exposures. In the line of previous reasoning, the gradual development of the error would suggest that the visual system puts increasingly more emphasis on the extraretinal signal as time progresses. In the first few hundred milliseconds either the visual system ignores extraretinal eye movement information, or the extraretinal signal is not yet fully developped. Support for this hypothesis comes from a study of the Filehne illusion, which is supposed to measure the strength of the eye movement reference signal during smooth pursuit. Wertheim (1987) described that for a brief exposure of the stimulus (300 msec) the eye movement reference signal was much smaller than for an exposure of the stimulus for 1000 msec. Therefore, in the case of the Filehne illusion, the visual system compensates for a pursuit eye movement and analyzes the visual input correctly only after a certain exposure time. In our case of simulated eye movement during actual fixation, the visual system may use a signal that no eye movement takes place and confuse the center of the radial structure in the flow field with the direction of heading also only after a certain duration.

But what would be the implications of that dynamical change in the importance of the extraretinal eye movement signal in natural movement conditions? For this question it is important to consider the possible origin of the eye rotations. Slow eye rotations like those simulated in our stimuli can result from either voluntary smooth pursuit eye movements usually generated when we want to track a moving object with the eyes. Or they might be generated by reflexive, involuntary mechanisms that are active during normal self motion. The vestibulo ocular, the optokinetic, and the ocular following reflexes induce eye movements that try to stabilize spatial vision during movements of the head.

Both types of eye movements, voluntary pursuit and stabilization reflexes, could lead to the eye rotation that was simulated in our experiments. However, the two types of eye movements typically have very different dynamical properties. Smooth pursuit is characterized by the tracking of a moving target over an extended period of time, i.e. as long as the subject is interested in following the target. Smooth pursuit eye movements can therefore last up to a few seconds. On the other hand, smooth pursuit takes about 100 msec to be established. In contrast, stabilization reflexes occur very fast (10–20 msec for the vestibulo–ocular reflex, 60 msec for ocular following) and without voluntary attention. During typical oculomotor behavior saccadic eye movements direct gaze to new targets about 2 or 3 times per second. This leaves about 300–500 msec time between two saccades. For a moving observer, after each saccade a new flow field occurs on the retina that differs from the one experienced before the saccade. Reflexive gaze stabilization mechanisms induce eye movements during the intersaccadic interval. Thus, in natural behavior eye movements during the first 300 msec after the saccade are most often generated by involuntary stabilization mechanisms. At later times, it is more likely that the smooth pursuit system is involved. A longer intersaccadic interval is evidence of a voluntary process that keeps gaze directed at a target of interest.

It is a long standing question whether an extraretinal eye movement signal is present during reflexive eye movements. It has been suggested (Post & Leibowitz, 1985) that only voluntary eye movements generate an extraretinal signal. A possible lack or even an incompleteness of the extraretinal signal during involuntary eye movements could explain the dynamic use of visual and extraretinal cues. During short stimulation periods that last about the length of a typical intersaccadic interval a potential eye rotation might be the result of an involuntary eye movement. Since the extraretinal signal might not be indicative in this case, the brain ought to rely more on visual cues. In contrast, long periods of uninterrupted flow stimulation are more likely to contain eye rotations that result from smooth pursuit. In this case, the extraretinal signal is reliable and can be used to compensate for the eye movement induced visual rotation.

10.4 Conclusion

Previous research has suggested that in the case of movement towards a wall human subjects tend to falsely associate their direction of heading with the pseudo–focus of expansion that is introduced by a simulated eye movement. Our experiments confirmed this tendency when the same parameters as in previous experiments were used for the stimuli. However, when the field of view was extended to $90° \times 90°$ and the presentation time of the stimulus was reduced to below 1 second, subjects were able to perform the task correctly. The larger field of view provides stronger texture and speed gradients that carry information about the slant of the wall. This information can be used to overrule the pseudo–focus and to correctly determine the direction of heading provided presentation time is small. With prolonged presentation duration the lack of extraretinal information about the involved eye movement causes the subjects to again confuse the heading direction with the fixation point. This possibly reflects the result of a dynamical use of the extraretinal eye movement signal by the visual system in the interpretation of retinal flow fields. This dynamical change might be related to different kinds of eye rotations occuring in natural situations.

Acknowledgments

This research was funded by the Human Frontier Science Program and by the DFG SFB–509.

References

Banks, M. S., Ehrlich, S. M., Backus, B. T., Crowell, J. A. (1996). "Estimating heading during real and simulated eye movements", *Vis. Res.*, 36: 431–443.

Cutting, J. E., Springer, K., Braren, P. A., Johnson, S. H. (1992). "Wayfinding on foot from information in retinal, not optical, flow", *J. Exp. Psych.: Gen.*, 121: 41–72.

Duffy, C. J., Wurtz, R. H. (1993). "An illusory transformation of optic flow fields", *Vis. Res.*, 33: 1481–1490.

Gibson, J. J. (1950). *The Perception of the Visual World*, Houghton Mifflin, Boston.

Gibson, J. J. (1966). *The Senses Considered As Perceptual Systems*, Houghton Mifflin, Boston.

Grigo, A., Lappe, M. (1998). "Interaction of stereo vision and optic flow processing revealed by an illusory stimulus" *Vis. Res.*, 38: 281-290.

Harris, L. R. (1994). "Visual motion caused by movements of the eye, head and body", in *Visual Detection of Motion*, A. T. Smith and R. Snowden Eds). Academic Press, London pp 397-436.

Johnston, I. R., White, G. R., Cumming, R. W. (1973). "The role of optical expansion patterns in locomotor control" *Am. J. Psychol.*, 86: 311–324.

Koenderink, J. J., van Doorn, A. J. (1981). "Exterospecific component of the motion parallax field" *J. Opt. Soc. Am.*, 71: 953–957.

Koenderink, J. J., van Doorn, A. J. (1987). "Facts on optic flow" *Biol. Cybern.*, 56: 247–254.

Lappe, M., Rauschecker, J. P. (1995). "An illusory transformation in a model of optic flow processing" *Vis. Res.*, 35: 1619–1631.

Llewellyn, K. R. (1971). "Visual guidance of locomotion" *J. Exp. Psych.*, 91: 245–261.

Longuet-Higgins, H. C., Prazdny, K. (1980). "The interpretation of a moving retinal image" *Proc. Royal. Soc. Lond. B*, 208: 385–397.

Post, R. B. & Leibowitz, H. W. (1985). "A revised analysis of the role of efference in motion perception", *Perception*, 14: 631–643.

Regan, D., Beverley, K. I. (1982). "How do we avoid confounding the direction we are looking and the direction we are moving?" *Science*, 215: 194–196.

Rieger, J. H., Lawton, D. T. (1985). "Processing differential image motion" *J. Opt. Soc. Am. A*, 2: 354–360.

Rieger, J. H., Toet, L. (1985). "Human visual navigation in the presence of 3–D rotations" *Biol. Cybern.*, 52: 377–381.

Royden, C. S., Banks, M. S., Crowell, J. A. (1992). "The perception of heading during eye movements" *Nature*, 360: 583–585.

Royden, C. S., Crowell, J. A., Banks, M. S. (1994). "Estimating heading during eye movements" *Vision Res.*, 34: 3197–3214.

van den Berg, A. V., Brenner, E. (1994). "Why two eyes are better than one for judgements of heading" *Nature*, 371: 700–702.

Vishton, P. M., Cutting, J. E. (1995). "Wayfinding, displacements and mental maps: Velocity fields are not typically used to determine ones's aimpoint", *J. Exp. Psychol.: Hum. Percept. Perform.*, 21: 978–995.

Warren Jr., W. H., Hannon, D. J. (1990). "Eye movements and optical flow", *J. Opt. Soc. Am. A*, 7: 160–169.

Warren Jr., W. H., Morris, M. W., Kalish, M. (1988). "Perception of translational heading from optical flow", *J. Exp. Psychol.: Hum. Percept. Perform.*, 14: 646–660.

Wertheim, A. H. (1987). "Retinal and extraretinal information in movement perception: How to invert the Filehne illusion", *Perception*, 16: 299–308.

11

Visual perception of 3D shape from motion: multisensory integration and cortical bases

V. Cornilleau-Pérès, A.L. Paradis and J. Droulez

11.1 Introduction

Optic flow is defined as the pattern of retinal velocities that is induced by a relative motion between the observer and a visual scene. Since the pioneering work by Gibson (1950), this visual input has been recognised as playing a critical role in motor control and equilibrium. It has been shown to be used by the visual system for the perception of 3D motion (self-motion or object-motion) but also of the 3D structure of objects. The ability to perceive the 3D structure of objects from optic flow has been evidenced by using distributions of dots randomly spread over a moving surface. Such stimuli are similar to random dot stereograms, in that they authorised depth perception while each individual static view of the dot distribution is devoid of any depth information. They have been widely used for the study of the processing of surface orientation or curvature from motion. However most experiments consisted in presenting such stimuli in small viewing angles (say smaller than 20 deg wide) to a static observer. The case of wide-field stimuli, and of the moving observer has almost never been addressed experimentally. In addition, the role of motion variables such as the translation and rotation components has never been assessed. Here we present the theoretical bases of the motion-structure dependency in the equations of 3D structure from motion. Then we report a series of studies that led us to understand the properties of the 3D processing of optic flow in the static and moving observer, in small field and wide field visual stimulation. Finally, for small field angles and static observer, we have questioned the specific structures that are involved in the processing of 3D structure from motion in an fMRI investigation.

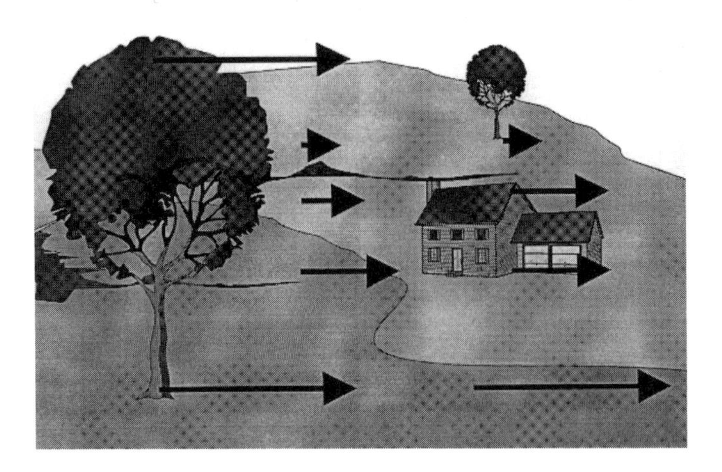

Fig. 11.1. The optic flow produced by a translation of the observer. The retinal image velocity depends on the distance of object points.

11.2 A non-linear relationship between optic flow, depth and 3D motion

During the relative movements between an observer and a visual scene, the pattern of optic flow created on the observer's retina can be described as a distribution of velocity vectors. This vector field usually presents complex variations over the observer's retina. Indeed if the observer's eye is not only rotating but also translating in space, the retinal velocity in each image point depends on the distance between the eye and the object point (Figure 11.1). In addition the relationship between image velocity and distance is *non-linear*. In a given image point, velocity is actually the sum of translation and rotation components, but the translation component depends on the depth map according to the following equation:

$$U = T.p + W \tag{11.1}$$

where T and W are linear combinations of the translation and rotation coordinates, respectively, and p is the proximity (inverse of the distance). Note that the center of rotation is chosen as being the observer's eye (which explains why $U = W$ if $T = 0$, meaning that rotations induce only a displacement of the image on the retina, with no information about 3D structure). Figure 11.2 presents a class of models proposing that 3D structure can be recovered from spatial derivatives of the optic flow. In the expression of these spatial derivatives (usually illustrated by patterns of expansion, rotation or shearing), translation is also multiplied by variables of surface orientation and curvature.

The multiplication between T and p has two major consequences. First depth and translation *cannot be recovered independently*, which means that

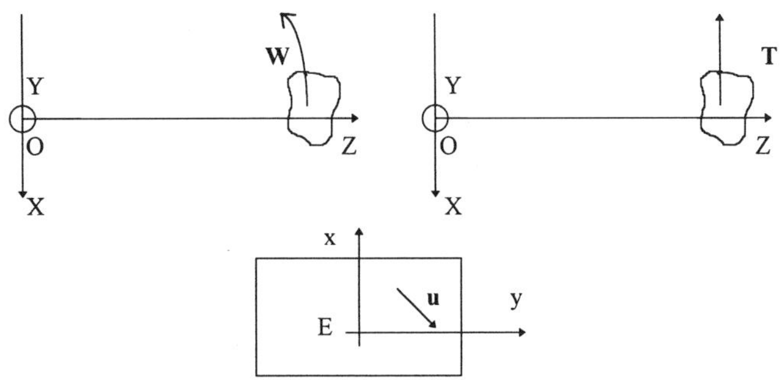

O represents the position of the observer's eye. E is the projection of an object point on the retina, u is the retinal velocity in an image point. W is the vector of object rotation around O, T the translation vector. (OXYZ) is a 3D coordinate system, with the axis Z pointing toward the surface to be analysed. (Exy) is the projected coordinate system in a plane approximating the retina. If the coordinates of T are (T_X, T_Y, T_Z) we defined τ as the vector representing the translation component, of coordinates $\tau_X = T_X/Z$ and $\tau_Y = T_Y/Z$ in (Exy).
The retinal velocity can then be expressed as the sum of τ and of a rotation component w:

$$u = \tau + w$$

In point O the spatial derivatives of the optic flow u can be expressed as a function of the surface orientation and curvature. Non linearities in this expression are due to the translational component. For instance:

$$\partial \tau_X / \partial x = -\tau_X \cdot p_X - \tau_Z$$

$$\partial^2 \tau_Y / \partial X^2 = -\tau_X \cdot C_{NX} \cdot Z \cdot (1 + p_X^2) \cdot (1 + p_X^2 + p_Y^2)^{1/2} + 2 \cdot \tau_Z \cdot P_X$$

where $p_X = \partial Z/\partial X$ and $p_X = \partial Z/\partial Y$ are the coordinates of the normal vector to the surface, and c_{NX} is the normal curvature to the surface in the direction X, and $\tau_Z = T_Z/Z$. From Koenderink and van Doorn (1975), Longuet-Higgins and Prazdny (1980) Droulez and Cornilleau-Pérès (1990)

Fig. 11.2. Non linearities in the problem of 3D structure from optic flow.

theoretically, the recovery of self-translation information from visual signals uniquely determine the depth distribution, and vice versa. This interdependency is particularly emphasised in the models proposed by Rieger and Lawton (1985), using depth discontinuities to compute T, or by Droulez and Cornilleau-Pérès (1990) accounting for the better performance in the detection of surface curvature in the direction orthogonal to T. We may not have as yet drawn out all the consequences of this interdependency. For instance, little is known about the influence of the 3D structure of the environment on heading perception or linear vection.

Second, this non-linearity renders the problem of recovering the 3D motion T and W, and the proximity map, p, difficult from a theoretical point of view (Koenderink & van Doorn, 1975; Longuet-Higgins & Prazdny, 1980; Droulez & Cornilleau-Pérès, 1990) and hard to solve in a robust fashion at the algorithmic level. In addition, the depth map (distribution of the distances between the eye and object points) is changing at each instant in the most general case. Hence the input (velocity field) and output (3D motion and structure) of the underlying process of recovering 3D information from optic flow are not stable over time which makes the problem still more difficult to solve using classical algorithms.

11.3 Multisensory integration in the 3D processing of optic flow

Due to the non-linearity of the relationship of retinal velocity and distance, several computer vision groups have proposed to use self-motion information to optimize the processing of moving images and to diminish the number of degrees of freedom (Aloimonos *et al.*, 1988; Ayache & Faugeras, 1989; Viéville, 1994; Luong & Faugeras, 1992; Ballard & Ozcanderli, 1988). For instance the equations in Figure 11.2 become linear if the frontal translation (T_X, T_Y) is known.

As far as human vision is concerned, many studies suggest that signals related to self-motion may indeed be used by the visual system to extract depth from moving images. In particular head translations can help in ther perception of visual absolute distances. Although the early work by Ferris (1972), Johansson (1973), and Gogel and Tietz (1980), indicates that absolute distances are not reliably estimated in this way, recent experiments do suggest that monocular viewing leads to an increase in head movements amplitude (Marotta *et al.*, 1995) suggesting a substitution of stereopsis by head movement. Similarly in the gerbil, Ellard *et al.* (1984) have demonstrated that head movements are spontaneously generated when the animal has to jump on distant platforms. The perception of absolute distances from self-motion could be due to the coupling between head translation, and the angle of eye rotation necessary to stabilize the target. Nonetheless, the facts that self-motion helps disambiguate the sign of depth order (Rogers & Rogers, 1992) and improves the adequacy between the actual and perceived depth extent of corrugated surfaces (Rogers & Graham, 1979; Ono & Steinbach, 1990) suggest an integration of self-motion signals in the 3D processing of a pattern of optic flow, and not only in the estimation of the distance of a target point.

Finally the recent exploration of the monkey posterior parietal cortex shows that cortical areas that seem to be involved in the coding of optic flow and its spatial derivatives, such as expansion-contraction or rotation are also areas of multisensory integration (see Colby & Duhamel, 1991; Lappe, 1997, for reviews) for instance areas MST and 7a.

From the theoretical and physiological arguments listed above, one can expect that the 3D processing of optic flow could be more reliable and precise when the observer, rather than the object, moves. In order to elucidate this question and to delineate the properties of 3D structure from motion perception in the moving observer, we performed a series of experiments in small field (8 deg visual angle) and large field (90 deg visual angle). We compared the ability of subjects to discriminate between surfaces that were either (i) planar (ii) concave spherical or (iii) convex spherical when the subject or the surface was moving. Computer-generated images represented these surfaces as made of random dot distributions. We made sure that motion parallax was the only depth cue available to subjects. We compared 3 conditions (Figure 11.3):

- self motion: the subject translated actively in front of a stationary surface,
- object translation: the surface translated in front of the stationary subject,
- object rotation: the surface rotated in depth about a frontoparallel axis.

During self-motion the subject translations were recorded and used to generate the transformations of the image (as in a virtual reality system) (see Cornilleau-Pérès and Droulez, 1994 and Dijkstra *et al.*, 1995 for methods). Subsequently these recordings were used to generate the translation component of object motion in the conditions "object translation" and "object rotation". Hence we insured that the potential depth information from motion parallax was identical in the three conditions.

11.3.1 Results of small field experiments

The results of small-field experiments are indicated in Figure 11.4B. The lower part of each bar indicates the percentage of errors in the sphere/plane discrimination, while the upper hashed part indicate the percentage of errors on the curvature sign (concave or convex).

First the detection of surface curvature appears to be optimal during object rotation, intermediate during self-translation, and the poorest during object translation. This result correlates well with the estimate of the retinal slip occurring in the 3 conditions. This retinal slip is null for object rotation in the image center. From Collewijn *et al.* (1981) and Ferman *et al.* (1987) retinal-slip was estimated to range between 1 and 6 deg/s for self motion. For object translation it was estimated to reach 15 deg/s (Lisberger *et al.*, 1981). Other studies showed that retinal slip is smaller during head movements than during target movements. This is probably due to the vestibularly-evoked stabilizing eye movements (Buizza *et al.*, 1981). Also, the fact that the perception of velocity gradients is impaired by a retinal slip as low as 1 deg/s (Nakayama, 1981) is consistent with the idea that eye movements are a major factor influencing subjects' performance in the detection of surface curvature.

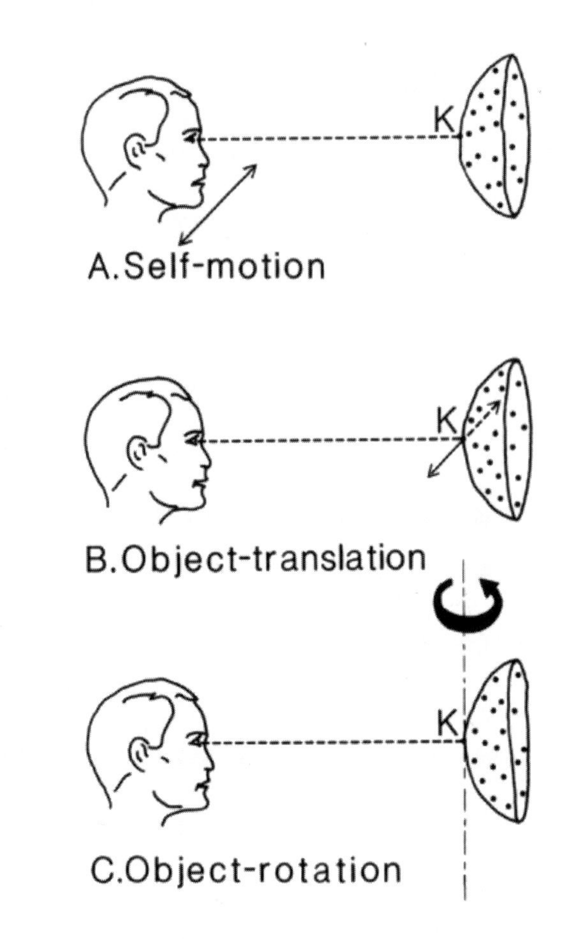

Fig. 11.3. Three experimental conditions for the study of the perception of 3D structure from motion. Condition "self-motion", the subject translate laterally his head, and this movement is used to produce images of a surface which is apparently stationary in space. Condition "object translation", the subject views a surface that translates laterally in space. Condition "object rotation", the surface is rotating about one of its frontoparallel tangent axes.

Hence we showed that rotation in depth is the optimal condition for the detection of surface curvature from optic flow, this being most probably due to the global stability of the motion pattern on the retina. Comparing the three conditions, we concluded that no major improvement of performance is provided by self-motion related signals, other than through a better retinal image stabilization, and that performance in the moving observer can be significantly degraded relative to performance that are usually measured for object-rotation in the static observer.

The relatively low refresh rate of our computer images (30 Hz) could induce a decrease of performance when retinal slip increases, because of

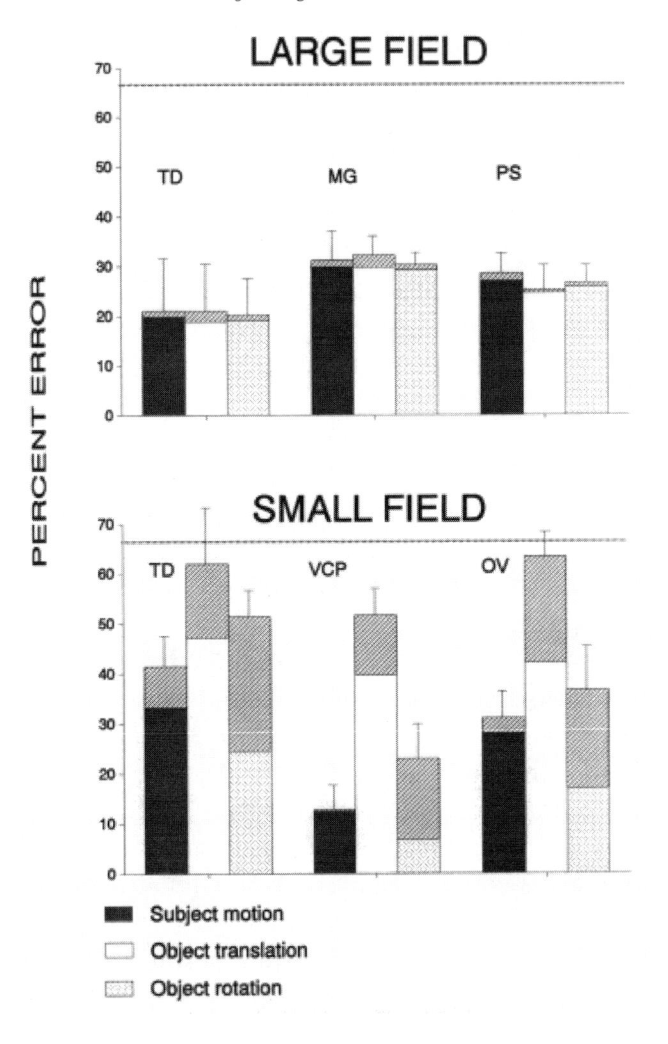

Fig. 11.4. Results of the wide-field and small-field experiments. The results are presented for 3 subjects in both field conditions. Planes, concave and convex spherical surfaces were presented in random order. The bars indicate the total percentage of errors. The lower parts of the bars indicate the percentage of errors relative to the sphere/plane discrimination. The upper hatched part indicates the percentage of errors on the curvature sign (errors between concave and convex spheres). The horizontal lines indicate chance level.

retinal persistence for instance. In order to check that this was not a critical factor, we performed a sphere/plane discrimination experiment with real objects. The results showed a similar variations of the performance across conditions (Cornilleau-Pérès & Droulez, 1996).

As far as the sign of curvature is concerned, Figure 11.4B confirms the finding by Rogers and Rogers (1992), that self-motion helps to disambiguate the sign of depth order. On the opposite, the curvature sign of our spherical surfaces was strongly ambiguous for object-rotation.

11.3.2 Results of large field experiments

Here the pattern of results is radically different (Figure 11.4A). The accuracy in detecting surface curvature does not vary across conditions, and also we observe no ambiguity relative to the sign of surface curvature.

11.3.2.1 Detecting surface curvature

These results confirm the small field experiment, in the sense that no increase of accuracy in the processing of spatial derivatives of optic flow is gained from self-motion related signals. Why the performance does not vary across conditions remains to be understood. The optokinetic nystagmus (OKN) gain and the dynamic of convergence eye movements improve as stimulus size increases (van den Berg & Collewijn, 1986; Erkelens *et al.*, 1989) and may explain why performance does not drop for object translation as it did in small field experiments. Also the subject is required to fixate as precisely as possible the central point of the display. Therefore the region which is visually stimulated on the retina does not vary a lot. In small field experiments, however, a 4 deg difference between the eye position and the stimulus centre reduces the area of central vision which is covered by the stimulus by more than 50%. Another interpretation is that the processing of large optic flow patterns involves neurons that are sensitive to spatial derivatives of optic flow over larger visual angles as are found in area MST of the monkey for instance. Finally it could be that the improvement due to self-motion signals is counterbalanced by the impairment due to retinal slip, but that reasoning is not supported by the fact that performance are also very good while retinal slip is the largest, i.e. during self-translation.

11.3.2.2 Detecting the curvature sign

Here we find that the curvature sign is unambiguously perceived in wide-field during subject or object motion. In our view, this result has two possible interpretations. First it may be that our large-field optic flow pattern was less ambiguous than the small field pattern, in that the difference between the velocity distribution for concave and convex surfaces is larger in wide field than in small field. This is probably not sufficient to explain our results however, because apparent 3D deformation was reported when the small field stimuli were perceived with the wrong curvature, suggesting that the difference between concave and convex stimuli is large enough to be detected by the visual system. Hence our second interpretation is that the rigidity of the visual scene is a stronger constraint for the processing of large field optic flow (which are generally consequent to the observer's motion in a rigid environment) than for the processing of small field optic flow, which is often created by deformable objects, or by objects moving with their own movements relative to a background (which is a case of non rigidity of the object relative to the background).

In summary our results demonstrate that for small field stimuli, the perception of 3D structure from motion does not only depend on the translation T, but also on the rotation W between the eye and the visual scene, as a factor conditioning retinal slip. Hence in small field the motion condition (object- or self-motion) as well as retinal image stabilization are critical factors. On the opposite for large field stimuli, the performance do not depend on these factors, which questions the possible existence of specific mechanisms dedicated to the processing of large field motion patterns. Finally, the hypothesis that self-motion signals improve the robustness of optic flow processing is strongly questioned. Our results suggest that the cooperation between self-motion signals and 3D optic flow processing is reduced to the disambiguation of the curvature sign (and perhaps of other variables) in small field. Finally the size of the visual stimulus appears to be a critical variable for the perception of 3D structure from motion.

11.4 Cortical imaging of the structures involved in the processing of 3D structure from motion

Although the cortical circuits involved in the processing of visual motion are now well documented, it remains unclear whether some areas of the brain are specifically active in the analysis of 3D shape from motion. In particular it is unknown whether this analysis takes place in the dorsal or in the ventral visual pathway. If, as now widely accepted in the literature, the dorsal pathway is specialised in the processing of motion and localisation in space, while the ventral pathway mediates object recognition (seeColly & Duhamel, 1991 for review), then both pathways seem to be required for the perception of 3D shape from motion. In particular, our working hypothesis is that the percept of a 3D shape emerging from a 3D spherical dotted surface in motion should elicit neuronal activities in the ventral pathway, as compared to random motion. Alternatively a coherent motion representing an expansion or contraction of a frontoparallel surface should not elicit such activities, because such stimulus does not elicit the percept of a 3D shape. From the literature related to the existence of neurons sensitive to expansion/contraction, or to the rotation in depth of objects, we also expect that both stimuli (the rotating sphere and the expanding/contracting pattern) should enhance the neuronal activity in the dorsal visual pathway, and particularly the posterior parietal cortex, as compared to random motion (see Lappe, 1997, for a review).

To look at which brain areas were involved we produced four types of

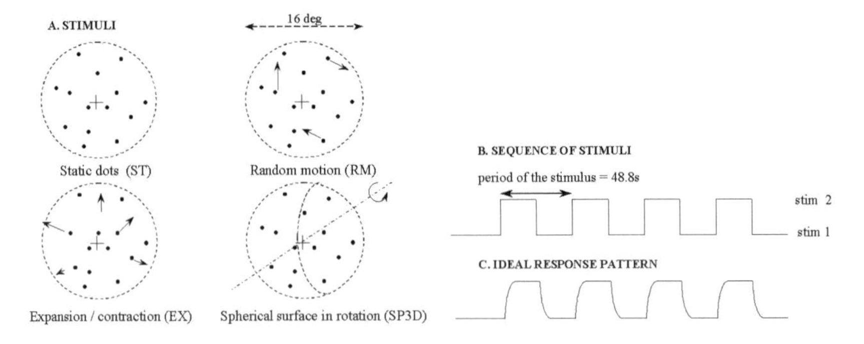

Fig. 11.5. A. The four stimuli used in the fMRI experiments are made of 300 dots randomly positioned over a disk of 16 deg diameter of visual angle. The stimuli were projected on a transluscent screen and viewed by the subjects lying in the magnet. The dots could be either stationary (ST), moving with a velocity that varied randomly in direction (RM), moving with alternated expansions and contractions (EX), or moving as if they belonged to a spherical surface rotating about one of its frontoparallel tangents. B. The temporal sequence of stimulation. Each stimulus was presented during 24.4 s. In each experiment, 2 stimuli were presented 4 times in alternation. C. The ideal response pattern of a voxel showing more neuronal activity for stimulus 2 than for stimulus 1. It is obtained by convoluting the experimental profile B with a function modelling the rise time of the MRI signal.

stimuli (Figure 11.5A), all made of 300 antialiased dots. Dots could be either

- stationary (stimulus ST),
- moving in random motion (RM) with velocities that varied randomly in direction over the stimulus,
- moving coherently in an expansion-contraction fashion (EX),
- moving as if they belonged to an opaque sphere oscillating about one of its tangential frontoparallel axis (SP3D). The sphere was larger than the display window, so that no apparent contour was visible.

The visual field of view was 16 deg wide. By adjusting the proportion of flickering dots (reappearing in a random position from one image frame to the next) we maintained the dot density uniform and constant over the visual field. We made sure that the number of flickering dots and their distribution, the mean dot velocity, the temporal frequency of the changes in velocity direction were equal or similar across conditions. Also, in all stimuli, the average velocity over any disc centred on the fixation cross was null, which was hoped to minimise eye movements. Subjects passively viewed the stimuli and were asked to fixate a cross in the center of the image.

We performed three experiments by alternating stimuli (1) ST and RM, (2) RM and EX, and (3) RM and SP3D according to a paradigm described on Figure 11.5B (Paradis *et al.*, 1998). Experiment 1 aimed at determining the areas involved in motion processing and compare our results to previously published description of these areas (Watson *et al.*, 1993; Tootell *et al.*, 1995). Experiment 2 aimed at finding which regions were activated by coherent motion (with no compelling perception of a 3D shape) as compared to stochastic motion. In experiment 3, the presence of a percept of a 3D shape, and a coherent motion field should both be responsible for cortical activity.

Functional MRI images were acquired in nine adult volunteers, using a 3T MRI scanner, with voxel size 4x4x5 mm^3. The results were processed with SPM96, with normalization of each brain in the Talairach coordinate system. The functional data were analysed for each individual, and mapped upon his anatomical MRI. A statistical group analysis was performed for the nine subjects. Due to some geometrical distortions in the lower functional slices, as compared to the anatomical slices for four subjects, a precise localisation of functional regions in lower slices ($z < 15$ mm) could not be performed for these subjects, and the Talairach coordinates of the activity foci in these low slices are subject to possible errors of up to 12 mm along the antero-posterior axis y, and 8 mm along the infero-superior axis z. In order to compensate for these effects, as well as for the inter-individual variations in anatomy, we performed a Gaussian spatial filtering of the signal for each subject (width 5 mm). The different steps of filtering and normalization led to functional volumes of voxel size 3x3x3 mm^3.

The activations regions were selected according to two criteria. The first criterion concerns the probability that the temporal signal of one voxel follows the modelled response course (figure 11.3C) by chance. The corresponding threshold could be strict ($p<0.001$) or permissive ($p<0.01$). The second criterion requires that a region is selected only if its size is larger than two adjacent voxels.

Each experiment was analysed for the two opposite contrasts between the two stimuli (for instance RM - ST and ST - RM in experiment 1). Figure 11.6 indicates the results of the group analysis for the three contrasts RM - ST, EX - RM, and SP3D - RM, for the strict criterion ($p<0.001$). Table 11.1 indicates the Talairach coordinates of the focus of maximum Z score, in each selected region, and the level of probability indicating the risk that this region has been selected by chance. The 2 contrasts between coherent (EX or SP3D) and random motion yield activity levels that are less significant than the contrast between random motion and stationary dots. Also, this activity is stronger when the rotating sphere, rather than the expansion/contraction pattern, was contrasted with the random motion.

In experiment (1) the contrast RM - ST yielded similar results for all subjects, with strong activations (up to 12%) over regions covering 800 to 1600 mm^3. Those "random motion areas" were

GROUP ANALYSIS 9 SUBJECTS			**Random motion - static**				
	BA		x	y	z	**Z score**	p
Lateral occipito-temporal junction (V5+)	19/37	left	-51	-72	3	7.61	<0.001
		right	45	-54	6	7.59	<0.001
Lingual and middle occipital gyri (V1/V2)	17/18	left	-12	-81	-3	6.19	<0.001
		right	27	-84	0	7.26	<0.001
Superior occipital gyrus (V3)	19	left	-27	-81	15	6.56	<0.001
Lingual/fusiform gyri, occipito-temporal junction (V2/VP?)	18/19	left	-24	-66	-12	5.42	<0.001
		right	21	-66	-6	5.93	<0.001
			Expansion/ contraction - random motion				
	BA		x	y	z	**Z score**	p
Superior occipital gyrus(V3)	19	left	-24	-84	27	4.17	<0.253
			3D sphere - random motion				
	BA		x	y	z	**Z score**	p
Superior occipital gyrus(V3)	19	left	-18	-87	21	5.32	<0.005
Lingual/fusiform gyri, occipito-temporal junction (V2/VP?)	18/19	left	-24	-69	-9	4.48	<0.094
Parieto-occipital junction	19/39/7	left	-33	-78	39	4.44	<0.109

Table 11.1. *BA is the Brodman area, (x, y, z) are the Talairach coordinates of the local maximum of the Z score distribution. The value of this local maximum is given, and the probability p that the region of interest is selected by chance.*

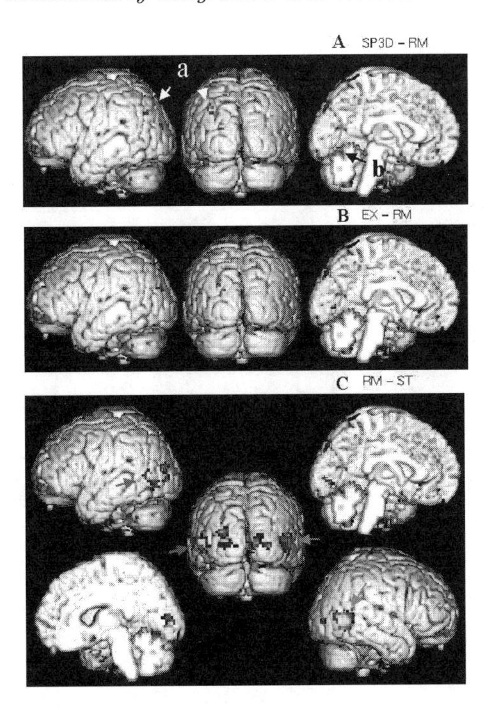

Fig. 11.6. Results of the group analysis (nine subjects) showing posterior views of the cortex (middle column) as well as lateral and medial views. A. The contrast SP3D - RM yields foci of ativity in the superior occipital gyrus, in the dorsal part of parietooccipital junction (a), and ventrally in the tempooccipital junction (b). B. The contrast EX - RM delineates a focus in the superior occipital gyrus. C. The contrast RM - ST allows to determine the position of motion areas, in particular the middle temporal V5 complex in the left (green arrow) and right (blue arrow) hemispheres.

- the bilateral lingual and middle occipital gyri (V1/V2). These regions extended ventrally to the occipito-temporal junction between the lingual and fusiform gyri (posterior part of the collateral sulcus),
- the bilateral middle-temporal (V5) complex, at the lateral occipito-temporal junction, in the anterior occipital sulcus (ascending branch of the inferior temporal sulcus),
- the left superior occipital gyrus (possibly dorsal V3). This region is at the border between the dorsal cuneus (medial occipital gyrus) and the lateral occipital cortex, in the intra-occipital sulcus.

Both contrasts between coherent (EX or SP3D) and incoherent motion yielded activations that were limited to the left hemisphere, at least at the strict threshold that we imposed in this first analysis. A first region of activity is located in the superior occipital gyrus and is also a "random motion area". Therefore this region is involved in the coding of motion, but is more activated by coherent than incoherent motion.

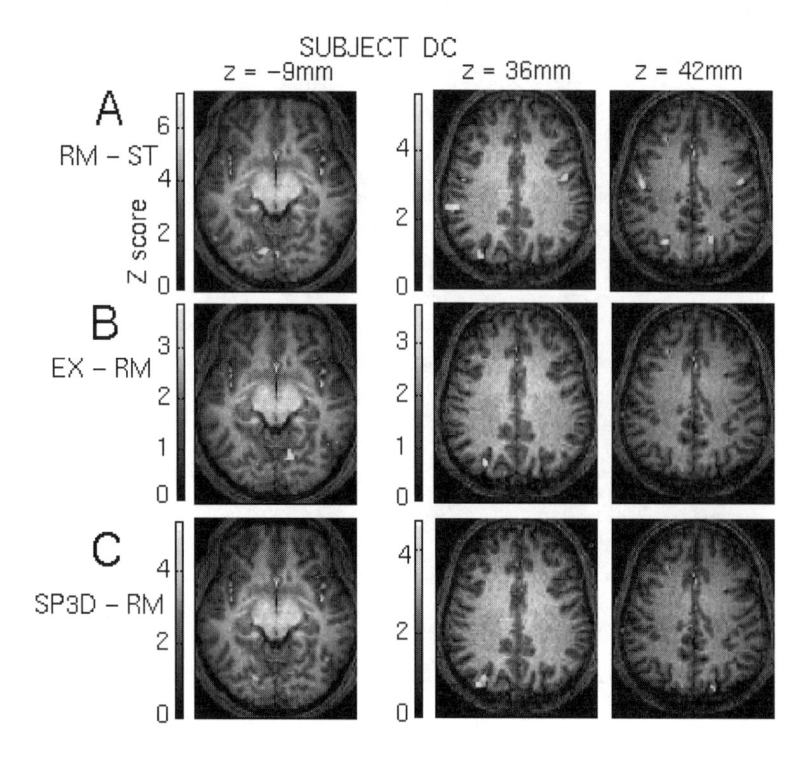

Fig. 11.7. Localisation of the activity foci on individual data for the comparison between random motion and static dots (A), between expansion/contraction and random motion (B), and between the sphere and random motion (C). Left column : the lower slices show the localisation of a ventral focus along the collateral sulcus in the left hemisphere, at the junction between the lingual and fusiform gyri, and between the occipital and temporal lobes. Such a focus was selected by the group analysis as statistically significant for the contrast SP3D - RM, but not for the contrast EX - RM. Right columns : these slices are located just above the parietooccipital junction. The posterior foci of activity are located in the intraparietal sulcus.

In addition, the contrast SP3D - RM induced two specific areas of activation :

- the parieto-occipital junction, at the border between the precuneus, angular gyrus and superior occipital gyrus. Individual analyses showed that these activations were often located in the intraparietal sulcus (Figure 11.7).
- the occipito-temporal junction, in the ventral part of the occipito-temporal junction. For two subjects we could locate these activations in the collateral sulcus, at the border between the fusiform and lingual gyri (Figure 11.7).

With the less selective threshold on the individual voxel signal ($p<0.01$)

we observed a similar spatial distribution of dorsal activity for contrasts EX -RM and SP3D - RM. In particular the parieto-occipital junction was weakly activated in the left and right hemisphere for the contrast EX - RM, and in the right hemisphere for contrast SP3D - RM (in addition to the strong left activity).

On the contrary, at the less selective threshold, the ventral region was seen bilaterally for the contrast SP3D - RM, but in none hemisphere for contrast EX - RM. Finally the two contrasts EX - RM and SP3D - RM did not yield any pattern of activation in V5+ and its vicinity, even at our most permissive threshold.

We conclude that

- the major regions involved in the processing of our three different motion patterns (RM, EX and SP3D) largely overlap, and consist mainly in occipital regions, the temporo-occipital V5 complex and the left superior occipital gyrus,
- as compared to random motion, the processing of coherent motion (expansion/contraction or rotating sphere) activates areas in the parieto-occipital junction, with a dominance of the left hemisphere. This dorsal region is more strongly activated by the rotating sphere than by the expansion/contraction stimulus.
- as compared to random motion, the rotating sphere stimulus yields a differential activity in the occipito-temporal junction, which supports a role of the ventral visual pathways in the perception of 3D shape from motion.

The results of our first experiments are in close agreement with previous studies. In particular, the coordinates of V5 complex are similar to what has been found in other fMRI or PET studies (reviewed by Mc Carthy *et al.*, 1995). Also, the activity of V5 complex is not enhanced by coherent motion as compared to random motion. This is in agreement with the PET study by McKeefry *et al.* (1997) who compared unidirectional and random motions, and might be explained by the existence of a population of neurons that have an increased activity when the stimulus velocity is different in the classical receptive field and its surround (Allman *et al.*, 1985).

As far as the comparison between coherent and incoherent motion is concerned, our results coincide partially with those obtained by De Jong *et al.* (1994) with PET. The basic stimulus consisted in an expansion flow of dots located on a horizontal plane (" heading" flow). This coherent motion pattern was alternated with an incoherent flow, obtained by randomising the directions of dot velocities. Comparing coherent with incoherent motion yields activity foci that are located in the lateral side of the cuneus and precuneus, and in the ventral occipital lobe. These foci correspond, in their description and Talairach coordinates, with the dorsal and ventral regions that we find in our sphere vs random motion comparison. The similarity between their results and our sphere vs random motion experiment can be explained by the fact that large heading stimuli (40 deg in their experiment) induce the percept of the ground surface. On the opposite, our small field (16 deg) expansion/contraction pattern do not yield a compelling percept of a 3D surface. Surprisingly, however, the location of dorsal activity was more consistent in the right than in the left hemisphere, which contrasts with our finding of a left dominance for dorsal areas selected by the SP3D - RM comparison.

A common finding in De Jong *et al.*, and in our work, is that in humans there does not seem to be a satellite of V5, within the V5 complex, that would be an analog of the monkey MSTd (V5A) in the sense that its neuronal activity would be enhanced by coherent motion patterns.

It is of interest to note that the intraparietal sulcus of the monkey is now known to contain neurons that are selective to object shape, object orientation, or binocular disparity. Some of these neurons, particularly in area AIP (anterior intraparietal) also have an hand-related activity (for a review see Sakata *et al.*, 1997). Alternatively, the neurons that are selective to object shape in ventral areas usually respond as well to the presentation of a 3D object, or of its 2D picture. Together with the imaging results, this suggests that in both monkey and human there exists a representation of object shape both in the dorsal and ventral visual pathway.

11.5 Conclusion

The studies described have shown that small-field visual motions seem to be processed in a way similar to the processing found in monkey brains, with an implication of the parietooccipital cortex for coherent motion, and of the ventral temporo-occipital junction when 3D structure emerges from a motion pattern.

This result is in agreement with the fact that the detection of surface curvature does not seem to be modified by the presence of self-motion information (i.e. the dorsal stream) in small field, other than through the involvement of eye-movements. Hence both our psychophysical and fMRI studies converge toward the conclusion that the coding of 3D variables related to object shape rather involves the ventral stream (although it passes first through motion areas classically seen as belonging to the dorsal stream).

Still the question arises of the cortical mechanisms underlying the disambiguation of the curvature sign through self-motion. In particular one can wonder whether this process involves the parieto-temporal connections studied by Seltzer and Pandya (1984) in the monkey. Also, it remains to be known which variables related to object shape are coded in the dorsal and ventral pathways respectively. One can wonder whether extrinsic variables such as object orientation are coded in similar areas as intrinsic variables such as surface curvature. Finally the monkey studies as well as the human imagery work by Corbetta and his collaborators (1991) show that the neuronal activity in different cortical areas can be strongly task dependent. Hence the recent physiological results reviewed by Sakata *et al.* (1997) lead to the hypothesis that the implication of the ventral and dorsal routes might depend very much on the context. In particular, the selectivity to object shape could be enhanced during object manipulation in dorsal areas, and during object recognition tasks in ventral areas.

Acknowledgments

This work was supported by Esprit BRA 3149 (MUCOM), by the ACC-SV program from the French Ministry of Research (no 951261/12), and by the Ecole Nationale Supérieure des Télécommunications.

References

Allman, J., Miezin, F. & McGuinness, E. (1985). "Direction and velocity-specific responses from beyond the classical receptive field in the middle temporal area (MT)", *Percept.*, 14: 105-126.

Aloimonos, J., Weiss, I. & Bandhopadhyay, A. (1988). "Active vision", *Int. J. Comp. Vis.*, 7: 333-356.

Ayache, N. & Faugeras, O. D. (1989). "Maintaining representations of the environment of a mobile robot", *IEEE Trans. Robot Automat.*, 5: 804-819.

Ballard, D. H. & Ozcanderli, A. (1988). "Eye fixation and early vision: kinetic depth", Proc. 2nd IEEE Int. Conf. Computer Vision, 524-531.

Buizza, A., Schmid, R. & Droulez, J. (1981). "Influence of Linear Acceleration on Oculomotor Control", in *Progress in Oculomotor Research*, Fuchs & Becker (Eds), Elsevier, North Holland, 517-524.

Colby, C. L. & Duhamel, J.-R. (1991). "Heterogeneity of extrastriate visual areas and multiple parietal areas in the macaque monkey", *Neuropsychologia*, 29: 517-537.

Collewijn, H., Martins, A. J. & Steinman, R. M. (1981). "Natural retinal image motion: origin and changes", *Ann. NY Acad. Sci.*, 374: 312-329.

Corbetta, M., Miezin, F. M., Shulman, G. L. & Petersen, S. E. (1991). "Selective and divided attention during visual discriminations of shape, color, and speed: functional anatomy by positron emission tomography", *J. Neurosci.*, 11: 2383-2402.

Cornilleau-Pérès, V. & Droulez, J. (1994). "The visual perception of 3D shape from self-motion and object-motion", *Vis. Res.*, 34: 2331-2336.

Cornilleau-Pérès, V. & Droulez, J. (1996). "Visual perception of the curvature of real objects from self-motion and object-motion", *Perception*, 25: 93.

De Jong, B. M., Shipp, S., Skidmore, B., Frackowiack, R. S. J. & Zeki, S. (1994). "The cerebral activity related to the visual perception of forward motion in depth", *Brain*, 117: 1039-1054.

Dijkstra, T. M. H., Cornilleau-Pérès, V., Gielen, C. C. A. M. & Droulez, J. (1995). "Perception of 3D shape from ego- and object-motion: comparison between small and large field stimuli", *Vis. Res.*, 35: 453-462.

Droulez, J. & Cornilleau-Pérès, V. (1990). "Visual perception of surface curvature. The spin variation and its physiological implications", *Biol. Cybern.*, 62: 211-224

Ellard, C. G., Goodale, M. A. & Timney, B. (1984). "Distance estimation in the mongolian gerbil: The role of dynamic depth cues", *Behav. Brain Res.*, 14: 29-39

Erkelens, C. J., Van der Steen, J., Steinman, R. M. & Collewijn, H. (1989). "Ocular vergence under natural conditions. I. Continuous changes of target distance along the median plane", *Proc. Roy. Soc. Lond.*, 236: 417-440

Ferman, L., Collewijn, H., Jansen, T. C. & Van den Berg, A. V. (1987). "Human gaze stability in the horizontal vertical and torsional direction during voluntary head movements, evaluated with a three-dimensional scleral induction coil", *Vis. Res.*, 27: 811-828

Ferris, S. H. (1972). "Motion parallax and absolute distance", *J. Exp. Psychol.*, 95: 258-263.

Gibson, J. J. (1950). *The perception of the visual world*, Houghton Mifflin Cx., Boston, MA.

Gogel, W. C. & Tietz, J. D. (1980). "Relative cues and absolute distance perception", *Percept. Psychophys.*, 28: 321-328.

Johansson, G. (1973). "Monocular movement parallax and near-space perception", *Perception*, 2: 135-146.

Koenderink, J. J. & Van Doorn, A. J. (1975). "Invariant properties of the motion parallax field due to the movement of rigid bodies relative to an observer", *Optica Acta*, 22: 773-791.

Lappe, M. (1997). "Analysis of self-motion by parietal neurons", in *Parietal Lobe contribution to orientation in 3D space*, P. Thier & H. O. Barnath (Eds), *Exp. Brain Res.*, 25: 597-618.

Lisberger, S. G., Evinger, C., Johanson, G. W. & Fuchs, A. F. (1981). "Relationship between eye acceleration and retinal image velocity during foveal smooth pursuit in man and monkey", *J. Neurophysiol.*, 46: 229-249.

Longuet-Higgins, H. C. & Prazdny, K. (1980). "The interpretation of a moving retinal image", *Proc. Roy. Soc. Lond. B*, 208: 385-397.

Luong, Q. T. & Faugeras, O. (1992). "Active head-movements help solve stereo correspondence?", Proc of the 10th Europ. Conf. on Artif. Intell., 800-802.

Marotta, J. J., Perrot, T. S., Nicolle, D., Servos, P. & Goodale, M. A. (1995). "Adapting to monocular vision : grasping with one eye".

McCarthy, G., Spicer, M., Adrignolo, A., Luby, M., Gore, J. & Allison, T. (1995). "Brain activation associated with visual motion studied by functional magnetic resonance imaging in humans", *Hum. Brain Mapping*, 2: 234-243.

McKeefry, D. J., Watson, J. D. G., Frackowiack, R. S. J., Fong, K. & Zeki, S. (1997). "The activity in human area V1/V2, V3 and V5 during the perception of coherent and incoherent motion", *Neuroimage*, 5: 1-12.

Nakayama, K. (1981). "Differential motion hyperacuity under conditions of common image motion", *Vis. Res.*, 21: 1475-1482.

Ono, H. & Steinbach, M. J. (1990). "Monocular stereopsis with and without head movement", *Percept. Psychophys.*, 48: 179-187.

Paradis, A. L., Cornilleau-Pérès, V., Droulez, J., van de Moortele, P. F., Berthoz, A. & Le Bihan, D. (1998). "The visual perception of 3D structure from motion: a fMRI study", Manuscript in preparation.

Rieger, J. H. & Lawton, D. T. (1985). "Processing differential image motion", *J. Opt. Soc. Am.*, 2: 354-359.

Rogers, B. J. & Graham, M. (1979). "Motion parallax as an independent cue for depth perception", *Perception*, 8: 125-134.

Rogers, S. & Rogers, B. J. (1992). "Visual and non-visual information disambiguate surfaces specified by motion parallax", *Percept. Psychophys.*, 52: 446-452.

Sakata, H., Taira, M., Kusunoki, M., Murata, A. & Tanaka, Y. (1997). "The parietal association cortex in depth perception and visual control of hand action", *Trends in Neurosci.*, 20: 350-357.

Seltzer, B. & Pandya, D. N. (1984). "Further observations on parieto-occipital connections in the rhesus monkey", *Exp. Brain. Res.*, 55: 301-312.

Tootell, R. B. H., Reppas, J. B., Kwong, K. K., Malach, R., Born, R. T., Brady, T. J., Rosen, B. R. & Belliveau, J. W. (1995). "Functional analysis of human MT and related visual cortical areas using magnetic resonance imaging", *J. Neurosci.*, 15: 3215-3230.

Van den Berg, A.V. & Collewijn, H. (1986). "Human smooth pursuit: effects of stimulus extent and of spatial and temporal constraints of the pursuit trajectory", *Vis. Res.*, 26:1209-1222

Viéville, T. (1994). *A few steps towards 3-D Active Vision*, Springer Verlag.

Watson, J. D. G., Myers, F., Frackowiack, R. S. J., Hajnal, J. V., Woods, R. P., Mazziotta, J. C., Shipp, S. & Zeki, S. (1993). "Area V5 of the human brain: evidence from a combined study using positron emission tomography and magnetic resonance imaging", *Cerebral Cortex*, 3: 79-94.

12

Vision and action in artificial animals

Demetri Terzopoulos

12.1 Introduction

Advances in the emerging field of artificial life (ALife) make possible a new approach to the computational study of perception and action in living systems.* A major theme in ALife research is the synthesis of artificial animals, or "animats", a term coined by Wilson (1991). Wilson's original animat, a point marker in an idealized 2D grid world that could move between squares containing food or obstacles, was a highly abstracted animal model (or "vehicle" in the sense of Braitenberg, 1984) proposed to study the acquisition of simple behavior rules. Current animat theory encompasses the physics of the animal and its world, its ability to locomote, its adaptive, sensorimotor behavior, and its cognitive faculties, including learning (Terzopoulos *et al.*, 1994).

The recently proposed *animat vision* paradigm prescribes the use of artificial animals as autonomous virtual robots for active vision research, which includes the computational modeling of perception and action (Terzopoulos & Rabie, 1997; Rabie & Terzopoulos, 1996)†. Active vision became popular within mainstream computer vision following the seminal papers by Bajcsy (1988) and Ballard (1991), and it is now a prevalent theme (see, e.g., Ballard, 1992; Blake & Yuille, 1992; Swain & Ballard, 1993). Note that *animat* vision should not be confused with Ballard's *animate* vision; the latter does not involve artificial animals. The idea in a nutshell is to implement, entirely in software, realistic artificial animals and to give them the ability to locomote, perceive, and in some sense understand the realistic virtual worlds they inhabit so that they may achieve individual and social functionality within these worlds. The artificial animal is an autonomous agent possessing a muscle-actuated body capable of locomotion and a mind with

* For an engaging introduction to the ALife field, see, e.g., S. Levy, *Artificial Life* (Pantheon, 1992).

† Active vision has its roots in the work of the late psychologist J. J. Gibson, who stressed in pre-computational terms the importance of understanding an active observer situated in a dynamic environment (Gibson, 1979)

Fig. 12.1. Artificial fishes in their physics-based virtual world as it appears to an underwater observer. The 3 reddish fish (center) are engaged in mating behavior, the greenish fish (upper right) is a predator hunting for small prey, the remaining 3 fishes are feeding on plankton (white dots). Seaweeds grow from the ocean bed and sway in the current.

perception, motor, and behavior centers. It is endowed with functional eyes that can image the dynamic 3D virtual world onto 2D virtual retinas. The perceptual center of the animat's brain exploits active vision algorithms to continually process the incoming stream of dynamic retinal images in order to make sense of what the artificial animal sees and, hence, to purposefully navigate its world.

An artificial animal has been implemented that emulates natural animals as complex as teleost fishes in their marine habitats (Terzopoulos & Rabie, 1994). Imagine a virtual marine world inhabited by a variety of realistic fishes (Figure 12.1).* In the presence of underwater currents, the fishes employ their muscles and fins to swim gracefully around immobile obstacles and among moving aquatic plants and other fishes. They autonomously explore their dynamic world in search of food. Large, hungry predator fishes stalk smaller prey fishes in the deceptively peaceful habitat. The sight of predators compels prey fishes to take evasive action. When a dangerous predator appears, similar species of prey form schools to improve their chances of survival. As the predator nears a school, the fishes scatter in terror. A chase ensues in which the predator selects victims and consumes them until satiated. Some species of fishes seem untroubled by predators. They find comfortable niches and feed on floating plankton when they get hungry. Driven by healthy libidos, they perform elaborate courtship rituals to attract mates.

* Note that the figures in this chapter depict monochrome versions of original color images.

In this chapter I describe our work on synthesizing an active vision system for the artificial fish which is based solely on retinal image analysis. We have endeavored to make the vision system extensible so that it will eventually support the broad repertoire of individual and group behaviors described above. Rather than attempt to model accurately any particular form of piscine perception within the great diversity of possibilities (Fernald, 1993), we have found it a fascinating and challenging problem to endow the artificial fish with a generic biomimetic vision system coupled tightly to motor control, hence driving purposive action. Its vision system enables the artificial animal to be a functional, active observer of its dynamic world. Furthermore, I describe our work on implementing generic yet biologically plausible learning mechanisms that enable the artificial fish to acquire locomotion and some more complex motor skills through perceptual feedback.

In particular, the basic functionality of the active vision system starts with binocular perspective projection of the 3D world onto the retinas of the artificial animal. Retinal imaging is accomplished by photorealistic, color graphics rendering of the world from the animat's viewpoint. This projection respects occlusion relationships among objects. It forms spatially variant visual fields with high resolution foveas and low resolution peripheries. The analysis of the incoming color retinal image stream is coupled closely to motor mechanisms in the body of the artificial animal. The visual center of the animat's brain supplies saccade control signals to the eyes in order to stabilize the visual fields during locomotion through compensatory eye movements (an optokinetic reflex), to attend to interesting colored targets, and to keep these dynamic targets fixated. Through the visually guided coordination of multiple muscle actuators, the artificial fish is thus able to approach and track other artificial fishes. Motor learning algorithms enable an artificial dolphin to discover through practice how to execute stunts not unlike those performed by trained marine mammals to the delight of spectators at aquatic theme parks. It demonstrates that realistic, biomechanical models of animals situated in physics-based virtual worlds are fertile ground for learning sensorimotor control strategies.

12.2 The artificial fish

I will now review the artificial fish model rather superficially, yet with sufficient detail to comprehend the remainder of the chapter (see Terzopoulos & Rabie, 1994; Tu, 1996 for the full details). Each artificial fish is an autonomous agent comprising a graphical display model, a body model, and a brain model.

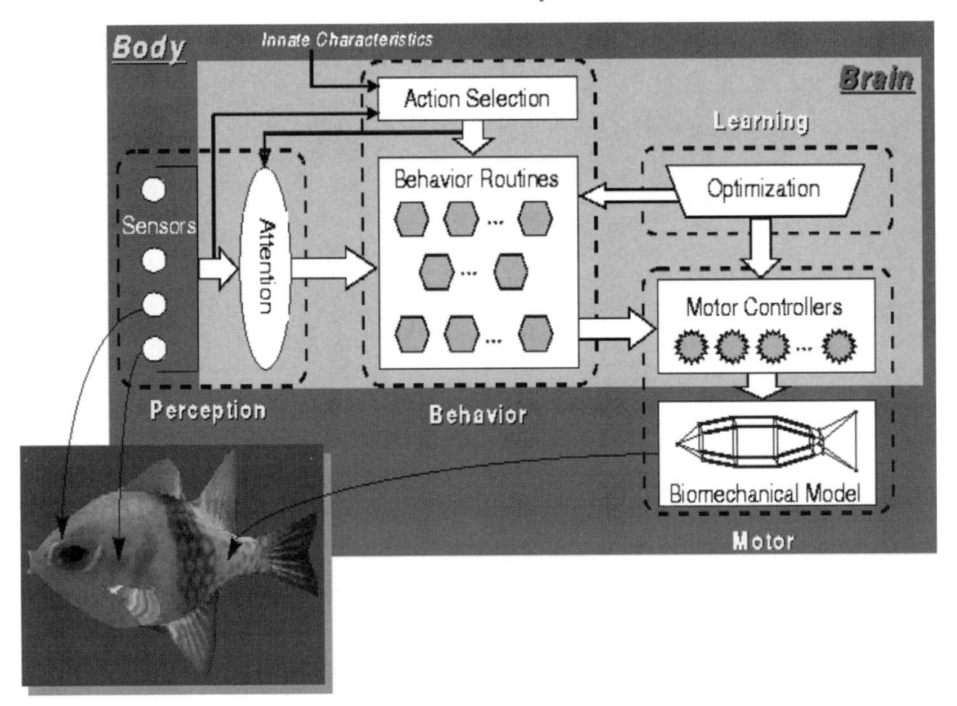

Fig. 12.2. The body of an artificial fish comprises a muscle-actuated biomechanical model, perceptual sensors, and a brain with motor, perception, behavior, and learning centers. To the lower left is an artificial fish graphical display model.

12.2.1 Body, brain, and graphical display models

As Figure 12.2 illustrates, the body model comprises a biomechanical model, eyes (among other on-board sensors), and a brain with motor, perception, behavior, and learning centers. Through controlled muscle actions, artificial fishes are able to swim in simulated water in accordance with simplified hydrodynamics. Their functional fins enable them to locomote, maintain balance, and maneuver in the water. Thus the artificial fish model captures not just 3D form and appearance, but also the basic physics of the animal and its environment. Though rudimentary compared to real animals, the minds of artificial fishes are nonetheless able to learn some basic motor functions and carry out perceptually guided motor tasks within a repertoire of piscine relevant behaviors, including collision avoidance, foraging, preying, schooling, and mating.

Animat vision requires that artificial animals capture the form and appearance of natural animals with considerable visual fidelity. To this end, using interactive image analysis tools (Terzopoulos & Rabie, 1994), color photographs of real fishes, such as the one shown in Figure 12.3(a), are converted into 3D spline (NURBS) surface body models (Figure 12.3(b)) that are mapped with RGB textures extracted from the original images to produce the 3D graphical display models of the fishes (Figure 12.3(c)).

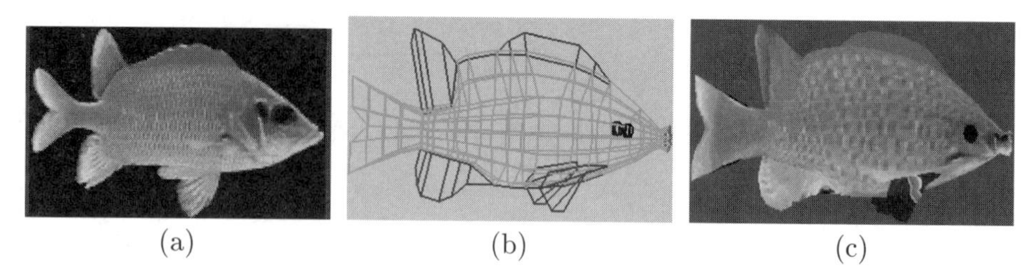

Fig. 12.3. Capturing form and appearance in the graphical display model. (a) Digitized image of a fish photo. (b) 3D NURBS surface fish body. (c) Color texture mapped 3D graphical display model of fish.

12.2.2 Motor system

The motor system (see Figure 12.2) comprises the fish biomechanical model, including muscle actuators and a set of motor controllers (`MC`s). Figure 12.4(a) illustrates the mechanical body model which produces realistic piscine locomotion using only 23 lumped masses and 91 uniaxial viscoelastic elements, 12 of which are actively contractile muscle elements. These mechanical components are interconnected so as to maintain the structural integrity of the body as it flexes due to the muscle actions.

Artificial fishes locomote like real fishes, by autonomously contracting their muscles in a coordinated fashion. As the body flexes it displaces virtual fluid which induces local reaction forces normal to the body. These hydrodynamic forces generate thrust that propels the fish forward. The model mechanics are governed by Lagrange equations of motion driven by the hydrodynamic forces. The system of coupled second-order ordinary differential equations is continually integrated through time by a numerical simulator. The artificial fish mechanical model achieves a good compromise between realism and computational efficiency.

The model is sufficiently rich to enable the design of motor controllers by gleaning information from the fish biomechanics literature. The motor controllers coordinate muscle actions to carry out specific motor functions, such as swimming forward (`swim-MC`), turning left (`left-turn-MC`), and turning right (`right-turn-MC`). They translate natural control parameters such as the forward speed or angle of the turn into detailed muscle actions that execute the function. The artificial fish is neutrally buoyant in the virtual water and has a pair of pectoral fins which enable it to navigate freely in its 3D world by pitching, rolling, and yawing its body. Additional motor controllers coordinate the fin actions.

12.2.3 Behavioral modeling

Artificial fishes gain awareness of their world through sensory perception. As Figure 12.4(b) suggests, it is necessary to model not only the abilities

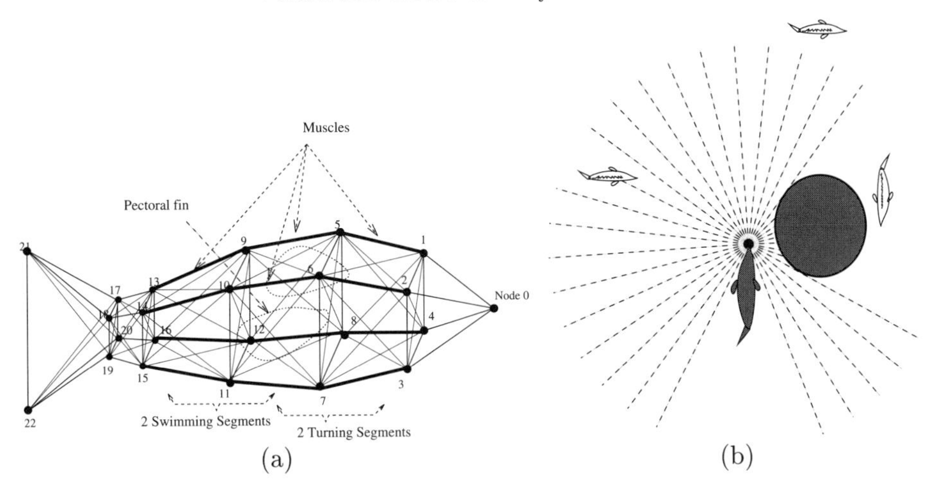

Fig. 12.4. Biomechanical fish model (a). Nodes denote lumped masses. Lines indicate springs (shown at their natural lengths). Bold lines indicate muscle springs. Artificial fishes perceive objects (b) within a limited field view if objects are close enough and not occluded by other opaque objects (only the fish towards the left is visible to the animat at the center).

but also the limitations of animal perception systems in order to achieve natural sensorimotor behaviors. Hence, the artificial fish has a limited field of view extending frontally and laterally to an effective radius consistent with visibility in the translucent water. An object may be detected only if some visible portion of it (i.e., not occluded behind some other opaque object) enters the fish's field of view (Figure 12.4(b)).

The behavior center of the artificial fish's brain mediates between its perception system and its motor system (Figure 12.2). A set of innate characteristics determines the (static) genetic legacy, dictating whether the fish is male or female, predator or prey, etc. A (dynamic) mental state comprises variables representing hunger, fear, and libido, whose values depend on sensory inputs. The fish's cognitive faculty resides in part in the action selection component of its behavior center. At each simulation time step, action selection entails combining the innate characteristics, the mental state, and the incoming stream of sensory information to generate sensible, survival sustaining goals for the fish, such as to avoid an obstacle, to avoid predators, to hunt and feed on prey, or to court a potential mate. The action selector en sures that goals have some persistence by exploiting a single-item memory. The behavior memory reduces dithering, thereby improving the robustness of prolonged behaviors such as foraging, schooling, and mating. The action selector also controls the perceptual attention mechanism. At every simulation time step, the action selector activates behavior routines that attend to sensory information and compute the appropriate motor control parameters to carry the fish a step closer to fulfilling its immediate goals. The behavioral repertoire of the artificial fish includes primitive, reflexive behavior routines, such as obstacle avoidance, as well as more sophisticated motivational behavior routines such as schooling and mating whose activation is dependent on the mental state.

12.3 Active vision system

The basic functionality of the active vision system starts with binocular perspective projection of the color 3D world onto the 2D retinas of the artificial fish. Retinal imaging is accomplished by photorealistic graphics rendering of the world from the animal's point of view. This projection respects occlusion relationships among objects. It forms spatially nonuniform visual fields with high resolution foveas and low resolution peripheries. Based on an analysis of the incoming color retinal image stream, the perception center of the artificial fish's brain supplies saccade control signals to its eyes and stabilize the visual fields during locomotion, to attend to interesting targets based on color, and to keep targets fixated. The artificial fish is thus able to approach and track other artificial fishes using sensorimotor control.

Figure 12.5 is a block diagram of the active vision system showing two main modules that control foveation of the eyes and retinal image stabilization.

12.3.1 Eyes and foveated retinal imaging

The artificial fish is capable of binocular vision and possesses an ocular motor system that controls eye movements (Goldberg *et al.*, 1991). The movements of each eye are controlled through two gaze angles (θ, ϕ) which specify the horizontal and vertical rotation of the eyeball, respectively, with respect to the head coordinate frame (when $\theta = \phi = 0°$, the eye looks straight ahead).

Each eye is implemented as four coaxial virtual cameras to approximate the spatially nonuniform, foveal/peripheral imaging capabilities typical of biological eyes. Figure 12.6(a) shows an example of the 64×64 images that are rendered (using the GL library and SGI graphics pipeline) by the four coaxial cameras of the left and right eye. The level $l = 0$ camera has the widest field of view (about 120°). The field of view decreases with increasing l. The highest resolution image at level $l = 3$ is the fovea and the other images form the visual periphery. Figure 12.6(b) shows the 512×512 binocular retinal images composited from the coaxial images at the top of the figure (the component images are expanded by factors 2^{l-3}). To reveal the retinal image structure in the figure, we have placed a white border around each magnified component image. Significant computational

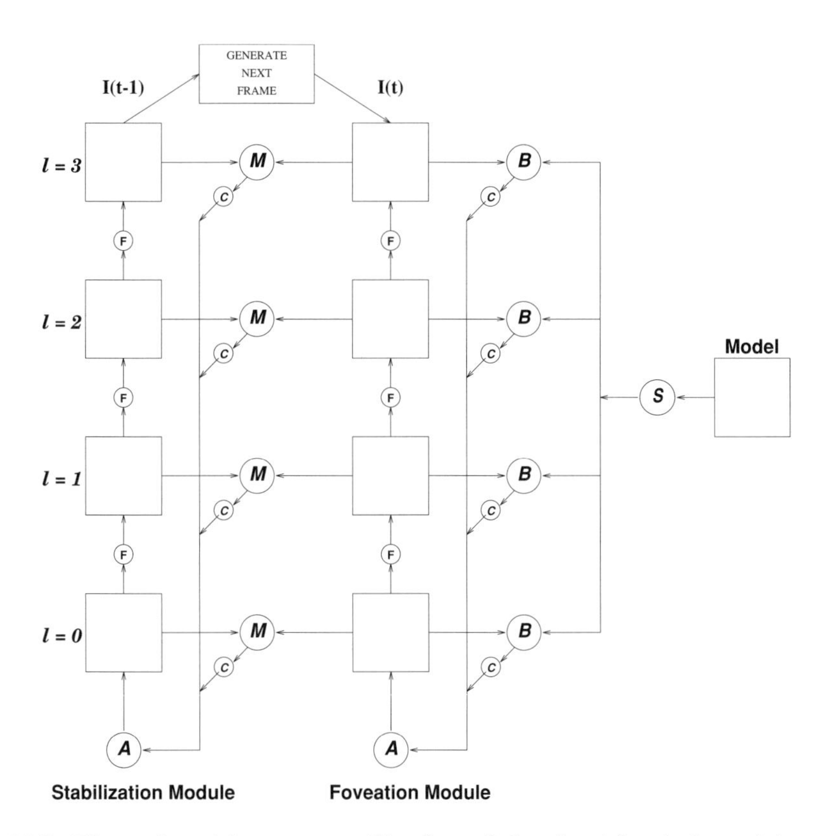

Fig. 12.5. The active vision system. The flow of the algorithm is from right to left. A: Update gaze angles (θ, ϕ) and saccade using these angles, B: Search current level for model target and if found localize it, else search lower level, C: Select level to be processed (see text), F: Reduce field of view for next level and render, M: Compute a general translational displacement vector (u, v) between images $I(t-1)$ and $I(t)$, S: Scale the color histogram of the model for use by the current level.

efficiency accrues from processing four 64×64 component images rather than a uniform 512×512 retinal image.

12.3.2 Foveation by color object detection

The brain of the fish stores a set of color models of objects that are of interest to it. For instance, if the fish is a predator, it would possess mental models of prey fish. The models are stored as a list of 64×64 RGB color images in the fish's visual memory.

To detect and localize any target that may be imaged in the low resolution periphery of its retinas, the active vision system of the fish employs an improved version of a color indexing algorithm proposed by Swain & Ballard (1991). Since each model object has a unique color histogram signature, it can be detected in the retinal image by histogram intersection and lo-

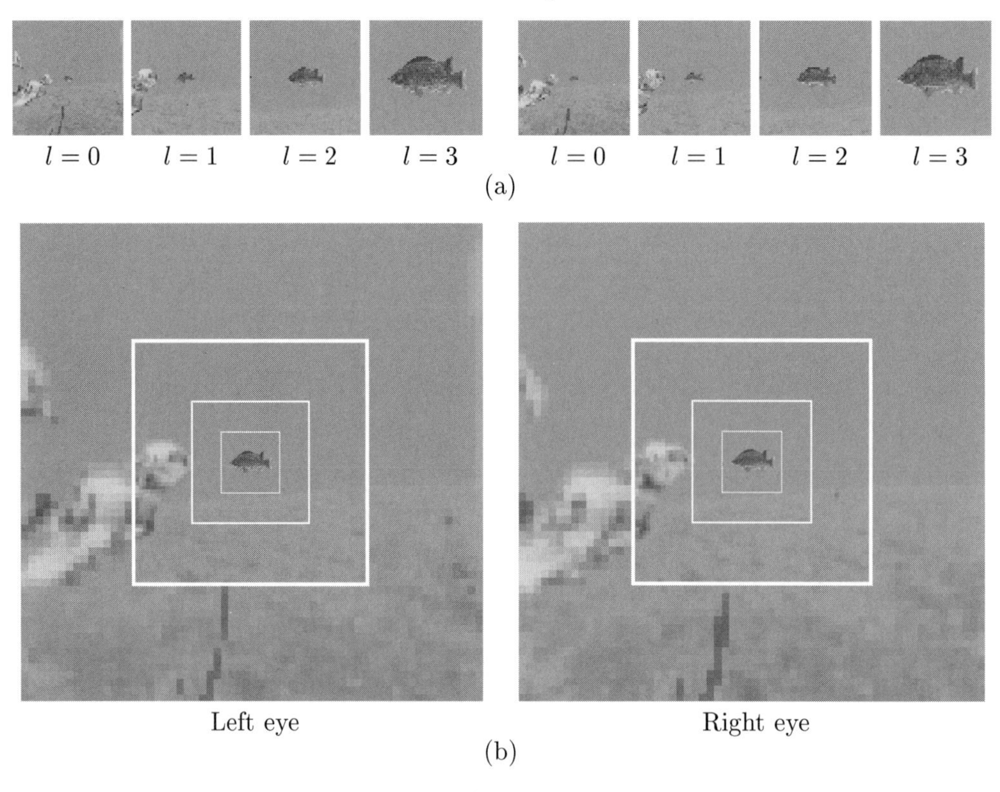

Fig. 12.6. Binocular retinal imaging (monochrome versions of original color images). (a) 4 component images; $l = 0, 1, 2$, are peripheral images; $l = 3$ is foveal image. (b) Composited retinal images (borders of composited component images are shown in white).

calized by histogram backprojection. Our algorithms are explained fully in Terzopoulos & Rabie (1997).

12.3.3 Saccadic eye movements

When a target is detected in the visual periphery, the eyes will saccade to the angular offset of the object to bring it within the fovea. With the object in the high resolution fovea, a more accurate foveation is obtained by a second pass of histogram backprojection. A second saccade typically centers the object accurately in both left and right foveas, thus achieving vergence.

Module A in Figure 12.5 performs the saccades by incrementing the gaze angles (θ, ϕ) in order to rotate the eyes to achieve the required gaze direction.

12.3.4 Visual field stabilization using optical flow

It is necessary to stabilize the visual field of the artificial fish because its body undulates as it swims. Once a target is verged in both foveas, the stabilization process (Figure 12.5) assumes the task of keeping the target foveated as the fish locomotes. Thus, it emulates the optokinetic reflex in animals.

Stabilization is achieved by computing the overall translational displacement (u, v) of light patterns between the current foveal image and that from the previous time instant, and updating the gaze angles to compensate. The displacement is computed as a translational offset in the retinotopic coordinate system by a least squares minimization of the optical flow between image frames at times t and $t-1$ (Horn, 1986).

The optical flow stabilization method is robust only for small displacements between frames. Consequently, when the displacement of the target between frames is large enough that the method is likely to produce bad estimates, the foveation module is invoked to re-detect and re-foveate the target as described earlier.

Each eye is controlled independently during foveation and stabilization of a target. Hence, the two retinal images must be correlated to keep them verged accurately on the target. Referring to Figure 12.7, the vergence angle is $\theta_V = (\theta_R - \theta_L)$ and its magnitude increases as the fish comes closer to the target. Therefore, once the eyes are verged on a target, it is straightforward for the fish vision system to estimate the range to the target by triangulation using the gaze angles.

12.4 Vision-guided navigation

The fish can use the gaze direction for the purposes of navigation in its world. In particular, it is natural to use the gaze angles as the eyes are fixated on a target to navigate towards the target. The θ angles are used to compute the left/right turn angle θ_P shown in Figure 12.7, and the ϕ angles are similarly used to compute an up/down turn angle ϕ_P. The fish's turn motor controllers (see Section 12.2) are invoked to execute a left/right turn—left-turn-MC for an above-threshold positive θ_P and right-turn-MC for negative θ_P—with $|\theta_P|$ as parameter. Up/down turn motor commands are issued to the fish's pectoral fins, with an above-threshold positive ϕ_P interpreted as "up" and negative as "down".

The problem of pursuing a moving target that has been fixated in the foveas of the fish's eyes is simplified by the gaze control mechanism described above. The fish can robustly track a target in its fovea and locomote to follow it around the environment by using the turn angles (θ_P, ϕ_P) computed from the gaze angles that are continuously updated by the foveation/stabilization algorithms.

We have carried out numerous experiments in which the moving target is a reddish prey fish whose color histogram model is stored in the memory of a predator fish equipped with the active vision system. Figure 12.8 shows plots of the gaze angles and the turn angles generated over the course of

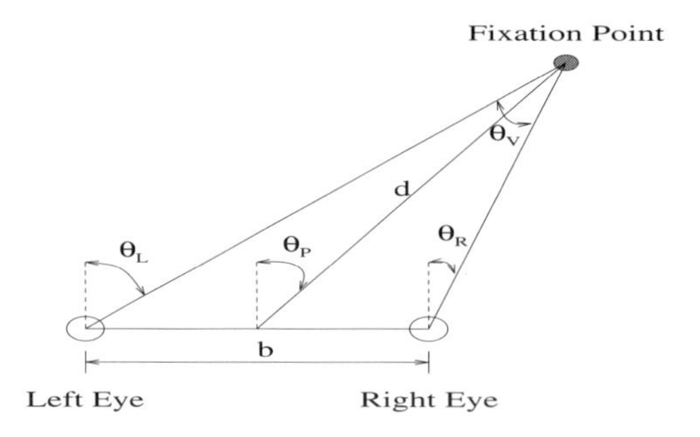

Fig. 12.7. Gaze angles and range to target geometry.

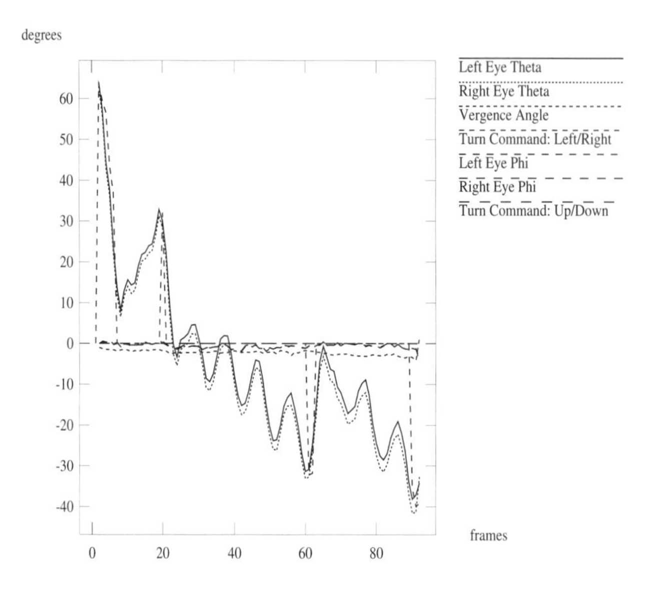

Fig. 12.8. Gaze angles generated while pursuing a moving target fish.

100 frames in a typical experiment as the predator is fixated upon and actively pursuing a prey target. Figure 12.9 shows a sequence of image frames acquired by the fish during its navigation (monochrome versions of only the left retinal images are shown). Frame 0 shows the target visible in the low resolution periphery of the fish's eyes (middle right). Frame 1 shows the view after the target has been detected and the eyes have performed a saccade to foveate the target (the scale difference of the target after foveation is due to perspective distortion). The subsequent frames show the target remaining fixated in the fovea despite the side-to-side motion of the fish's body as it swims towards the target.

The saccade signals that keep the predator's eyes fixated on its prey as both are swimming are reflected by the undulatory responses of the gaze angles in Figure 12.8. The figure also shows that the vergence angle increases

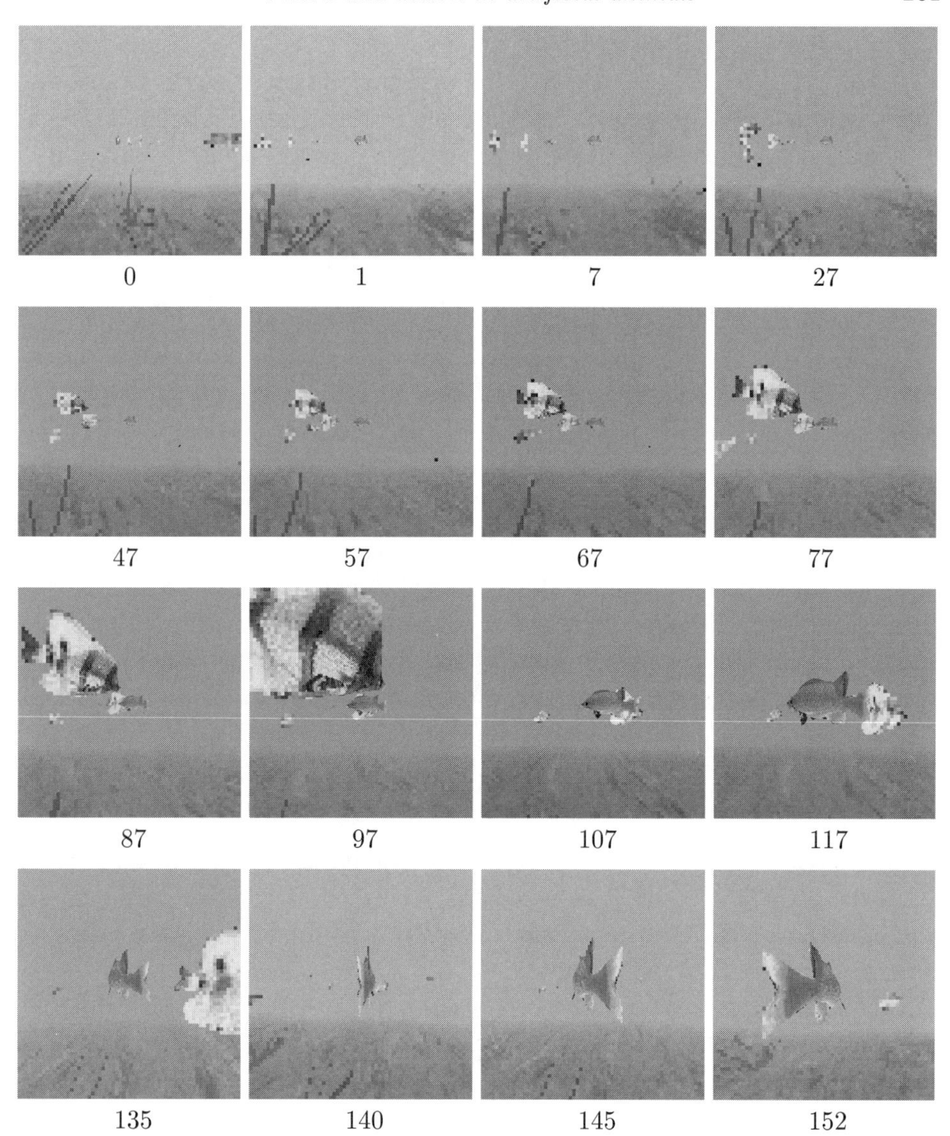

Fig. 12.9. Retinal image sequence from the left eye of the active vision fish as it detects and foveates on a reddish fish target and swims in pursuit of the target (monochrome versions of original color images). The target appears in the periphery (middle right) in frame 0 and is foveated in frame 1. The target remains fixated in the center of the fovea as the fish uses the gaze direction to swim towards it (frames 7–117). The target fish turns and swims away with the observer fish in visually guided pursuit (frames 135–152).

as the predator approaches its target (near frame 100). In comparison to the θ angles, the ϕ angles show little variation, because the fish does not undulate vertically very much as it swims forward. It is apparent from the graphs that the gaze directions of the two eyes are well correlated.

Note that in frames 87–117 of Figure 12.9, a yellow fish whose size is similar to the target fish passes behind the target. In this experiment the predator was programmed to be totally disinterested in and not bother to foveate any non-reddish objects. Because of the color difference, the yellowish object does not distract the fish's gaze from its reddish target. This demonstrates the robustness of the color-based fixation algorithm.

12.5 Perception-based learning of action

The learning center of its brain (see Figure 12.2) enables the artificial fish to acquire effective locomotion skills through practice and sensory reinforcement. Our second challenge has been to enhance the algorithms comprising the artificial fish's learning center so that it can learn more complex motor skills than those we demonstrated in reference (Terzopoulos *et al.*, 1994).

12.5.1 Low-level motor learning

Recall that some of the deformable elements in the biomechanical model (Figure 12.4(a)) play the role of contractile *muscles* whose natural length decreases under the autonomous control of the motor center of the artificial animal's brain. To dynamically contract a muscle, the brain must supply an *activation function* $a(t)$ to the muscle. This continuous time function has range $[0, 1]$, with 0 corresponding to a fully relaxed muscle and 1 to a fully contracted muscle. Typically, individual muscles form muscle groups, called *actuators*, that are activated in unison. Referring to Figure 12.4(a), the artificial fish has 12 muscles which are grouped pairwise in each segment to form 3 left actuators and 3 right actuators. Each actuator i is activated by a scalar *actuation function* $u_i(t)$, whose range is again normalized to $[0, 1]$, thus translating straightforwardly into activation functions for each muscle in the actuator. Thus, to control the fish's body we must specify the actuation functions $\vec{u}(t) = [u_1(t), \ldots, u_i(t), \ldots, u_N(t)]'$, where $N = 6$. The continuous vector-valued function of time $\vec{u}(t)$ is called the *controller* and its job is to produce locomotion. Learned controllers may be stored within the artificial animal's motor control center.

A continuous *objective functional* E provides a quantitative measure of the progress of the locomotion learning process. The functional is the weighted sum of a term E_u that evaluates the controller $\vec{u}(t)$ and a term E_v that evaluates the motion $\vec{v}(t)$ that the controller produces in a time interval

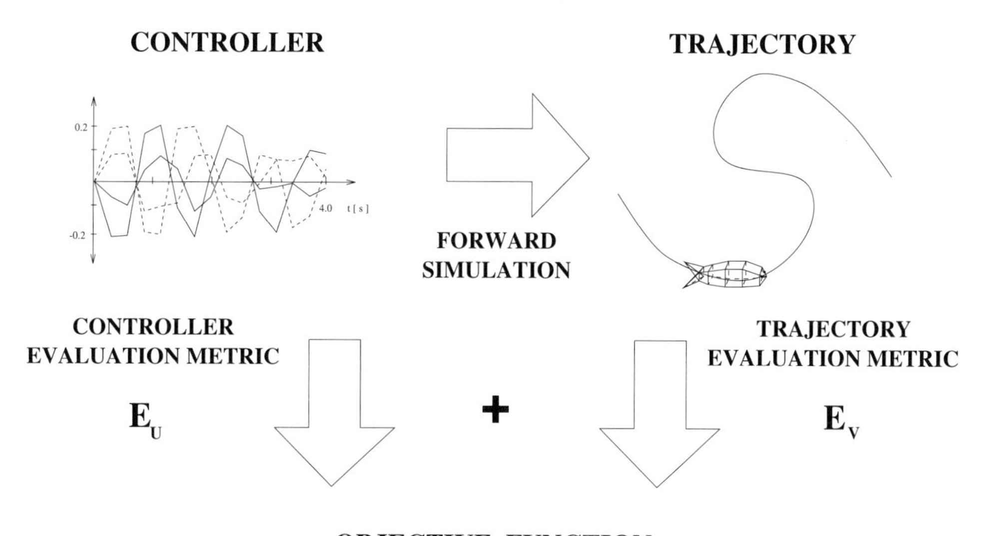

Fig. 12.10. The objective function that guides the learning process is a weighted sum of terms that evaluate the controller and the trajectory.

$t_0 \leq t \leq t_1$, with smaller values of E indicating better controllers $\vec{u}$. Mathematically,

$$E(\vec{u}(t)) = \int_{t_0}^{t_1} (\mu_1 E_u(\vec{u}(t)) + \mu_2 E_v(\vec{v}(t)))\, dt, \tag{12.1}$$

where μ_1 and μ_2 are scalar weights. Figure 12.10 illustrates this schematically.

It is important to note that the complexity of our models precludes the closed-form evaluation of E. As Figure 12.10 indicates, to compute E, the artificial animal must first invoke a controller $\vec{u}(t)$ to produce a motion $\vec{v}(t)$ with its body (in order to evaluate term E_v). This is done through forward simulation of the biomechanical model over the time interval $t_0 \leq t \leq t_1$ with controller $\vec{u}(t)$.

We may want to promote a preference for controllers with certain qualities via the controller evaluation term E_u. For example, we can guide the optimization of E by discouraging large, rapid fluctuations of $\vec{u}$, since chaotic actuations are usually energetically inefficient. We encourage lower amplitude, smoother controllers through the function $E_u = (\nu_1 |d\vec{u}/dt|^2 + \nu_2 |d^2\vec{u}/dt^2|^2)/2$, where the weighting factors ν_1 and ν_2 penalize actuation amplitudes and actuation variation, respectively. The distinction between good and bad controllers also depends on the goals that the animal must accomplish. In our learning experiments we used trajectory criteria E_v such as the final distance to the goal, the deviation from a desired speed, etc. These and other criteria will be discussed shortly in conjunction with specific experiments.

The low level motor learning problem optimizes the objective functional (12.1). This cannot be done analytically. We convert the continuous optimization problem to an algebraic parameter optimization problem (Goh & Teo, 1988) by parameterizing the controller through discretization using basis functions. Mathematically, we express $u_i(t) = \sum_{j=1}^{M} u_i^j B^j(t)$, where the u_i^j are scalar parameters and the $B^j(t)$, $1 \leq j \leq M$ are (vector-valued) temporal basis functions. The simplest case is when the u_i^j are evenly distributed in the time interval and the $B^j(t)$ are tent functions centered on the nodes with support extending to nearest neighbor nodes, so that $\vec{u}(t)$ is the linear interpolation of the nodal variables.

Since $\vec{u}(t)$ has N basis functions, the discretized controller is represented using NM parameters. Substituting the above equation into the continuous objective functional (12.1), we approximate it by the discrete *objective function* $E([u_1^1, \ldots, u_N^M]')$. Learning low level motor control amounts to using an optimization algorithm to iteratively update the parameters so as to optimize the discrete objective function and produce increasingly better locomotion.

We use the simulated annealing method to optimize the objective function (Press *et al.*, 1992). Simulated annealing has three features that make it particularly suitable for our application. First, it is applicable to problems with a large number of variables yielding search spaces large enough to make exhaustive search prohibitive. Second, it does not require gradient information about the objective function. Analytic gradients are not directly attainable in our situation since evaluating E requires a forward dynamic simulation. Third, it avoids getting trapped in local suboptima of E. In fact, given a sufficiently slow annealing schedule, it will find a global optimum of the objective functional. Robustness against local suboptima can be important in obtaining muscle control functions that produce realistic motion.

In summary, the motor learning algorithms discover muscle controllers that produce efficient locomotion through optimization. Muscle contractions that produce forward movements are "remembered". These partial successes then form the basis for the fish's subsequent improvement in its swimming technique. Their brain's learning center also enable these artificial animals to train themselves to accomplish higher level sensorimotor tasks, such as maneuvering to reach a visible target (see Terzopoulos & Rabie, 1994 for the details).

12.5.2 Learning complex motor skills

12.5.2.1 Abstracting controllers

It is time consuming to learn a good solution for a low level controller because of the high dimensionality of the problem (large NM), the lack of gradient information to accelerate the optimization of the objective functional, and the presence of suboptimal traps that must be avoided. For tractability, the learning procedure must be able to abstract compact higher level controllers from the low level controllers that have been learned, retain the abstracted controllers, and apply them to future locomotion tasks.

The process of abstraction takes the form of a dimensionality reducing change of representation. More specifically, it seeks to compress the many parameters of the discrete controllers to a compact form in terms of a handful of basis functions. Natural, steady-state locomotion patterns tend to be quasi-periodic and they can be abstracted very effectively without substantial loss. A natural approach to abstracting low-level motor controllers is to apply the fast Fourier transform (FFT) (Press *et al.*, 1992) to the parameters of the controller and then suppress the below-threshold amplitudes.

Typically, our artificial animals are put through a "basic training" regimen of primitive motor tasks that it must learn, such as locomoting at different speeds and executing turns of different radii. They learn effective low level controllers for each task and retain compact representations of these controllers through controller abstraction. The animals subsequently put the abstractions that they have learned into practice to accomplish higher level tasks, such as target tracking or leaping through the air. To this end, abstracted controllers are concatenated in sequence, with each controller slightly overlapping the next. To eliminate discontinuities, temporally adjacent controllers are smoothly blended together by linearly fading and summing them over a small, fixed region of overlap, approximately 5% of each controller (Figure 12.11).

12.5.2.2 Composing macro controllers

Next the learning process discovers composite abstracted controllers that can accomplish complex locomotion tasks. Consider the spectacular stunts performed by marine mammals that elicit applause at theme parks like "Sea-World". We can treat a leap through the air as a complex task that can be achieved using simpler tasks; e.g., diving deep beneath a suitable leap point, surfacing vigorously to gain momentum, maintaining balance during the ballistic flight through the air, and splashing down dramatically with a belly flop.

We have developed an automatic learning technique that constructs a macro jump controller of this sort as an optimized sequence of basic ab-

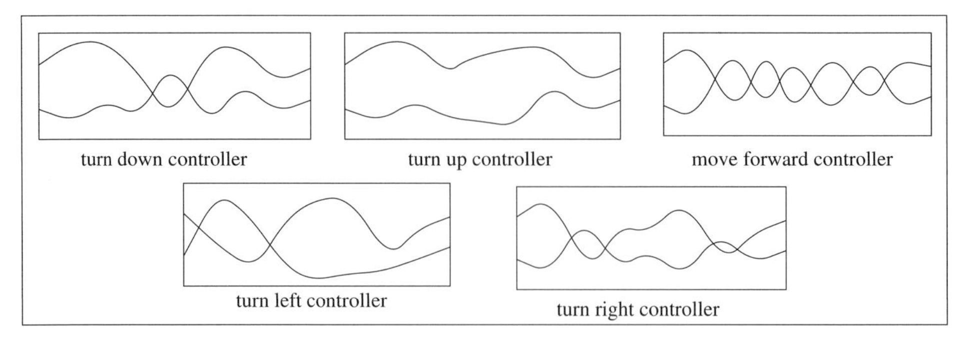

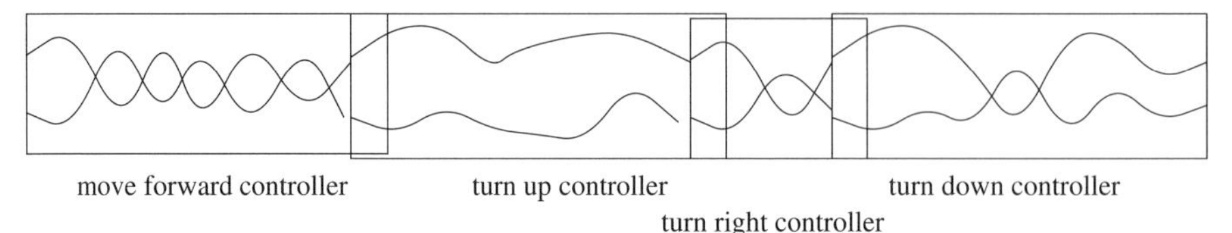

Fig. 12.11. Higher level controller for jumping out of water is constructed from a set of abstracted basic controllers.

stracted controllers. The optimization process is, in principle, similar to the one in low level learning. It uses simulated annealing for optimization, but rather than optimizing over nodal parameters or frequency parameters, it optimizes over the selection, ordering, and duration of abstracted controllers. Thus the artificial animal applying this method learns effective macro controllers of the type shown at the bottom of Figure 12.11 by optimizing over a learned repertoire of basic abstracted controllers illustrated at the top of the figure.

We have trained an artificial dolphin to learn effective controllers for 5 basic motor tasks: turn-down, turn-up, turn-left, turn-right, and move-forward. We then give it the task of performing a stunt like the one described above and the dolphin discovers a combination of controllers that accomplishes the stunt. In particular, it discovers that it must build up momentum by thrusting from deep in the virtual pool of water up towards the surface and it must exploit this momentum to leap out of the water. Figure 12.12(a) shows a frame as the dolphin exits the water. The dolphin can also learn to perform tricks while in the air. Figure 12.12(b) shows it using its nose to bounce a large beach-ball off a support. The dolphin can learn to control the angular momentum of its body while exiting the water and while in ballistic flight so that it can perform aerial spins and somersaults. Figure 12.12(c) shows it in the midst of a somersault in which it has just bounced the ball with its tail instead of its nose. Figure 12.12(d) shows

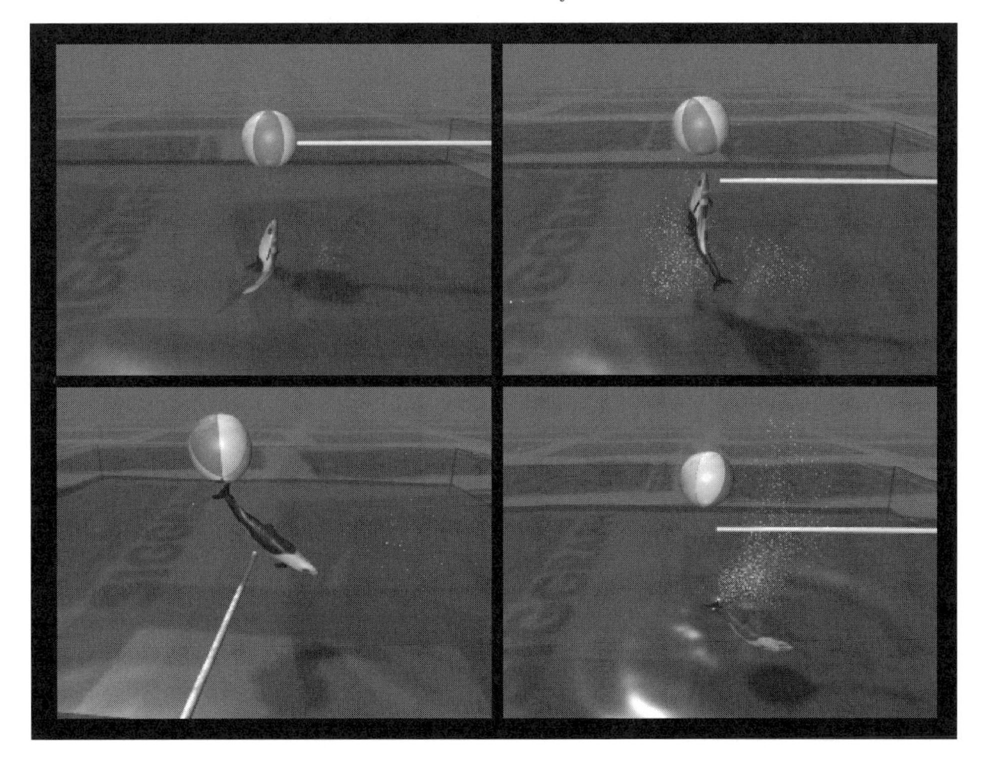

Fig. 12.12. "SeaWorld" skills learned by an artificial dolphin.

the dolphin right after splashdown. In this instance it has made a dramatic bellyflop splash.

12.6 Conclusion

We have demonstrated a new approach to the computational study of vision and action that draws upon recent advances in the fields of artificial life and computer graphics. The approach prescribes artificial animals situated in physics-based virtual worlds as autonomous virtual robots with active perception systems.

In particular, we have described life-like artificial fishes inhabiting a physics-based, virtual marine world. For purposeful, visually-guided navigation in its world, the artificial fish incorporates a sensorimotor control loop that directs eye movements and coordinates the actions of its muscle-actuated body, thus tying vision to motor control. The active vision system implements foveated retinal imaging, computes saccade signals to stabilize the spatially nonuniform visual fields, and directs the gaze at interesting objects based on color analysis and recognition. This enables artificial fishes to pursue moving colored targets, such as other artificial fishes. We have also illustrated sensorimotor learning, enabling artificial animals to acquire basic locomotion plus some more complex motor skills through practice and perceptual reinforcement. In particular, our learning algorithms enable an artificial dolphin to acquire a repertoire of motor skills supporting basic aquatic locomotion and more complex aerial leaps.

The use of realistic, physics-based virtual worlds inhabited by biomimetic autonomous agents appears to be a fruitful strategy for exploring biological information processing and control. Artificial animals can also be useful for developing, implementing, and evaluating computational models that profess sensorimotor competence for biological or robotic agents.

Acknowledgments

I thank my students Xiaoyuan Tu, Radek Grzeszczuk, and Tamer Rabie for their outstanding contributions to the research reviewed herein. The research has been made possible in part by grants from the Natural Sciences and Engineering Research Council of Canada.

References

Bajcsy, R. (1988). "Active perception", *Proceedings of the IEEE*, 76: 996–1005.

Ballard, D. H. (1991). "Animate vision", *Artificial Intelligence*, 48: 57–86.

Ballard, D. H. & Brown, C. M. (1992). "Principles of animate vision", *CVGIP: Image Understanding*, 56: 3–21.

Blake, A. & Yuille, A. (Editors) (1992). *Active Vision.* MIT Press, Cambridge, MA.

Braitenberg, V. (1984). *Vehicles, Experiments in Synthetic Psychology.* MIT Press, Cambridge, MA.

Fernald, R. D. (1993). "Vision", in *The Physiology of Fishes*, D. H. Evan (Ed.), 161–189, CRC Press, Boca Raton, FL.

Gibson, J. J. (1979). *The Ecological Approach to Visual Perception*, Houghton Mifflin, Boston, MA.

Goh, C. J. & Teo, K. L. (1988). "Control parameterization: A unified approach to optimal control problems with general constraints", *Automatica*, 24: 3–18.

Goldberg, M. E., Eggers, H. M., & Gouras, P. (1991). "The ocular motor system", In *Principles of Neural Science*, E. R. Kandel, J. H. Schwartz, & T. M. Jessel (Eds.), 660–678. Elsevier, New York, NY.

Horn, B. K. P. (1986). *Robot Vision*, MIT Press, Cambridge, MA.

Press, W. H., Flannery, B. P., Teukolsky, S. A., & Vetterling, W. T. (1992). *Numerical Recipes: The Art of Scientific Computing, Second Edition.* Cambridge University Press.

Rabie, T. F. & Terzopoulos, D. (1996). "Motion and color analysis for animat perception", In *Proceedings of the 13th National Conference on Artificial Intelligence*, 1090–1097, Portland, OR.

Swain, M. & Ballard, D. (1991). "Color indexing" *Int. J. Comp. Vis.*, 7: 11–32.

Swain, M. J. & Stricker, M. A. (1993). "Promising directions in active vision", *Int. J. Comp. Vis.*, 11: 109–126.

Terzopoulos, D. & Rabie, T. F. (1997). "Animat vision: Active vision in artificial animals", *Videre: J. Comp. Vis. Res.*, 1: 2–19, 1997. Earlier version in *Proc. Fifth International Conference on Computer Vision*, 801–808, Cambridge, MA, 1995. IEEE Computer Society Press.

Terzopoulos, D., Tu, X. & Grzeszczuk, R. (1994). "Artificial fishes: Autonomous locomotion, perception, behavior, and learning in a simulated physical world", *Artificial Life*, 1: 327–351.

Tu, X. (1996). *Artificial animals for computer animation*, PhD Thesis. Department of Computer Science, University of Toronto. Toronto, Canada.

Wilson, S. W. (1991). "The animat path to AI", in *From Animals to Animats*, J.-A. Meyer & S. Wilson (Eds.), 15–21. MIT Press, Cambridge, MA.

13

When vision is not sight: dissociations between perception and action in human vision

Melvyn A. Goodale and Angela Haffenden

13.1 Introduction

Think about the following scenario for a moment. You and your partner are cleaning up after a big dinner party. You pick up the dirty plates and glasses and carry them through to the kitchen and place them one by one in the dishwasher. Your partner, who is busy washing the pots and pans, throws you a tea towel to dry the pots that are already piling up in the dish rack. Throughout all of this activity, the two of you are engaged in a rather heated discussion about an issue that was raised at dinner. You scarcely notice the fact that you are reaching out and picking up plates and glasses with skill and economy of action – or that you catch the tea towel with one hand and pick up a pot with the other. All of this happens quite "automatically" while you continue to talk and listen to your partner. Of course, you could attend to the task at hand – and perhaps you would be able to perform some of the required actions even more efficiently. But – and this is the main thesis of this chapter – you would never really experience the actual visual information that is being used to program and control the movements of your fingers, hand, and limb as you reach out and pick up a plate or catch the tea towel flung towards you. It is not simply a question of attention; no matter how much you attend to the plate, you will never perceive what it is about the plate that determines the posture and trajectory of your hand and fingers as you reach out and lift it from the table. This does not mean that you are unaware of the size and shape of the plate or its position on the table, but rather that the computations that use various object parameters to generate the appropriate movements of your hand and fingers are quite different from those that generate your conscious perception of the plate. Indeed, as we shall see in this chapter, the main reason you remain unaware of the visual information controlling the movements of your hand is that the visuomotor systems mediating this control operate quite independently from those that allow us to perceive the objects and events that make up our experience of the visual world*.

The assertion that the visual control of skilled actions and the visual perception of objects depend on different mechanisms might appear counter-intuitive. After all, it seems self-evident that the actions we direct at objects are controlled by the perceptions that we have of those same objects. Evidence from neurological patients tells us otherwise however. Consider the case of DF, a young woman who developed a profound visual form agnosia following carbon monoxide-induced anoxia (Milner *et al.*, 1991). Even though DF's "low-level" visual abilities are reasonably intact, she can no longer recognize common objects on the basis of their form or even the faces of her friends and relatives; nor can she identify even the simplest of geometric shapes. If an object is placed in her hand she has no trouble identifying it by touch, and she can recognize people from their voices. Remarkably, however, DF shows strikingly accurate guidance of her hand and finger movements when she attempts to pick up the very objects she cannot identify. Thus, when she reaches out to grasp objects of different sizes, her hand opens wider mid-flight for larger objects than it does for smaller ones, just as it does in people with normal vision (Goodale *et al.*, 1991). Similarly, she rotates her hand and wrist quite normally when she reaches out to grasp objects in different orientations, and she places her fingers correctly on the boundaries of objects of different shapes (Goodale *et al.*, 1991; Goodale *et al.*, 1994). At the same time, she is quite unable to describe or distinguish between any of these objects when they are presented to her in discrimination tests. In other words, DF's visual system is no longer able to deliver any perceptual information about the size, shape, and orientation of objects in the world. Yet at the same time, the visuomotor systems in DF's brain that control the programming and execution of visually guided actions remain sensitive to these same object features. Although the damage in DF's brain is quite diffuse, the ventrolateral regions of her occipital lobe are particularly compromised; primary visual cortex, however, appears to be largely spared.

There is evidence that patients with damage to other visual areas in the cerebral cortex, such as the superior regions of the posterior parietal cortex, show a pattern of visual behaviour that is essentially the mirror image of that of DF. Such patients cannot use visual information to rotate their hand or scale the opening of their fingers when reaching out to pick up an object, even though they have no difficulty describing the size or orientation of objects in that part of the visual field (Goodale *et al.*, 1994; Jakobson *et al.*, 1991; Perenin & Vighetto, 1988).

These neurological dissociations suggest that the human visual system

* This chapter is largely drawn from material that appeared in an earlier article by Goodale & Haffenden (1998).

does not construct a single representation of the world for both visual perception and the visual control of action. Instead, *vision for perception* and *vision for action* appear to depend on separate neural mechanisms that can be differentially affected by neurological damage. Goodale & Milner (1992) have proposed that this distinction can be mapped onto the two prominent pathways or "streams" of visual projections that have been identified in the cerebral cortex of the monkey: a ventral stream, which arises from primary visual cortex and projects to inferotemporal cortex, and a dorsal stream, which also arises from primary visual cortex but projects instead to the posterior parietal cortex (Ungerleider & Mishkin, 1982). Additional support for this proposal comes from electrophysiological and behavioural studies in the monkey (for review, see Milner & Goodale, 1993, 1995). Although some caution must be exercised in generalizing from monkey to human (Crick & Jones, 1993), it seems likely that the visual projections from primary visual cortex to the temporal and parietal lobes in the human brain may involve a separation into ventral and dorsal streams similar to that seen in the monkey.

Ungerleider and Mishkin (1982) originally proposed that the ventral stream plays a special role in the identification of objects, whereas the dorsal stream is responsible for localizing objects in visual space. Goodale and Milner's (1992) re-interpretation of this story places less emphasis on the differences in the visual information that is received by the two streams (object features versus spatial location) than it does on the differences in the transformations that the streams perform upon that information (Goodale, 1993; Goodale & Milner, 1992; Milner & Goodale, 1993, 1995). According to their account, both streams process information about object features and about their spatial locations, but each stream uses this visual information in different ways. In the ventral stream, the transformations deliver the enduring characteristics of objects and their relations, permitting the formation of long-term perceptual representations. Such representations play an essential role in the identification of objects and enable us to classify objects and events, attach meaning and significance to them, and establish their causal relations. Such operations are essential for accumulating a knowledge-base about the world. In contrast, the transformations carried out by the dorsal stream deal with moment-to-moment information about the location and disposition of objects in egocentric coordinates and thereby mediate the visual control of skilled actions, such as manual prehension, directed at those objects. In some ways then, the dorsal stream can be regarded as a cortical extension of the subcortical visual structures, such as the superior colliculus, that mediate visually guided movements in all vertebrates. Of course, the two systems work together in controlling the rich stream of behaviour that we produce as we live our complex lives. Their respective roles in this control differ however. The perceptual representations constructed by the ventral stream are part of a high-level cognitive network that enables an organism to select a particular course of action with respect to objects in the world; the visuomotor networks in the dorsal stream (and associated cortical and subcortical pathways) are responsible for the programming and on-line control of the particular movements that the selected action entails.

This division of labour in visual processing requires that different transformations be carried out on incoming visual information by the two streams. Consider first the task of the perceptual mechanisms in the ventral stream. To generate long-term representations of objects and their relations, perceptual mechanisms must be 'object-based' so that constancies of size, shape, color, lightness, and relative location can be maintained across different viewing conditions. Some of these mechanisms might use an array of viewer-centered representations of the same object (e.g., Bülthoff & Edelman, 1992); others might use a set of canonical representations (e.g., Palmer, Rosch, & Chase, 1981); still others might generate representations that are truly "object-centered" (Marr, 1982). But whatever the particular coding might be, it is the identity of the object, not its disposition with respect to the observer that is of primary concern to the perceptual system. This is not the case for the visuomotor mechanisms in the dorsal stream, and other related structures, that support actions directed at that object. Here the underlying visuomotor transformations must be viewer-centered; in other words, both the location of the object and its disposition and motion must be encoded relative to the observer in egocentric coordinates, that is in retinocentric, head-centered, torso-centered, or shoulder-centered coordinates. Some object-based computations, such as those related to size, must be carried out, but even here the computations must reflect nature of the effector system to be used. Finally, because the position and disposition of a goal object in the action space of an observer is rarely constant, such computations must be carried out on each occasion an action is performed (for a discussion of this issue, see Goodale, Jakobson, and Keillor, 1994). To use a computer metaphor, the action systems of the dorsal stream do most of their work *on-line*; only the perception systems of the ventral stream can afford to work *off-line*. To summarize then, while similar (but not identical) visual information about object shape, size, local orientation, and location is available to both systems, the transformational algorithms that are applied to these inputs are uniquely tailored to the function of each system. According to the Goodale and Milner (1992) proposal, it is the nature of the functional requirements of perception and action that lies at the root of the division of labour in the ventral and dorsal visual projection systems of the primate cerebral cortex.

The differences in the visual abilities of the neurological patients described earlier are certainly consistent with these ideas. Thus, DF who has relatively intact visuomotor abilities despite her compromised visual perception has a pattern of brain damage that suggests that her ventral stream is not functioning properly. In contrast, patients with posterior parietal damage, which would interfere with dorsal stream function, show normal visuoperceptual performance but poor visuomotor control. But if visual perception and the visual control of action depend on different neural mechanisms in the human cerebral cortex, then it should be possible to demonstrate a dissociation between these two kinds of visual processing even in neurologically-intact individuals with the right kind of task. In other words, the visual information underlying the calibration and control of a skilled motor act directed at an object might not always match the perceptual judgements made about that object. In this chapter, we will explore the evidence for this kind of dissociation in normal subjects and discuss some of the ecological reasons why such dissociations might arise.

13.2 Responding to unseen visual events

If two visual stimuli of unequal intensity occur at the same time, the more intense stimulus often "masks" the less intense one. In other words, the less intense stimulus is more difficult to detect than when presented alone. This masking effect can be observed even if the more intense stimulus precedes ("forward masking") or follows ("backward-masking") the less intense stimulus –provided the interval between the two is short enough (see Fox, 1978). Although the degree of masking varies inversely with the interstimulus interval, some backward-masking has been reported with intervals as long as 100 ms.

But even though the perception of a visual event might be masked by such procedures, a variety of voluntary motor outputs can still be driven by the very stimulus that subjects fail to detect. For example, Fehrer and Biederman (1962) found that detection of a test stimulus (a 5-ms blanking of a steady dim light) could be masked by a bright flash presented some 50 ms later. Nevertheless, subjects continued to show a voluntary motor response to the test stimulus; moreover, their reaction times were little different than those obtained on trials in which the test stimulus was presented by itself. This result has been replicated in a number of other studies that used a somewhat different interference paradigm, metacontrast, in which the contours of the test stimulus are adjacent to those of the mask (Fehrer & Raab, 1962; Harrison & Fox, 1966; Schiller & Smith, 1966). In all these studies, subjects initiated a motor response to a visual event that they did not perceive.

Motor responses to unseen stimuli can be relatively complex. In one experiment, Taylor and McCloskey (1990) required subjects to initiate simultaneous or sequential movements of the two hands when a small light was turned on. The muscle groups required to make the actions in each hand were quite different and subjects had to pay close attention to the instructions. Despite the complexity of the movements, subjects continued to initiate them in response to the small light even when it was presented in a backward masking paradigm that left it unavailable to perceptual report. In other words, subjects could generate relatively complex motor acts in response to unseen visual stimuli.

Although the mechanisms of backward masking are not well understood (for review, see Breitmeyer, 1984), the results of the experiments just described suggest that the neural processing giving rise to the perception of a visual event may not be complete before a motor act triggered by that event has been initiated. Thus, the subsequent presentation of a masking stimulus may interfere only with the perception of the earlier visual stimulus and not with the programming of the movements constituting the motor act. Libet (1966, 1981) has argued that conscious perception of a sensory stimulus might require as much as 500 ms of processing time – far longer than the typical reaction time of most motor responses to that same sensory stimulus. But it may not be simply a question of length of processing time. As was suggested earlier, it could be the case visual perception and visuomotor control are mediated by independent visual pathways – the ventral and dorsal stream – with different information transmission times and different processing requirements.

Even when a visual event *is* perceived, there is evidence that the time required to deliver that perception is quite a bit longer than the time required for a motor reaction to the same stimulus. A hint of this difference in processing time was observed in a study by Paulignan, McKenzie, Marteniuk, and Jeannerod (1990). Subjects in this experiment reported that they did not notice a sudden and unpredictable displacement in the position of an object which they were attempting to pick up until their hand, whose trajectory they had already adjusted, was nearly at the target. This temporal dissociation between the motor adjustment and perceptual report was studied more formally by Castiello, Paulignan, and Jeannerod (1991). In this study, subjects were asked to indicate (using a vocal response) when they perceived the displacement in the position of the object they were attempting to grasp. On displacement trials, in which the object was displaced at the onset of the grasping movement, the vocal response was emitted 420 ms after the onset of the movement. In contrast, adjustments to the trajectory of the grasping movement could be seen in the kinematic records as little as 100 ms after movement onset, i.e. more than 300 ms earlier than the vocal response. The authors concluded that the delay between motor correction and the vocal response reflects the fact that perceptual awareness of a visual stimulus takes far longer to achieve than adjustments to the ongoing motor activity provoked by the visual stimulus. Processing time is likely to be a much more critical constraint for visuomotor networks than for the perceptual networks supporting conscious awareness. To catch a rapidly moving insect or to block the blow of a competitor, for example, an animal would have to be able to make extremely rapid adjustments in the trajectory of its limb movements in response to sudden and unpredictable changes in the position of the target. Such temporal demands would not be placed on the perception of these changes; indeed, in many instances, the changes might not have to be perceived at all – provided the requisite adjustments were made in the animal's motor responses.

Although objects, particularly animate ones, often change their location as we reach out to pick them up, they rarely (if ever) change size. Thus, we might expect that, if an object were suddenly to shrink or expand in size as we attempted to pick it up, any adjustments that we made to grasp would be quite sluggish. In fact, Castiello & Jeannerod (1991) found exactly that in an experiment in which a dowel that subjects were attempting to grasp suddenly and unpredictably increased or decreased in width as they began their grasping movement. In this situation, subjects took nearly 300ms to make adjustments in the size of their grasp – a response time that was much longer than the 100 ms needed respond to a shift in the location of the goal object. It is possible, as Castiello and Jeannerod suggest, that the slower responses to perturbations in object size arise because the visuomotor pathways mediating grasping are more complex than those mediating movements of the limb. It is also possible that there is little need for the ability to make rapid adjustments to abrupt changes in object size because such changes are quite uncommon in the real world. It is likely, however, that the ability to make rapid adjustments to sudden changes in the location of objects has been highly selected in primate evolution. Of course, there are situations where grip aperture and distal components of the grasp might also have to be adjusted on-line; a sudden change in the orientation of an object, for example, could require an adjustment in the posture of the hand and fingers. In any case, even though the responses to changes in object size were much slower than those to changes in object location, they were still significantly faster than the perceptual reports of the change in size. Again, subjects appeared to be adjusting their motor response to a visual event before they actually perceived it.

13.3 Different spatial coding for perception and action

The dissociation between motor adjustment and perceptual report has also been demonstrated in experiments in which the position of a target is moved unpredictably during a saccadic eye movement (Bridgeman *et al.*, 1979; Bridgeman, Kirch & Sperling, 1981; Goodale, Pélisson, & Prablanc, 1986; Hansen & Skavenski, 1985). In these experiments, subjects typically failed to report the change in the position of the target even though a later correction saccade and even a manual aiming movement directed at the target accurately accommodated the shift in position.

To understand how these experiments have been carried out, it is necessary to consider what happens when we reach out and place our finger on a target that suddenly appears in the peripheral visual field. Not only does our arm and finger extend toward the target, but the eyes, head, and body also move in such a way that the image of the target brought to the fovea. Even though the motor signals reach the ocular and brachial musculature at roughly the same time (Biguer, Jeannerod, & Prablanc, 1982), the saccadic eye movements directed at the target are typically completed while the hand is still moving*. The first and largest of the target-directed saccades is often completed before (or shortly after) the hand has begun to move. A second saccade – the so-called 'correction saccade' – puts the image of the target on the fovea. This means that during the execution of the aiming movement, the target, which was originally located in the peripheral visual field, is now located either on or near the fovea. Theoretically then, the more precise information about the position of the target provided by central vision (coupled with extra-retinal information about the relative position of the eye, head, body, and hand) can be used to update the trajectory of the hand as it moves toward the target.

Support for this idea came from an experiment by Goodale *et al.* (1986) in which the position of a target was changed unpredictably during the execution of the movement. Subjects were asked to move their finger from a central target to a new target (a small light) that appeared suddenly in their peripheral visual field. On half the trials, the peripheral target stayed in position until the subject had completed the aiming movement. On the remainder of the trials, however, the target was displaced to a new position 10% further out from where it had originally appeared. This sudden displacement of the target occurred just after the first saccadic eye movement had reached its peak velocity. The two kinds of trials were presented in random order, and the subjects were not told that the target would sometimes change position during the first eye movement.

The effects of this manipulation on the aiming movements and correction saccades were clear and unambiguous. As Figure 13.1 illustrates, the final position of the finger on trials in which the target was displaced during the saccadic eye movement was shifted (relative to 'normal' trials) by an amount equivalent to the size of the target displacement†. In other words, subjects corrected the trajectory of their aiming movement to accommodate the dis-

* One reason for this, of course, is that the eye is much less influenced by inertial and gravitational forces than the limb.

† Similarly, the correction saccade, which followed the first saccade, always brought the target, displaced or not, onto the fovea.

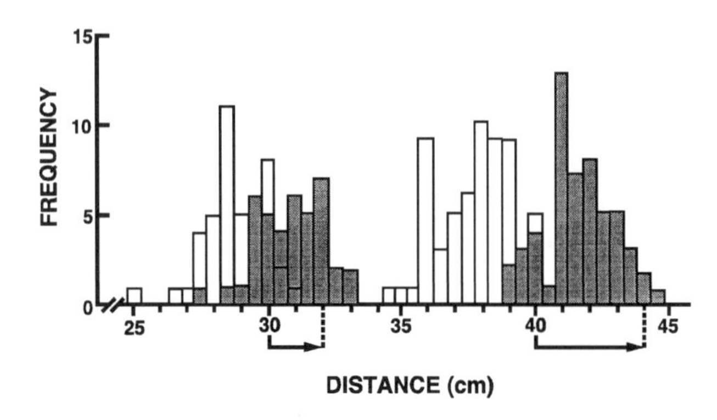

Fig. 13.1. Frequency distributions of the final positions of the index finger for the four subjects in single- and double-step trials for the 30- and 32 cm and the 40- and 44-cm targets. Unfilled bars indicate the final position for single-step trials. Filled bars indicate the final position for double-step trials. The extent of unfilled bars overlapped by filled bars is indicated by a horizontal line. The extents of the second displacements are indicated by the small arrows beneath the abscissa. Adapted with permission from Goodale, M.A., Pélisson, D., and Prablanc, C. (1986). "Large adjustments in visually guided reaching do not depend on vision of the hand or perception of target displacement", *Nature*, 320, 748-750.

placement of the target that occurred on some of the trials. In addition, the duration of limb movements made to a displaced target corresponded to the duration of the movement that would have been made had the target been presented at that location right from the start of the trial. Thus no additional processing time was required on displaced- target trials.

These particular findings confirm the suggestion that aiming movements are typically updated during their execution – even when the target does not change position. Although the initial programming of the movements is made on the basis of information about target location from the peripheral retina (combined with extra-retinal information about eye and head position), the trajectory of the moving limb and the amplitude of the final correction saccade is updated at the end of the first saccade – presumably on the basis of the more accurate information about target position that it is available from central vision. This explains why the duration of a limb movement is the same for a given position of the target, independent of whether the target appeared there initially or was moved to that position from another during the first saccade. In either case, the same post-saccadic information would be used to update motor signals controlling the trajectory of the hand (provided the target displacement during the saccade was not too large). In other words, the apparent 'correction' in the trajectory that occurred on displaced-target trials was nothing more than the normal updating of the motor programming that occurs at the end of the first saccade on an ordinary trial.

But the most remarkable thing about these experiments was the dissociation between what the subjects did when the target was displaced and what they actually perceived was happening. Even though the subjects' motor output was exquisitely sensitive to changes in the position of the target, the subjects remained quite unaware of the fact that the target had jumped to a new location while they were reaching towards it. Indeed, even when subjects were told that the target would be displaced during the eye movement on some trials and were asked to indicate after each trial whether or not such a displacement had occurred, their performance was no better than chance. Of course, the failure to perceive stimulus change, particularly position, during saccadic eye movements is well-known and is sometimes termed 'saccadic suppression' (e.g. Volkman, Schick & Riggs, 1968; Bridgeman, 1983). But in this case subjects could not even detect that the target had changed position at the end of their first saccade. These and other experiments using similar paradigms (for review, see Goodale 1988) have consistently shown that subjects fail to perceive changes in target position even though they modify their visuomotor output to accommodate the new position of the target.

In fact, had the subjects perceived the change in target position that occurred on displaced-target trials and, on the basis of this perception, consciously altered the trajectory of their reach, then one might have expected the duration of the movements on these trials to have fallen outside the amplitude-duration curve observed on normal trials where no displacement occurred. The fact that this did not occur is additional evidence that the subjects were treating normal trials and displaced-target trials in an identical fashion and suggests that the 'correction' observed on displaced-target trials is nothing more than the fine-tuning of the trajectory that occurs at the end of the first saccade in the normal course of events. Later experiments by Denis Pélisson (pers. comm.) have shown that if the displacement of the target during the first saccade is made large enough, subjects will perceive the change in target position. When this happens, the subjects show a large re-adjustment in the trajectory of their reach that falls well outside the normal amplitude-duration curve for movements to targets that are initially presented in those locations. In other words, if the change in target position is perceived, then subjects fail to make efficient on-line adjustments in their motor output. Efficient on-line adjustments would appear to occur only if the subjects fail to perceive the displacement of the target.

Complementary dissociations between perception and action have also been observed in which the perception of a visual stimulus is manipulated while visuomotor output remains unaffected. Bridgeman *et al.* (1981) have shown that even though a fixed visual target surrounded by a moving frame appears to drift in a direction opposite to that of the frame, subjects persist in pointing to the veridical location of the target. Similar observations have been made using saccadic eye movements rather than pointing. In an experiment by Wong and Mack (1981), for example, a small target was presented within a surrounding frame and after a 500 ms blank, the frame and target re-appeared, but now with the frame displaced a few degrees to the left or right. The target itself was presented at exactly the same location as before. In this case, however, instead of perceiving the frame as having changed position, subjects had the strong illusion that it was the target that had moved, in a direction opposite to that of the actual displacement of the frame. This illusion was maintained even when the target was displaced in the same direction as the frame, but by only one third of the distance. In this latter condition, the perceived change in target position following the blank period and the actual change of target position on the retina were in opposite directions. Yet despite the presence of this strong illusory displacement of the target, subjects consistently directed their saccades to the true location of the target. In short, an illusory displacement of target position occurred without any change in the spatial organization of the target-directed motor output.

Taken together, these findings suggest that the mechanisms mediating the perception of object location operate largely in allocentric coordinates whereas those mediating the control of object-directed actions operate in egocentric coordinates. In other words, perception uses a coordinate system that is world-based in which objects are seen as changing location relative to a stable or constant world; the systems controlling action systems, however, cannot afford these kinds of constancies and must compute the location of the object with respect to the effector that is directed at that target. Thus, in the experiments by Bridgeman *et al.* (1981) and Wong and Mack (1981), the target within the moving or displaced frame was perceived as moving relative to the frame while the frame itself, which was the only large visible feature in the field of view, was perceived as a stable background. The visuomotor systems computing the saccadic eye movements (or the aiming movements) directed at the target simply ignored the movement of the frame and computed the actual position of the target in retinocentric (and perhaps also in head and/or shoulder-centered) coordinates. In the experiments by Goodale *et al.* (1986) in which the position of the target was changed unpredictably during a saccade, the subjects' failure to perceive the displacement of the target was probably a reflection of the broad tuning of perceptual constancy mechanisms – mechanisms which preserve the identity of a target as its position is shifted on the retina during an eye movement. When no other reference points are available in the field of view, the perceptual system assumes that the position of the target (which was stable at the beginning of the saccade) has not changed. Such an assumption has little consequence for perception and is computationally efficient. The visuomotor systems controlling saccadic eye movements and manual aiming movements cannot afford this luxury. At the end of the first saccade, they must recompute the position of the target (within egocentric frames of reference) so that the appropriate correction saccade and amendment to the trajectory of the moving hand can be made. In short, visuomotor control demands different kinds of visual computations than visual perception. As we suggested earlier, it is this difference in the computational requirements of the visual control of action and visual perception that explains the division of labour between the two streams of visual processing in the primate cerebral cortex.

13.4 Perceptually driven motor outputs

Until now, we have emphasized the fact that motor outputs, such as saccadic eye movements and aiming movements with the limb, are directed to the location of targets within egocentric frames of reference. These egocentric locations may or may not be coincident with the perceived location of the target. There are special circumstances, however, where motor output can be driven by the perceived position of a target rather than by its actual location in egocentric coordinates. One of the clearest demonstrations of the effect of perception on motor output comes from work by Wong and Mack (1981) in which they employed the same paradigm they had used to demonstrate a dissociation between perception and motor control. In that paradigm, the target was displaced in the same direction as the surrounding frame but by only one third of the distance. Subjects moved their eyes correctly in the direction of the target displacement but perceived the target as having moved in the opposite direction. In Wong and Mack's second experiment, subjects were asked to look back to the 'original' location of the target after making their saccade to its new location. These memory-driven saccades were made to a location in perceptual, not retinal, coordinates. In other words, the 'look-back' saccade was made to a location that corresponded with the subjects' perceptual judgement about the previous location of the target - a judgement that was determined by the relative location of the target in the frame rather than by its absolute location in egocentric space.

The suggestion that memory-driven saccades may be driven by visual information that is different from that driving normal target-directed saccades (and by inference may depend on separate neural pathways) is consistent with a number of other observations showing that the kinematic profiles and trajectories of saccades to remembered targets can be very different from those of target-directed saccades. Thus, saccades made 1-3 s following the offset of a target light were 10-15 ms longer and achieved lower peak velocities than saccades directed towards an identical target that remained illuminated (Becker & Fuchs, 1969). In addition, the coupling between peak velocity and movement amplitude also appears to be much more variable in saccades made to remembered rather than to visible targets (Smit, Van Gisbergen, & Cools, 1987). Memory-driven saccades also appear to be less accurate than those made to visible targets (Gnadt, Bracewell, & Andersen, 1991): these errors were seen even when the delay between target offset and the initiation of the saccade was as short as 100 ms. Moreover the amplitude of the errors increased as the delay was lengthened systematically up to 2 s. This observation led Gnadt *et al.* to propose that a transition occurs in the oculomotor system between visually-linked and memory-linked representations during the first 800-1000 ms following offset of a visual target and that "the 'memory' of intended eye movement targets does not retain accurate retinotopic registration" (p. 710). In summary, these differences in saccadic dynamics and accuracy suggest that the neural subsystems for the generation of saccades to remembered locations are to some degree independent from those subserving normal target-driven saccades (see, for example, Guitton, Buchtel, & Douglas 1985; Wurtz & Hikosaka 1986). Indeed, if, as the work by Wong and Mack (1981) described above suggests, memory-driven saccades are computed in allocentric rather than egocentric coordinates, then the visuomotor transformations mediating those saccades may use the same visual information that underlies perceptual reports of spatial location.

Like eye movements, manual aiming movements are most accurate when directed to still-visible targets, and errors accumulate as the length of the delay period between target viewing and movement initiation increases (Holding, 1968; Elliot & Mandalena, 1987). Indeed, Elliot and Mandalena (1987) have suggested that a visual representation of the environment which would be useful for guiding manual aiming movements does not persist for longer than 2 s. It seems likely that the action systems underlying the control of pointing, like those underlying the control of saccadic eye movements, work optimally in 'real time' with visible targets. Once the target is gone and some time has elapsed (2 s or more), the production of an aiming movement, like the saccadic eye movements discussed earlier, may have to be referred to earlier perceptual representations of the target – representations that depend on neural circuitry quite separate from that mediating aiming movements to visible targets.

13.5 Differential size coding for perception and action

Just as the perception of object location appears to operate within relative or allocentric frames of reference, so does the perception of object size.

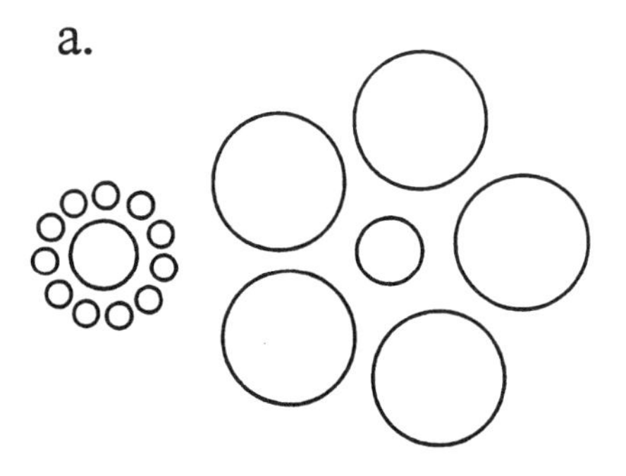

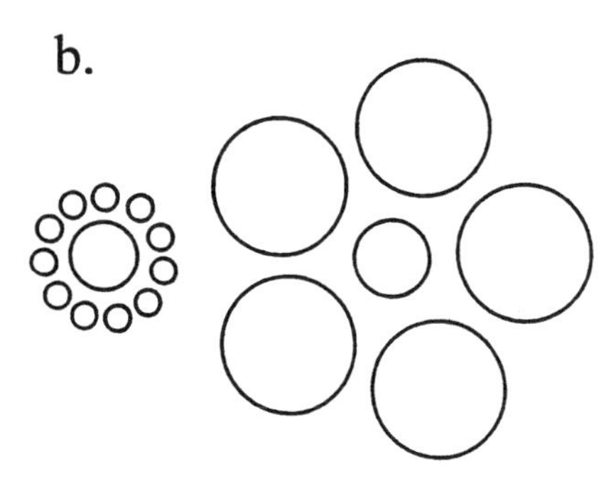

Fig. 13.2. The 'Ebbinghaus' illusion. Panel A shows the standard version of the illusion in which physically identical target circles appear perceptually different. Most people judge the target circle surrounded by the annulus of smaller circles to be larger than the other target circle. Panel B shows a version of the illusion in which the target circle surrounded by the annulus of larger circles has been made physically larger than the other target, compensating for the effect of the illusion. Most people now see the two target circles as equivalent in size. (Adapted with permission from Aglioti, S., DeSouza, J.F.X., & Goodale, M.A. (1995). "Size-contrast illusions deceive the eye but not the hand" *Current Biology*, 5, 679-685.)

Although we often make subtle judgements of the relative sizes of objects, we rarely make judgements of their absolute size. Indeed, our judgements of size appear to be so inherently relative that we can sometimes be fooled by visual displays in which visual stimuli of the same size are positioned next to comparison stimuli which are either much smaller or much larger than the target stimuli. Such size-contrast illusions are a popular demonstration in many introductory textbooks in psychology and perception. One such illusion is the so-called Ebbinghaus Illusion (or Titchener Circles Illusion) in which two target circles of equal size, each surrounded by a circular array of either smaller or larger circles, are presented side by side (Figure 13.2). Subjects typically report that the target circle surrounded by the array of smaller circles appears larger than the one surrounded by the array of larger circles, presumably because of the difference in the contrast in size between the target circles and the surrounding circles. Although the illusion is usually depicted in the way just described, it is also possible to make the two target circles appear identical in size by increasing the actual size of the target circle surrounded by the array of larger circles (Figure 13.2).

While perception is clearly affected by these manipulations of the stimulus array, there is good reason to believe that the calibration of size-dependent motor outputs, such as grip aperture during grasping, would not be. After all, when we reach out to pick up an object, particularly one we have not seen before, our visuomotor system must compute the object's size accurately if we are to pick it up efficiently. It is not enough to know that the target object is larger or smaller than surrounding objects; the visuomotor systems controlling hand aperture must compute its real size. Moreover, the visual control of motor output cannot afford to be fooled by an accidental conjunction of contours in the visual array that might lead to illusions of size or location. For these reasons therefore, one might expect grip scaling to be insensitive to size-contrast illusions.

To see just how insensitive grip scaling might be to such an illusion, Aglioti, DeSouza, and Goodale (1995) developed a three dimensional version of the Ebbinghaus illusion in which two thin "poker-chip" disks were used as the target circles. In this experiment, trials in which the two disks appeared perceptually identical but were physically different in size were randomly alternated with trials in which the disks appeared perceptually different but were physically identical. Subjects were given the following instructions: if you think the two disks are the same size, pick up the one on the left; if you think they are different in size, pick up the one on the right. (Half the subjects were given the opposite instructions.) The aperture between their index finger and thumb was monitored using opto-electronic recording.

All the subjects remained sensitive to the size-contrast illusion throughout testing. On trials in which the disks were physically identical, their choice indicated that they thought the disks were different in size; on trials in which the disks were physically different, they behaved as though they were the same size. Remarkably, however, the scaling of their grasp was affected very little by these beliefs. Instead, the maximum grip aperture, which was achieved approximately 70% of the way through the reach towards the disk, was almost entirely determined by the true size of that disk. As Figure 13.3 illustrates, the difference in grip aperture for large and small disks was the same for trials in which the subject believed the two disks were equivalent in size (even though they were different) as it was for trials in which the subject believed the two disks were different in size (even though they were identical). In short, the calibration of grip size was largely impervious to the effects of the size-contrast illusion.

As we saw earlier, however, the control of skilled movements is not com-

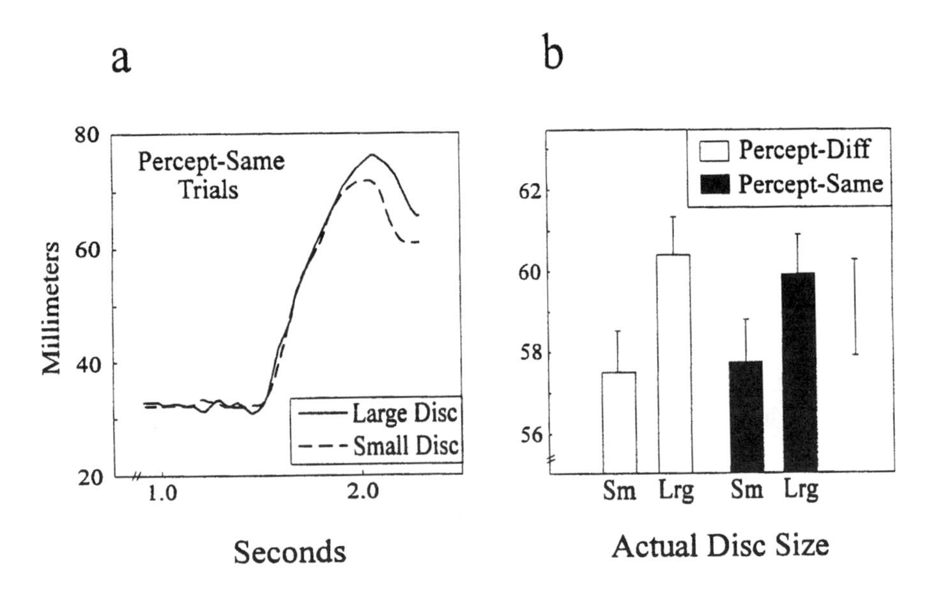

Fig. 13.3. Panel A shows representative grip aperture profiles for one subject picking up a large disk and a small disk on separate trials in which he judged the two disks to be identical in size (even though, of course, the two disks were physically different). In both cases, the disk was located on the left hand side of the display. Panel B shows the mean maximum grip aperture for the 14 subjects in different testing conditions. The two solid bars on the right indicate the maximum aperture on trials in which the two disks were judged to be perceptually the same, even though they were physically different in size. The two open bars on the left indicate the mean maximum aperture on trials in which the two disks were judged to be perceptually different, even though the were physically the same size (either two small disks presented together or two large disks presented together). The line on the extreme right of the graph indicates the average difference in disk size that was required to achieve perceptual equivalence as determined in a pretest. The difference between the maximum grip aperture achieved for large disks was significantly greater than the maximum grip aperture for small disks independent of whether or not the subject perceived the disks to be the same or different sizes ($p < .001$). No other comparisons were significant. Error bars indicate the standard error of the mean. (Adapted with permission from Aglioti, S., DeSouza, J.F.X., and Goodale, M.A. (1995). "Size-contrast illusions deceive the eye but not the hand", *Current Biology*, 5, 679-685.)

pletely isolated from perception. There is no doubt, for example, that the perceived function of an object can affect the nature of the grasp that we adopt when we pick it up. Moreover, we often pick up the same object differently depending on what we plan to do with it. Thus, the grasp we use to pick up a glass of wine we are going to drink from is usually quite different from the grasp we use when placing the glass on a serving tray. It is perhaps not surprising therefore that perception had some effect on grip scaling even in the Ebbinghaus illusion task – at least on those trials in which subjects perceived the two disks to be different in size when in fact they were identical. On some of these trials, some subjects opened their fingers slightly more for the disk surrounded by the small circles than they did for the disk surrounded by the large circles. Nevertheless, this perceptual effect on grip aperture circles was quite variable and was significantly smaller than the size difference that was required to achieve perceptual equivalence between the two disks in judgement tasks. In other words, the effect of the illusion on grip size was much smaller and more variable than the effect of the illusion on perceptual judgements of size. In contrast, as we have already seen, all subjects showed a strong effect of real object size on the calibration of their grip – independent of whether or not it was on perceptually different or perceptually identical trials.

It is remotely possible of course that the accurate scaling of the grasp was simply due to the fact that the subjects were somehow comparing their grip aperture with the diameter of the target disk in flight. This is unlikely for several reasons however. For one thing, as can be seen in Figure 13.3 the maximum grip aperture was actually much larger than the diameter of the disk – more than twice as large on most trials. But just as has been observed in many previous studies of grasping (Jeannerod, 1988; Jakobson and Goodale, 1991) maximum grip aperture was still well correlated with the size of the target. Second, subjects remained susceptible to the illusion throughout the experiment, despite the fact that they were scaling their grasp to the real size of the target disk. If they were adjusting their grasp on the basis of information delivered by the same visual networks that they used to make their perceptual judgements, then one might expect to see some weakening of the illusion over the course of the experiment. No such weakening was observed. Third, the calibration of maximum grip aperture is largely determined by motor programming that is carried out before the hand has actually left the table (Jakobson & Goodale, 1991), and as we saw earlier, can be modified by new information only late in the trajectory (Castiello & Jeannerod, 1991). In fact, on a few occasions, subjects expressed surprise when they handled the disk after they had picked it up – claiming that it seemed larger or smaller than they had thought it was – even though their grasp had been calibrated accurately in flight.

Nevertheless, it seemed important to eliminate the possibility of on-line modulation, since adjustments to the grasp were undoubtedly occurring at least in the final portion of the reach trajectory. In addition, there was another, perhaps more serious, problem with the Aglioti *et al.* (1991) study: the measure of the effect of the illusion on perception was essentially dichotomous, while the measure of its effect on the grasp was continuous. For these reasons, in a subsequent experiment (Haffenden & Goodale, 1998), we replicated the Aglioti *et al.* study, but with two modifications. First, the entire experiment was run in “visual open-loop”. In other words, as soon

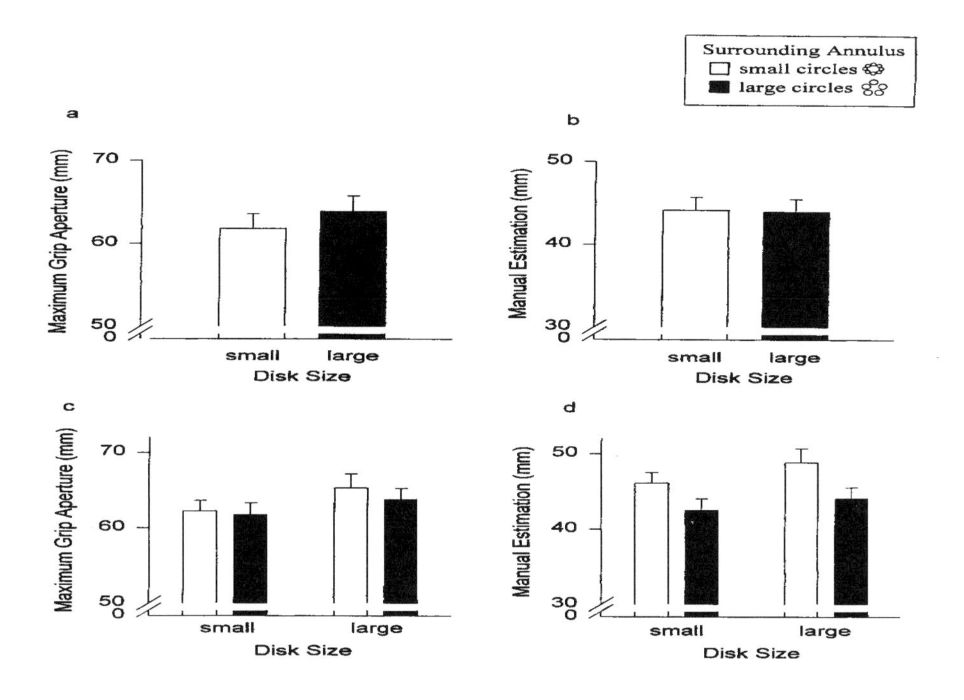

Fig. 13.4. Calibration of the grasp (on left) and manual estimations (on right) for disks surrounded by the illusory annuli: (a and b) perceptually identical conditions, and (c and d) perceptually different conditions. The difference between the maximum grip aperture achieved during a grasping movement was significantly greater for large disks than the maximum grip aperture for small disks independent of whether or not the subject perceived the disks to be the same or different sizes ($p < .05$). Manual estimations were influenced by the illusory display. The difference between manual estimations of the large and small disks in the perceptually identical condition was not significant ($p > .05$). Perceptually different, but physically identical disk pairs produced significantly different manual estimations. The small disk surrounded by the small circle annulus was estimated to be larger than the small disk surrounded by the large circle annulus ($p < .01$). Manual estimations of the pair of large disks produced a similar result. The large disk surrounded by the small circle annulus was estimated to be larger than the large disk surrounded by the large circle annulus ($p < .05$). Error bars indicate the standard error of the mean averaged within each condition for all subjects. Adapted with permission from Haffenden, A. M. and Goodale, M. A. (1998) "The effect of pictorial illusion on prehension and perception", *J. Cog. Neurosci.*, 10: 122-136.

as subjects had made their choice about whether the disks were the same or different in size and had begun to move their hand toward one of the disks, the lights were turned off, leaving them with no opportunity to use vision to modify their grip aperture on-line. Second, an additional condition was added to the experiment in which subjects were asked not to pick up the disk after making their choice but rather to estimate the diameter of the disk by opening their index finger and thumb a matching amount. This response, which was also run in open loop, gave a continuous measure of the subjects' perception of disk size that could be more directly compared to the calibration of the grasp. The results of this study were clear and unambiguous. As Figure 13.4 shows, the scaling of grip aperture was again correlated with the true size of the disk, even though the subjects were now grasping the disk without the benefit of any visual feedback. The subjects' manual estimations of disk size, however, were strongly biased in the direction of the illusion.

But why should perception be so susceptible to this illusion while the calibration of grasp is not? The mechanisms underlying the size-contrast illusion are not well understood. It is possible that it arises from a straightforward relative size judgement, whereby an object that is smaller than its immediate neighbors is assumed to be smaller than a similar object that is larger than its immediate neighbors. It is also possible that some sort of image-distance equation is contributing to the illusion in which the array of smaller circles is assumed to be more distant than the array of larger circles; as a consequence, the target circle within the array of smaller circles will also be perceived as more distant (and therefore larger) than the target circle of equivalent retinal image size within the array of larger circles. In other words, the illusion may be simply a consequence of the perceptual system's attempt to make size constancy judgments on the basis of an analysis of the entire visual array (Coren, 1971; Gregory 1963).

Mechanisms such as these, in which the relations between objects in the visual array play a crucial role in scene interpretation, are clearly central to perception. In contrast, the execution of a goal-directed act like prehension depends on metrical computations that are centered on the target itself. Moreover, the visual mechanisms underlying the control of the grasping movements must compute the real distance of the object (presumably on the basis of reliable cues such as stereopsis and retinal motion). As a consequence, computation of the retinal image size of the object, coupled with an accurate estimate of distance, will deliver the true size of the object for calibrating the grasp. There is increasing evidence that binocular vision is particularly critical in this computation (for review, see Goodale & Servos, 1996). In a recent study, Marotta, DeSouza, Haffenden, and Goodale (in press) showed that removing binocular vision made the scaling of the grasp more sensitive to the illusory display. Of course, when binocular vision was available, subjects continued to scale accurately for the real size of the object on the illusory background.

The relative insensitivity of reaching and grasping to pictorial illusions has been demonstrated in several other recent experiments employing a number of different classical illusions, including the Ponzo illusion (Ian Whishaw, pers. comm.), and the Müller-Lyer illusion (Gentilucci *et al.*, 1996). What is particularly interesting about the last of these studies is that the introduction of a 5 s delay between viewing the display and responding significantly increased the effect of the illusion on the calibration of the final motor response. As was discussed earlier, motor responses to remembered stimuli appear to rely more on a perceptual than an action-based frame of reference (for a discussion of a similar effect of delays on grasp scaling, see Goodale, Jakobson, & Keillor, 1994).

13.6 Perception and Visuomotor Control in the Periphery

There can be little doubt that the best way to see something is to look at it. Indeed, our ability to identify objects falls off dramatically as our view becomes more and more peripheral (for review, see Strasburger, Harvey, & Rentschler, 1991). But why are our visual fields so extensive if most of our perception of the world comes from central vision? Part of the answer, of course, is related to the need to detect biologically relevant stimuli across a wide region of visual space. Humans, like most highly visual vertebrates, typically orient their gaze (and thus their fovea) towards objects and events that suddenly appear in the visual periphery. But this is not the only reason why our visual fields are so large. Many of the movements that humans make through the world require processing of information far into the periphery. Thus, when we walk from one place to another in our immediate environment we use the optic flow that is generated by our movement to control both our heading and our rate of locomotion (for review, see Warren, 1995). We also use optic flow to modulate our posture, particularly on uneven terrain or in situations, such as balancing on a narrow beam, where proprioceptive information is unreliable (for review see, Lee, 1993). We even seem able to pick up objects, such as coffee cups and pens, without looking at them directly.

Most of the time we remain quite unaware of the fact that we are using information from the visual periphery to control our actions. But is this simply a question of attention? The evidence we have reviewed throughout this chapter would suggest that visuomotor control is mediated by networks that are quite separate from those delivering our perception of the world. Perhaps these visuomotor systems have privileged access to information from the visual periphery. To put it another way, even if we were *directed* to attend to stimuli in the visual periphery, our ability to discriminate between these stimuli would never be as good as our ability use them to direct our actions. Recent experiments in our laboratory suggest that this is indeed the case (Murphy and Goodale, 1994; Goodale & Murphy, 1997).

In these experiments, subjects were presented with one of five different rectangular blocks for 100 ms at various positions in the peripheral visual field along the horizontal meridian, from 5° to 70°. The change in eccentric viewing was accomplished by having the subjects direct their gaze increasingly leftward at fixation points located at different distances from the target

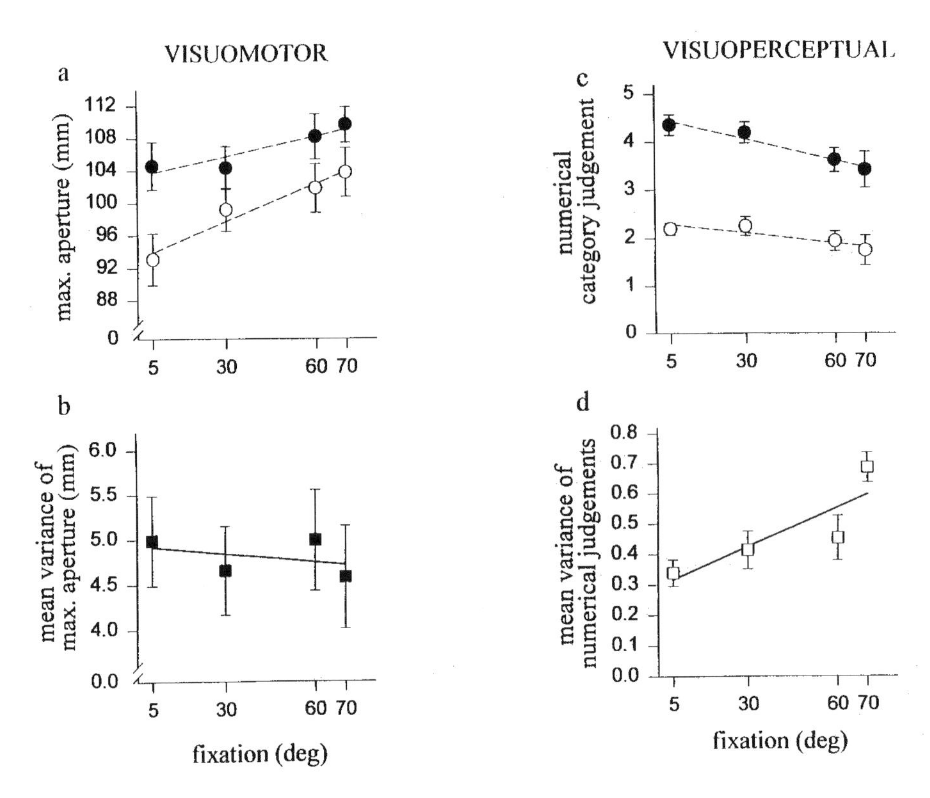

Fig. 13.5. Reaching to and estimating two different sizes of blocks: (a) Maximum grip aperture increases with increasing retinal eccentricity, but (b) the variance seen in grip scaling remains constant. (c) In contrast to the visuomotor results, estimations of block size decrease with increasing retinal eccentricity and (d) the variance of the estimations increases. (Adapted with permission from Goodale, M.A., and Murphy, K.J. (1997). "Action and perception in the visual periphery". In P. Thier and H. Karnath (Eds.), "Parietal lobe contributions to orientation in 3D space", *Exp. Brain Res.*, supplements. Springer-Verlag, Berlin.)

object along the same horizontal plane. Thus, although the retinal location of the target block varied from trial to trial, the location of the block with respect to the subject's hand and body remained constant. Sensitivity to the width of the blocks (which varied in their dimensions but not in their overall area) was measured under two response conditions: a visuomotor condition in which subjects were required to reach out and grasp the block, and a visuoperceptual condition in which subjects were required to categorize the different blocks on the basis of their width.

When subjects reached out and grasped the blocks, the maximum aperture of their grasp increased with the eccentricity of the view but continued to be well-tuned to the actual width of the target block. Moreover, as Figure 13.5 illustrates, trial-to-trial variability in grip aperture did not change as a function of eccentricity. In contrast, when subjects were asked to make perceptual judgements of the width of the objects, they progressively underestimated the width of the blocks as retinal eccentricity increased. In addition, even though the subjects continued to discriminate between the blocks in the peripheral visual field, the variability of their judgements increased significantly the further away from central vision the block was viewed (Figure 13.5). These and other findings (for review, see Goodale and Murphy, 1997) suggest that visuomotor systems may be more sensitive to stimuli in the visual periphery than the perceptual systems mediating our experience of the visual world beyond the fovea.

These behavioural observations in humans are consistent with anatomical and electrophysiological studies in the monkey showing that areas in the dorsal stream receive extensive inputs from the peripheral visual fields while inputs to the ventral stream are largely from more central regions of the visual field (for review, see (Baizer, Ungerleider, & Desimone, 1991; Colby *et al.*, 1988). The difference in representation of the visual field might explain why the visual control of grasping movements directed at targets in the visual periphery are so much more reliable than perceptual judgements about those same objects; in short, the dorsal action is simply better connected with the visual periphery than the ventral perception system.

13.7 Concluding Remarks

The various examples of dissociations between perceptual judgement and visuomotor control that we have reviewed in this chapter provide powerful evidence for the parallel operation, within our everyday life, of two types of visual processing systems. Each system has evolved to transform visual inputs for quite different functional outputs, and, as a consequence, is characterized by quite different transformational properties. The nature of the dissociations lend further support to the proposal by Goodale and Milner (1992; Milner & Goodale, 1993, 1995) that these two types of processing are mediated by separate visual pathways in the primate cerebral cortex. The parallel operation of these two systems lies at the heart of the paradox that what we think we 'see' is not always what guides our actions.

References

Aglioti, S., DeSouza, J.F.X., & Goodale, M.A. (1995). "Size-contrast illusions deceive the eye but not the hand", *Current Biology*, 5: 679-685.

Baizer, J.S., Ungerleider, L.G., & Desimone, R. (1991). "Organization of visual inputs to the inferior temporal and posterior parietal cortex in macaques", *J. Neurosci.*, 11: 168-190.

Becker, W., & Fuchs, A.F. (1969). "Further properties of the human saccadic system: eye movements and correction saccades with and without visual fixation points", *Vis. Res.*, 9: 1247-1258.

Biguer, B., Jeannerod, M., & Prablanc, C. (1982). "The coordination of eye, head, and arm movements during reaching at a single visual target", *Exp. Brain Res.*, 46: 301-304.

Breitmeyer, B.G. (1984). *Visual Masking: An Integrative Approach*, Oxford University Press, New York.

Bridgeman, B. (1983). "Mechanisms of space constancy", in *Spatially Oriented Behavior*, A. Hein & M. Jeannerod (Eds.), 263-279, Springer-Verlag, New York, NY.

Bridgeman, B., Kirch, M., & Sperling, A. (1981). "Segregation of cognitive and motor aspects of visual function using induced motion", *Perception and Psychophysics*, 29: 336-342.

Bridgeman, B., Lewis, S., Heit, G., & Nagle, M. (1979). "Relation between cognitive and motor-oriented systems of visual position perception", *J. Exp. Psych.: Human Perception and Performance*, 5: 692-700.

Bülthoff, H.H., & Edelman, S. (1992). "Psychophysical support for a two-dimensional view interpolation theory of object recognition", *Proceedings of the National Academy of Science*, 89: 60-64.

Castiello, U., & Jeannerod, M. (1991). "Measuring time to awareness", *Neuroreport*, 2: 797-800.

Castiello, U., Paulignan, Y., & Jeannerod, M. (1991). "Temporal dissociation of motor responses and subjective awareness", *Brain*, 114: 2639-2655.

Colby, C.L., Gattas, R., Olson, C.R., Gross, C.G. (1988). "Topographic organization of cortical afferents to extrastriate visual are PO in the macaque: a dual tracer study", *Journal of Comprehensive Neurology*, 269: 392-413.

Coren, S. (1971). "A size contrast illusion without physical size differences", *American Journal of Psychology*, 84: 565-566.

Crick, F., & Jones, E. (1993). "Backwardness of human neuroanatomy", *Nature*, 361: 109-110.

Elliot, D., & Madalena, J. (1987). "The influence of premovement visual information on manual aiming", *Quarterly Journal of Experimental Psychology*, 39A: 541-559.

Fehrer, E., & Biederman, I. (1962). "A comparison of reaction time and verbal report in the detection of masked stimuli", *J. Exp. Psych.*, 64: 126-130.

Fehrer, E., & Raab, D. (1962). "Reaction time to stimuli masked by metacontrast", *J. Exp. Psych.*, 63: 143-147.

Fox, R. (1978). "Visual Masking", in *Perception. Handbook of Sensory Physiology VIII*, R. Held, H. W. Leibowitz & H. L. Teuber (Eds.) 629-653, Springer-Verlag, Berlin.

Gentilucci, M., Chieffi, S., Daprati, E., Saetti, M.C., & Toni, I. (1996). "Visual illusion and action", *Neuropsychologia*, 34: 369-376.

Gnadt, J. W., Bracewell, R. M., & Andersen, R. A. (1991). "Sensorimotor transformation during eye movements to remembered visual targets", *Vis. Res.*, 31: 693-715.

Goodale, M. A. (1988). "Modularity in visuomotor control: From input to output", in *Computational Processes in Human Vision: An Interdisciplinary Perspective*, Z. Pylyshyn (Ed.), 262-285. Ablex Press, Norwood, NJ.

Goodale, M. A. (1993). "Visual pathways supporting perception and action in the primate cerebral cortex", *Current Opinion in Neurobiology*, 3: 578-585.

Goodale, M. A., & Haffenden, A. (In press). "Frames of reference for perception and action in the human visual system", *Neurosci. and Biobehavioral Reviews*.

Goodale, M. A., Jakobson, L. S., & Keillor, J. M. (1994). "Differences in the visual control of pantomimed and natural grasping movements", *Neuropsychologia*, 32: 1159-1178.

Goodale, M. A., Meenan, J. P., Bülthoff, H. H., Nicolle, D. A., Murphy, K. J., & Racicot, C. I. (1994). "Separate neural pathways for the visual analysis of object shape in perception and prehension", *Current Biology*, 4: 604-610.

Goodale, M. A., & Milner, A. D. (1992). "Separate visual pathways for perception and action", *Trends In Neuroscience*, 15: 20-25.

Goodale, M. A., Milner, A. D., Jakobson, L. S. & Carey, D. P. (1991). "A neurological dissociation between perceiving objects and grasping them", *Nature*, 349: 154-156.

Goodale, M.A., & Murphy, K. J. (1996). "Action and perception in the visual periphery", in *Parietal lobe contributions to orientation in 3D space. Experimental Brain Research Supplements*, P. Thier & H. Karnath (Eds.) 447-461, Springer-Verlag, Berlin.

Goodale, M.A., Pélisson, D., & Prablanc, C. (1986). "Large adjustments in visually guided reaching do not depend on vision of the hand or perception of target displacement", *Nature*, 320: 748-750.

Goodale, M. A., & Servos, P. (1996). "The visual control of prehension", in *Advances in motor learning and control*, H. Zelaznik (Ed.) 87-121, Human Kinetics Publishers, Champaign, IL.

Gregory, R. L. (1963). "Distortion of visual space as inappropriate constancy scaling", *Nature*, 199: 678-680.

Guitton, D., Buchtel, H. A., & Douglas, R. M. (1985). "Frontal lobe lesions in man cause difficulties in suppressing reflexive glances and in generating goal-directed saccades", *Exp. Brain Res.*, 58: 455-472.

Haffenden, A.M., & Goodale, M.A. (1998). "The effect of pictorial illusion on prehension and perception", *J, Cog. Neurosci.*, 10: 122-136.

Hansen, R. M., & Skavenski, A. A. (1985). "Accuracy of spatial localization near the time of a saccadic eye movement", *Vis. Res.*, 25: 1077-1082.

Harrison, K., & Fox, R. (1966). "Replication of reaction time to stimuli masked by metacontrast", *J. Exp. Psych.*, 71: 162-163.

Holding, D. H. (1968). "Accuracy of delayed aiming responses", *Psychonomic Science*, 12: 125-126.

Jakobson, L.S., Archibald, Y.M., Carey, D.P. & Goodale, M.A. (1991). "A kinematic analysis of reaching and grasping movements in a patient recovering from optic ataxia", *Neuropsychologia*, 29: 803-809.

Jakobson, L.S., & Goodale, M. A. (1991). "Factors affecting higher-order movement planning: a kinematic analysis of human prehension", *Exp. Brain Res.*, 86: 199-208.

Jeannerod, M. (1988). "The neural and behavioral organization of goal directed movements", Oxford University Press, Oxford, UK.

Johansson, R. S. (1991). "How is grasping modified by somatosensory input?", in *Motor Control: concepts and issues*, D. R. Humphrey & H.-J. Feund (Eds.), 331-355. Chichester, UK. Wiley.

Lee, D. (1993). "Body-environment coupling", in *The Perceived Self: Ecological and Interpersonal Sources of Self-Knowledge*, U. Neisser (Eds.), 43-67, Cambridge University Press, New York, NY.

Libet, B. (1966). "Brain stimulation and the threshold of conscious experience", in *Brain and Conscious Experience*, J.C. Eccles (Ed.), 165-181, Springer-Verlag, New York, NY.

Libet, B. (1981). "The experimental evidence for subjective referral of a sensory experience backwards in time: reply to P.S. Churchland", *Philosophy of Science*, 48: 182-197.

Marotta, J. J., DeSouza, J. F. X., Haffenden, A. M., & Goodale, M.A. (In press). "Does a monocularly presented size-contrast illusion influence grip aperture?", *Neuropsychologia*.

Marr, D. (1982). *Vision*, Freeman Press, San Francisco.

Milner, A. D. & Goodale, M. A. (1993). "Visual pathways to perception and action", in *The visually responsive neuron: From basic physiology to behavior. Progress in Brain Research*, T.P. Hicks, S. Molotchnikoff & T. Ono (Eds.), 317-338, Elsevier, Amsterdam.

Milner, A.D., and Goodale, M.A. (1995). *The visual brain in action*, Oxford University Press, Oxford.

Milner, A. D., Perrett, D. I., Johnston, R. S., Benson, P. J., Jordan, T. R., Heeley, D. W., Bettucci, D., Mortara, F., Mutani, R., Terazzi, E., and Davidson, D. L. W. (1991). "Perception and action in 'visual form agnosia"', *Brain*, 114: 405-428.

Murphy, K. J. & Goodale, M. A. (1994). "The effect of retinal eccentricity on implicit and explicit judgments of object width", *Society For Neuroscience Abstracts*, 20: 1580.

Palmer, S., Rosch, E. & Chase, P. (1981). "Canonical perspective and the perception of objects", in *Attention and Performance IX*, J. Long & A. Baddeley (Eds.) 135-151, L. Earlbaum, Hillsdale, NJ.

Paulignan, Y., MacKenzie, C., Marteniuk, R., & Jeannerod, M. (1990). "The coupling of arm and finger movements during prehension", *Exp. Brain Res.*, 79: 431-435.

Perenin, M.-T. & Vighetto, A. (1988). "Optic ataxia: A specific disruption in visuomotor mechanisms. I.Different aspects of the deficit in reaching for objects", *Brain*, 111: 643-674.

Schiller, P. H., & Smith, M. C. (1966). "Detection in metacontrast", *J. Exp. Psych.*, 71: 32-39.

Smit, A. C., Van Gisbergen, J. A. M. & Cools, A. R. (1987). "A parametric analysis of human saccades in different experimental paradigms", *Vis. Res.*, 27: 1745-1762.

Strasburger, H., Harvey, L. O. Jr. & Rentschler, I. (1991). "Contrast thresholds for identification of numeric characters in direct and eccentric view", *Perception and Psychophysics*, 49: 495-508.

Taylor, J. L. & McCloskey, D. I. (1990). "Triggering of preprogrammed movements as reactions to masked stimuli", *J. Neurophysiol.*, 63: 439-446.

Ungerleider, L.G. & Mishkin, M. (1982). "Two cortical visual systems", in *Analysis of Visual Behavior*, D. J. Ingle, M. A. Goodale & R. J. W. Mansfield (Eds.) 549-586, MIT Press, Cambridge, MA.

Volkman, F. C., Schick, A. M. L. & Riggs, L. A. (1968). "Time course of visual inhibition during voluntary saccades", *J. Opt. Soc. Am.*, 58: 562-569.

Warren, W. H. (1995). "Self-motion: Visual perception and visual control", in *Handbook of Perception and Cognition v. 5: Perception of Space and Motion*, W. Epstein & S. Rogers (Eds.), 263-325. Academic Press, San Diego, CA.

Wong, E. & Mack, A. (1981). "Saccadic programming and perceived location", *Acta Psychologica*, 48: 123-131.

Wurtz, R. H. & Hikosaka, O. (1986). "Role of the basal ganglia in the initiation of saccadic eye movements", in *Progress in Brain Research, 64*, H.-J. Freund, U. Buttner, B. Cohen, & J. North (Eds.), 175-190, Elsevier, Amsterdam.

14

Exocentric pointing

Jan J. Koenderink and Andrea J. van Doorn

Abstract

We have conducted experiments in visual space perception in free field, daylight conditions. Observers were constrained to particular locations and had to maintain eye height, but were otherwise free to make eye and head movements, twist at the hips, change feet position, etc. Observers had to point a pointing device at a location different from the vantage point, to a target at yet another location ("exocentric pointing task"). The task was done by radio control, thus the observer could simply look at pointer and target and change the pointer's attitude by moving a slider on a portable control box. In this paper we report only on experiments in which the observer was at the barycenter of equilateral triangles of variable size with pointer and target at vertices. The angle sum is an angular *excess* in near space and an angular *deficit* in far space. We analyze the results in terms of various models current in the literature and speculate on the use and nature of the concept of "visual space".

14.1 The concept of "Visual Space"

We intuitively take it for granted that the stage of our daily environment is "Galilean space–time". That is to say, that time and space are independent (we will completely ignore the temporal dimension in this paper) and that space is three–dimensional Euclidean space. The Euclidean structure used to be taken for granted even by scientists and philosophers until Gauss in the early nineteenth century compared it empirically with homogeneous, curved spaces (Gauss, 1880). In curved spaces one expects the sum of angles of triangles to add up to either *less* (the "hyperbolic" case) or *more* (the "elliptic" case) than the Euclidean 180°. Gauss measured angle sums for very large triangles using precise geodesic instruments and could detect no significant deviations from 180°. This settles the question for experiments concerning visual perception in the natural environment: the stage is Euclidean.

When we consider judgements of geometrical quantities (coincidence, collinearity, paralellism, angle, bisection, length, area, ...) made by human observers, we often note striking deviations from the corresponding geometrical measurements. One possible reason is immediately obvious: all human perception is necessarily bound to the present *perspective**. Granting considerable simplification and overstatement, one tends to roughly agree with the view that there is an essential difference between the visual *field* and the visual *world*, especially relating to the depth dimension of the latter. Indeed, considerable efforts have been spent on the understanding of random and systematic deviations of the geometrical judgements of human observers from the geometrical "ground truth".

The wealth of theories can be ordered according to various viewpoints, but one aspect is certainly whether the theory distinguishes between "geometrical" (or objective) and "visual" spaces.

Theories in which the notion of *visual space* doesn't occur are for instance theories that relate human (mis–)judgements to erroneous range estimates (we use "range" for the geometrical distance between some fiducial object and the observer, and "depth" for the observer's judgement of this range). We refer to such theories (*e.g.*, Gilinsky, 1951, 1955) collectively as the "sloppy rangefinder model". In such theories one often deals with "depth functions". Theories differ in many possible ways, for instance in whether a depth function is suggested from first principles or whether it is an empirical entity, whether the depth function is determined once and for all or depends on the perspective (for instance, to many observers the world appears to shrink when fog sets in), whether observers estimate absolute range or only range ratios, and so forth.

When a *fixed depth function* is assumed the sloppy rangefinder model can be framed in a slightly different way: for a given vantage point there exists a one–to–one relation (isomorphism) between geometrical space and another space (let's call it "visual space") in which the radial distance from the observer (the range) has been "corrected for the depth function". With a slight (but essential!) change of emphasis one might say that the observer's judgements are taken from the "visual space" ("in the head") *instead* of from geometrical space. One assumes some automatic mapping of geometrical on visual space, apparently due to the idiosyncratic structure of the human visual system. At first blush it might seem that this is mere word play. However, such a view invites novel questions. For instance, when the depth function is not the identity (that is to say, if the range *is* 10m the observer calls out something else than "10m"), then visual space is a locally *deformed* image of geometrical space: Should we expect that an object not only is judged to be at the (systematically) wrong distance, but also is judged to have the (systematically) wrong shape (notice that distant mountains look flat)? Moreover, are such shape errors *related* to the depth judgements via visual space?

* Here "perspective" means something like "setting", for instance, perceptions depend on vantage point, general illuminance level, whether we wear spectacles, are standing on our hands, are thirsty, etc..

Notice that the notion of a "visual space" of this type is *prima facie* an unlikely one: When there exists a fixed relation between visual and geometrical space, then why live with it? One could calibrate the visual system and operate as a more effective agent! It appears to be a definite evolutionary disadvantage.

The true "visual space models" go far beyond this. Typically they assume certain intrinsic properties of visual space, quite apart from the relation of visual space to geometrical space. For instance, in a well known report Luneburg (1947) argues that visual space *has to be* a Riemannian space of constant curvature. That visual space is Riemannian is suggested to Luneburg by the introspection that we cannot really imagine a space without an, at least locally, well-defined distance function and so forth and that the curvature is constant by the fact that we can imagine changes of position without changes in size or shape (isometries, or movements, are only possible if the curvature is constant). Such arguments may appear pretty weak, yet a considerable corpus of psychophysical effort has departed from these very assumptions (*e.g.*, Hardy, 1949; Hardy *et al.*, 1953; Zajaczkowska, 1956a, b; Blank, 1953, 1957, 1958, 1959, 1961), with only a few (notable) exceptions (*e.g.*, Foley, 1964; Indow *et al.*, 1962a, b, 1963; Indow, 1968).

Yet the notion of a "visual space" does have some attractive sides to it when taken in a more operational sense. Suppose we succeeded in bringing (at least partial) order in the confusing empirical data concerning human geometrical judgements. Such an order might well be described as an abstract geometrical entity (indeed, geometry *is* the study of possible order among the elements of sets). We may abstractly call it "visual space" without committing ourselves to the weird notion that we don't really see the world but only some deformed copy of it in our heads. Such a visual space would be quite structureless before empirical data are used to "shape" it. In building the structure the logical approach is by *adding* structure. One starts with the simplest, global features and adds structure till the systematic contents of the data are accounted for. This is much the reverse of the Luneburg program: as Blank saw it, visual space *was* Riemannian of constant curvature to start with and the task of psychophysics was merely to *measure* that curvature. What one should really do, is start to work on the foundations, but then a (possible) Riemannian property is far beyond the horizon. Prior to any metric one has to establish topological properties, projective structure, and so forth.

In this contribution we study aspects of projective structure, quite independent of any metrical properties. We also attempt to address the notion of "visual space" as associated with the sloppy rangefinder model.

14.2 The paradigm of exocentric pointing

Consider the operation of pointing at a target. An agent (doing the pointing) rotates and translates a rigid object that carries two disjunct fiducial marks (the "sight") such that these marks and the target form a *collinear* triplet of points in geometrical space. In some cases the sight may be part of the agent, *e.g.*, in pointing the eye, the agent establishes collinearity between the *fovea centralis*, the optical center of the eye and the target. In such cases the pointer (eye–ball) has only to be rotated. When pointing a rifle (in the regular way) the pointer (rifle) has both to be rotated and translated ("brought to the eye"). In both cases the observer's eye is collinear with the marks of the sight and the target. We refer to such cases as "egocentric pointing" (or "taking aim"). When, at a party, you observe John looking at Sally, you make a judgement that we refer to as "exocentric pointing". Here the marks of the pointer (John's head) and the target (Sally) are indeed collinear, but not also collinear with your eye. Likewise, when you point a gun via remote control, you don't "take aim" in the conventional sense, you perform an act of *exocentric pointing.*

The essential difference between egocentric and exocentric pointing is that the former can be done completely in the *visual field* (the marks of the pointer and the target should appear *coincident* in the visual field) whereas the latter involves collinearity in the *visual world.* Indeed, even when three points appear collinear in the visual field, they (generically) fail to be collinear in the visual world. For instance, all objects appearing on your visual horizon are collinear, yet they are (typically) not lined up in your visual world. In other words, egocentric pointing is trivial in the sense that it doesn't involve the depth dimension whereas exocentric pointing involves depth in an essential way.

The basic geometry is simple enough (Figure 14.1). We consider the simple case of vantage point, target and pointer in the horizontal plane at eye height of the observer. Reckon all distances from the observer, r_t the range of the target, r_p that of the pointer. Define the veridical pointing angle φ such that zero means pointing at the observer, veridical pointing some value in the range $(0°, 180°)$. Let the apparent separation of target and pointer be ϑ. Then elementary trigonometry yields the relation:

$$\varphi = \arctan \frac{\sin\vartheta}{\mu - \cos\vartheta}, \tag{14.1}$$

where $\mu = r_p/r_t$ denotes the range ratio of pointer and target. When the observer performs the exocentric pointing, the pointer will be set to some angle ψ (say) that will—in general—differ from the veridical angle φ. The difference $\psi - \varphi$ will be referred to as the "mispointing" or "exocentric pointing error".

Notice that exocentric pointing addresses the issue of *collinearity*, or—

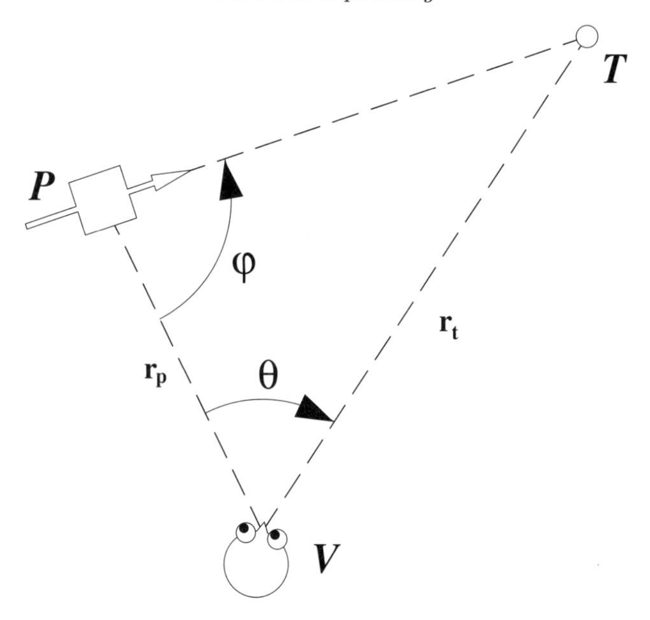

Fig. 14.1. The geometry of exocentric pointing. The observer defines the vantage point $\mathcal{V}$ and looks at the pointing device $\mathcal{P}$ and the target $\mathcal{T}$. The target is at egocentric distance r_t and the pointer at (different) egocentric distance r_p. Target and pointer are a visual angle ϑ apart. The task is to aim the arrow of the pointing device at the target, i.e., to adjust the angle φ to the correct value. Although the figure shows a veridical pointing, in general the observer will misadjust the angle φ and consequently take a wrong aim at the target.

equivalently—the relation of incidence of points and lines (notice that any two of the three points define a line). Thus it addresses purely *projective*, not *metrical* properties.

Because the pointer will usually be small with respect to the configuration as a whole (the triangle: observer–target–pointer), we may phrase the geometry somewhat differently: the pointing indicates the direction—at the location of the pointer—of the "straight line" (henceforth referred to as *pregeodesic*) between pointer and target. This may be taken as a working definition: it is up to empirical verification whether "visual space" admits of a nexus of pregeodesics such that (at least in some limited domain) any two points admit of a unique connecting pregeodesic arc at all. We speak of "pregeodesic" to distinguish these "straight lines" from the shortest (or extremal) paths of Riemannian geometry. The geodesics carry a distance scale, the pregeodesics do not.

14.3 Angular excess or deficit for equilateral triangles

Due to space limitations we discuss only a single experiment here: measuring the amount by which the angle sum of pregeodesic triangles of various sizes differs from 180°. Notice that this is the analogon of the fundamental experiment of Gauss in geometrical space.

Experiments were done outside in the open field in daylight. The terrain was flat, overgrown with weeds, at many places knee– or waist–, at places shoulder–height. Some buildings and trees appeared in the far distance. Observers were free to make body, head and eye movements in so far as the eye height and the location were not affected (thus they might turn on the feet or at the waist). They were encouraged to use both eyes. Target and pointer were placed at eye height and thus appeared on the visual horizon of the observer. Targets were spheres painted bright orange (thus contrasting well with the blue sky and green vegetation) placed on thin steel rods. Sizes were selected to keep their apparent size constant at $40'$. The pointer was a white cube with edge length of 25cm pierced by an orange arrow of total length 75cm. Although tip and tail of the arrow were defined as the fiducial marks of the pointer, the cube helps a lot to visually determine the way it is pointing. In fact, the major cues as to the pointing angle are probably monocular, rather than binocular. The subject controlled the pointing angle under remote radio control.

In this experiment pointer and target were placed at two of the vertices of equilateral triangles with the observer placed at the barycenter. This is a very special case, because the range ratio $\mu = r_p/r_t$ is unity, regardless of the size of the configuration. Thus the veridical pointing angle is simply 30° in all cases. The relevant parameter is simply the range (either r_p or r_t). We used sizes of the configuration such that the distance of the observer to the pointer (which equals that to the target) varied between 1.5–24 m (a 16–fold range). Notice that the visual angle between target and pointer is 120°, thus the observer *has* to move the head in order to be able to see both clearly. Three observers participated in the experiment.

A particularly intuitive representation of the results is obtained when we draw the configuration not with straight sides, but with the vertices connected with circular arcs (Figure 14.2). We determine the curvature of the arcs such that the initial directions at which the curves leave the vertices coincide with the average pointing angles. (One easily checks that this uniquely determines the arcs.) One can either regard this as merely one way to represent the raw data, or as an approximation to the pregeodesic arcs (revealed by the experiment) connecting the vertices. In any case, such figures most effectively summarize the result of the experiment.

Notice that the small triangles appear *inflated*, their angle sums *exceed* the Euclidean 180°, whereas the large triangles appear *deflated*, their angle sums *fall short of* the Euclidean 180° (see Figure 14.3). Since an angular excess indicates an elliptic curvature, and an angular deficit indicates a hyperbolic curvature, the results indicate in Luneburg's interpretation a *non–constant* curvature. The changeover from positive (elliptic) to negative (hyperbolic)

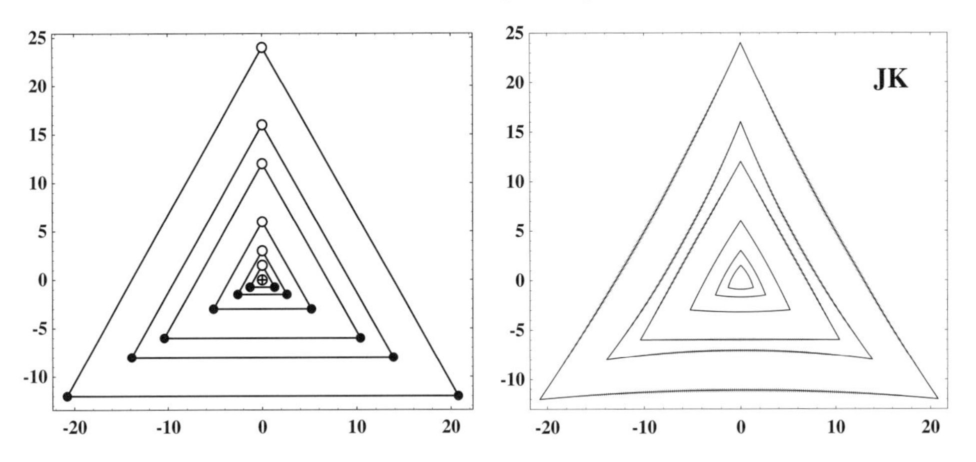

Fig. 14.2. Left: The configuration for the equilateral-triangles experiment. The targets are at distances of 1.5 m, 3 m, 6 m, 12 m, 16 m and 24 m from the origin. The pointer is at one of the vertices indicated by the open circles, targets are at the other vertices of the same triangle (filled circles). The location of the observer is at the barycenter (circle with plus sign). Right: Pregeodesic curvilinear arcs representation for the equilateral triangle results of observer JK. Dimensions are specified in meters. The curves are composed of a grey ribbon indicative of the standard error and a black curve indicating the mean. Results for the other observers are similar.

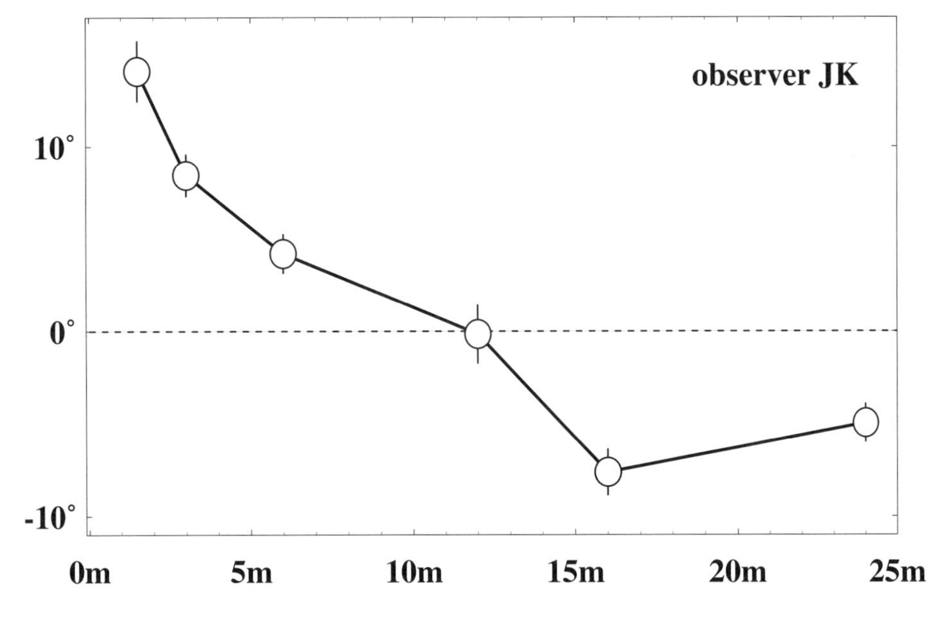

Fig. 14.3. The mispointing in degrees as a function of the distance to the pointing device (in meters) for the equilateral triangles configuration. Only results for observer JK are shown, but the results are similar for all observers. Pointings are typically nonveridical and are off by several degrees depending on distance. Observers change from overpointing in near space to underpointing in far space. The pointings are significantly non–veridical (the standard errors are indicated by the vertical line segments).

curvature violates Luneburg's assumption that visual space is necessarily a space of constant curvature.

14.4 Discussion of the experiment

Because the equilateral triangle configuration is so simple ($\mu = 1$) we have simple and very direct checks on the validity of the sloppy rangefinder model. When the observer misjudges the range and uses the correct formula to infer the pointing angle we expect

$$\psi = \arctan \frac{\sin \vartheta}{\xi - \cos \vartheta}, \tag{14.2}$$

where the *depth ratio* $\xi = d_p/d_t$. Let observers apply a depth function $d = F(r)$, then we expect $\xi = F(r_p)/F(r_t) = 1$ *regardless of* F because $r_p = r_t$. Thus the sloppy rangefinder model predicts veridical pointing throughout. When we assume that the observer estimates range *ratios* we expect a relation of the type $\xi = G(\mu)$. Such a relation has to satisfy $G(1) = 1$ in order to avoid pathologies that can be discarded offhand: again we predict veridical pointing. We conclude that sloppy rangefinder models simply fail to predict our results.

What happens when we consider the sophisticated sloppy rangefinder models that involve a *deformed visual space*? In such a case the deformation will change the shape of the pointer and thus the apparent pointing angle. A simple derivation yields:

$$\tan \psi = \frac{(r_p \frac{F'(r_p)}{F(r_p)}) \sin \vartheta}{\xi - \cos \vartheta} \tag{14.3}$$

in the general case, or, in the case of the equilateral triangles:

$$\tan \psi = \tan \varphi \, r \frac{F'(r)}{F(r)} \tag{14.4}$$

with

$$\tan \varphi = \frac{1}{3}\sqrt{3} \tag{14.5}$$

In this case we *can* predict our results: It is only necessary to select a depth function $d = F(r)$ such that:

$$\frac{d \log F(r)}{d \log r} = \sqrt{3} \tan \psi \tag{14.6}$$

Thus we easily find the required depth function by numerical integration of the empirical results, that is, up to an arbitrary scaling factor. Notice that this model is sufficiently general that *any* empirical result we might have found could be explained. Thus the model is rather weak. The issue is whether the depth function required for the fit is in general accord

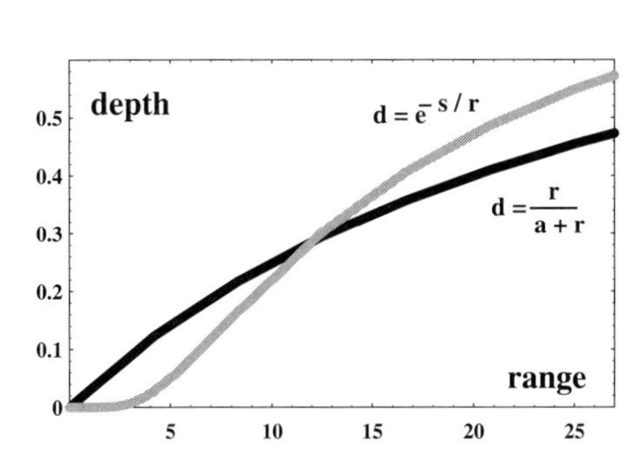

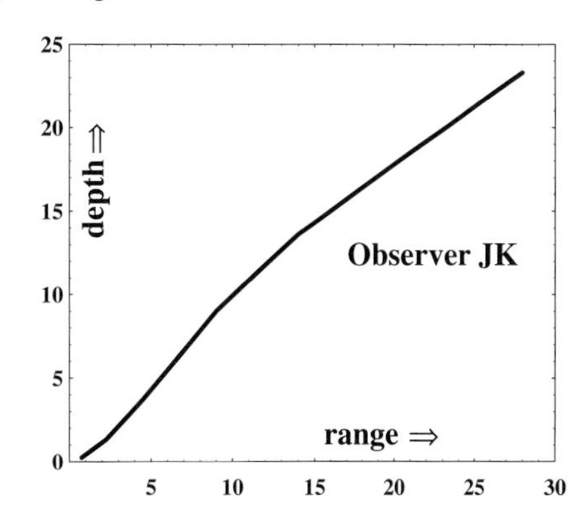

Fig. 14.4. Left: Two functions that have been used frequently in the literature to describe a functional relation between depth (subjective) and range (physical). We have adjusted the normalization of these functions such that both asymptotically (for infinite values of the range) approach unit depth. The relation $d(r) = r/(a+r)$ (usually written $d(r) = ar/(a+r)$) is compressive throughout and describes many empirical data remarkably well with $a = 30$ m. The relation $d(r) = \exp(-s/r)$ (usually written $d(r) = \exp(-\sigma\gamma)$, γ the vergence angle) is due to Luneburg. Instead of taking the value of s from the literature we picked a value ($s = 15$m) such that the two curves intersect in the middle of our experimental range. Right: Numerically integrated values of the logarithmic derivative of the depth function as calculated from the results of the equilateral triangles experiment for observer JK. These relations are only defined up to arbitrary scalings of the depth, thus the vertical units are arbitrary: They have (arbitrarily) been normalized to 9 at a distance of 9m. The range scale (horizontal) is in meters.

with findings from different experiments. Our empirical results reveal depth functions that are quite similar for the three subjects: the functions are accelerating in near space and decelerating in far space, much like the depth function suggested by Luneburg but qualitatively unlike the type of depth function suggested by many experiments and perhaps best characterized as Gilinsky's depth function (Gilinsky, 1951, 1955). The latter is decelerating throughout (see Figures 14.4 and 14.5).

In the sloppy rangefinder model, one maps geometrical into visual space via the depth function. The observer is assumed to employ straight line constructions in visual space and the results are backtranslated to geometrical space. Thus the "pregeodesics" predicted by this model are the preimages of the straight line in visual space as they appear in geometrical space. When the depth function is accelerating, equilateral triangles with the barycenter at the origin will appear inflated, when the depth function is decelerating, they will appear deflated.

Notice that we have been able to falsify the full class of sloppy rangefinder models although this particular experiment does not have the power to decide the issue on the "visual space" type of sloppy rangefinder models.

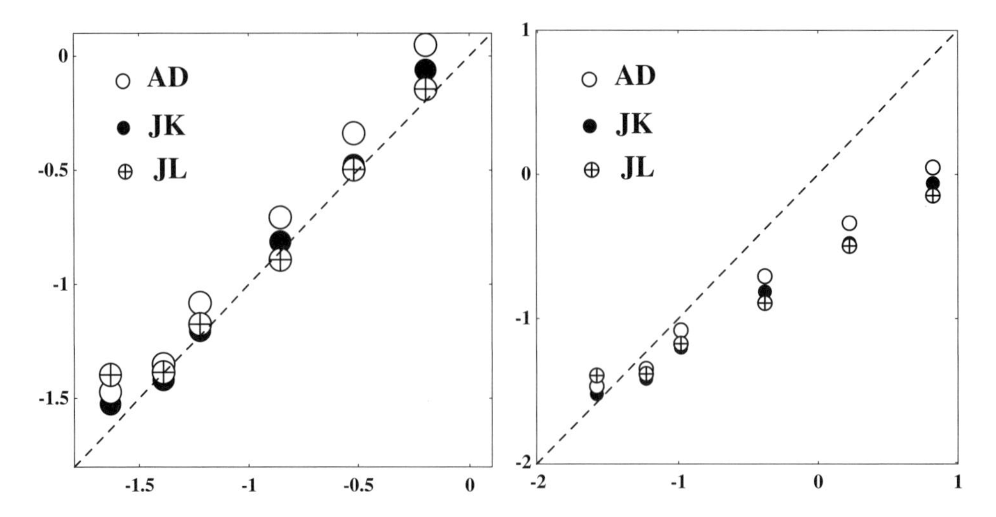

Fig. 14.5. Left: Log–log plot of the logarithmic derivative of the depth function as calculated from the results of the equilateral triangles experiment and those computed from Gilinsky's expression (Gilinsky, 1951, 1955); Right: Log–log plot of the logarithmic derivative of the depth function as calculated from the results of the equilateral triangles experiment and those computed from Luneburg's expression (Luneburg, 1947).

14.4.1 Is visual space Riemannian?

As already noted, our empirical results evidently falsify Luneburg's assumption of a *homogeneous* (constant curvature) visual space. Do the results bear on the issue of whether visual space is Riemannian? We think not, but this is also the case for many of the experiments described in the literature that purport to bear on the matter: many simply *assume* Riemannian structure but don't actually *test* it. For instance, one "measures" the curvature of visual space *under the assumption* that the space is Riemannian. The topic is of sufficient interest that we dedicate some space to it here.

With "space curvature" we mean the Gaussian (or intrinsic) curvature of the horizontal plane at eye height. According to Gauss (1880) the space curvature integrated over the area of the triangle (the so called *curvatura integra*) equals the angular excess. Thus one could calculate the average curvature over the triangle as the angular excess divided by the area. When we calculate the curvatures using the *areas in geometrical space* we may, by differencing, estimate the intrinsic curvature as a function of the range. Then, by numerical integration (Bour, 1862; Spivak, 1975; Struik, 1950), we may construct a surface (figure 14.6) with the following properties: the surface is defined by a height function over the horizontal plane at eye height; it is rotationally symmetric with the observer at the center; It is as flat as possible; and finally—and most importantly—it has geodesics that, when projected on the horizontal plane at eye height, reproduce the empirically

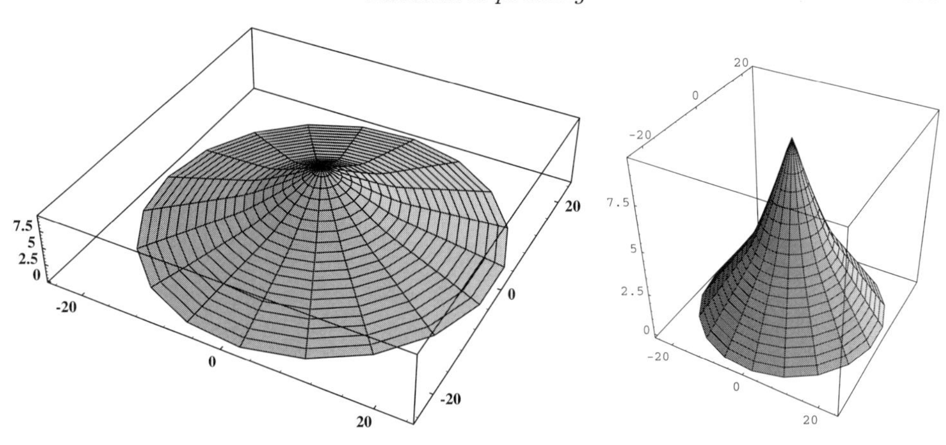

Fig. 14.6. Geodesic surface for observer JK. This surface has the Gaussian curvature as estimated from the equilateral triangle angle excesses. The surface is a surface of rotation by assumption. The range is measured as arclength along meridians from the axis of rotation. All distances are in meters. In the picture on the right the vertical scale has been expanded to yield a better rendering of the essential shape. Notice the conical singularity at the origin.

obtained pregeodesic arcs. This is feasible, and the result indeed reproduces the empirical data within the experimental scatter. Such surfaces are quite similar for the three observers. They have a conical singularity at the origin, have an elliptically curved region in near space changing into a hyperbolical region at far space, and tend to flatten out at the very large ranges. The curved plane also induces a metric: the geodesic distance, that is the distance measured over the curved plane. Thus we can calculate a depth function from the curvature representation. This depth function is again accelerating in near space and decelerating in far space.

The curved plane is indeed one possible way to summarize the results of the experiment. As summaries go it is about as efficient as the sloppy rangefinder visual space model: in both we have to fit an arbitrary function of range, intrinsic curvature in the former, depth in the latter. In both cases we deal with properties of observer's judgements in *geometrical space*, not with properties *intrinsic to visual space.* In the sloppy rangefinder model the range doesn't belong to visual space and in the curvature model we used the triangular areas in geometrical space.

The two models are remarkably similar in many respects. Both manage to predict a nexus of pregeodesics *overlaid on geometrical space* that summarizes the results of exocentric pointing experiments. They are essentially different *in principle* though. One can prove that such models can be equivalent *only* for the case of constant curvature. Thus the sloppy rangefinder model is less general than the curved space one. It is conceivable that future experiments might decide between these alternatives (though the present experiment doesn't).

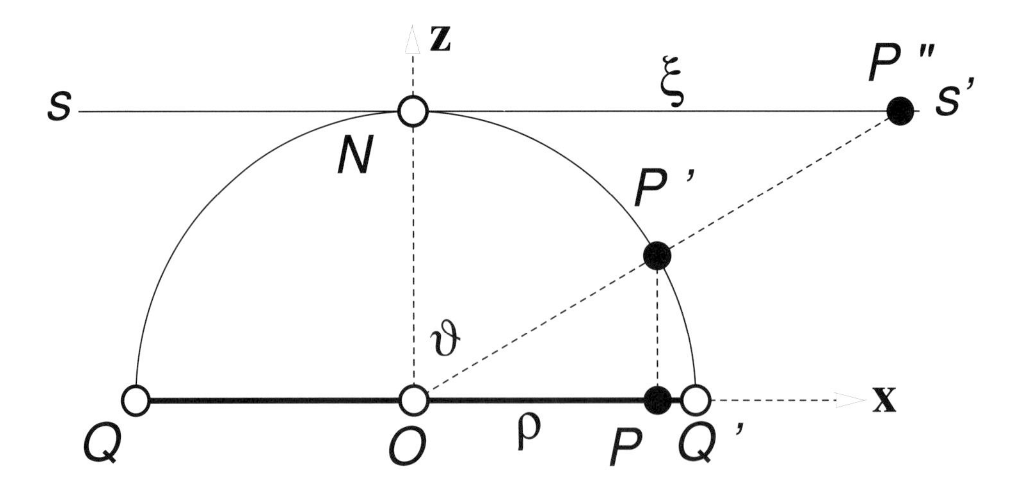

Fig. 14.7. The elliptic model. Shown is only a section (the xx–plane). The point $\mathcal{P}$ in the geometrical space $\mathcal{Q}\mathcal{Q}'$ is at a distance ρ from the origin. The curved space is the hemisphere $\mathcal{Q}\mathcal{N}\mathcal{Q}'$, the relation between the curved space and geometrical space is simply orthogonal projection: thus the point $\mathcal{P}'$ in the curved space corresponds to the point $\mathcal{P}$ in geometrical space. In the curved space the point $\mathcal{P}'$ is at geodesic distance $\vartheta = \arcsin \rho$ from the origin. The visual space of the sloppy rangefinder model is the plane ss', the point in this "visual space" $\mathcal{P}''$ is found via central projection (from the origin $\mathcal{O}$) of $\mathcal{P}'$. Thus the "depth" ξ as a function of the "range" ρ is $\xi = \tan \vartheta = \rho(1 - \rho^2)^{-1/2}$. Because of the central projection the geodesics of the hemisphere correspond to the straight lines of visual space and *vice versa*. However, straight lines in visual space correspond to the projections of great circles, that are elliptical arcs, in geometrical space: these are the "pregeodesics" in this model.

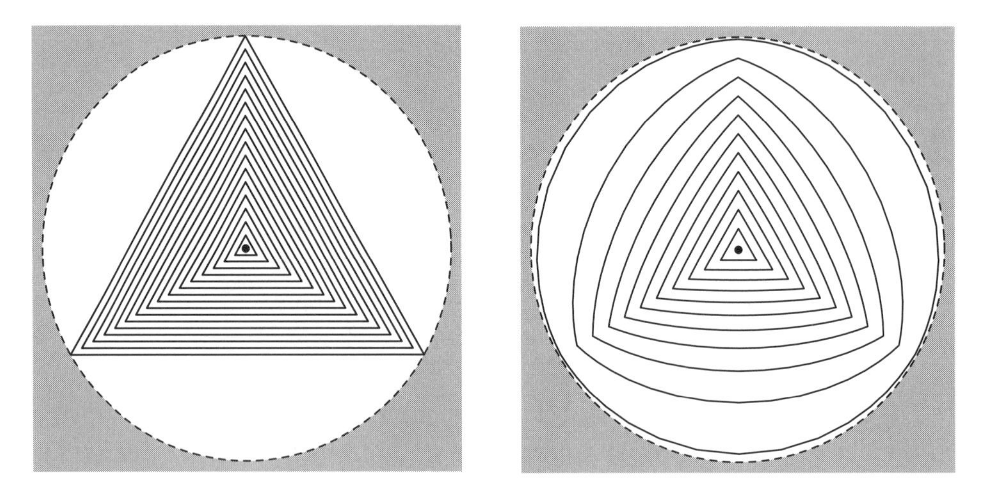

Fig. 14.8. On the left, some equilateral configurations of assorted sizes in geometrical space. On the right, the corresponding pregeodesic triangles in the simple elliptic model. The predictions of the curved space and the sloppy rangefinder models are identical.

In order to illustrate the relations between the models, we present here a simple example for the case of elliptical curvature. This model is not meant to be realistic at all, but merely to serve in developing an intuition for the essential structures. We take the Euclidean plane as a model for geometrical space. The location of the observer is taken as the origin. We locate points via Cartesian coordinates (x, y) or polar coordinates (ρ, φ) (where $x = \rho \cos\varphi$ and $y = \rho \sin\varphi$). In the curved space model we use the upper unit hemisphere as a model. It resides in three–dimensional space: we will parameterize it with Cartesian coordinates (x, y, z) where the first two coordinates are taken to coincide with those of geometrical space, thus the "polar distance" is the angle $\arcsin \rho$. The metric is the induced metric, that is to say, the geodesics are the great circles of the hemisphere (Figure 14.7). We will define the pregeodesics as the projections of these geodesics on the plane, stripped of their distance scales. Notice that we can only deal with the interior of the unit disk of geometrical space in this model, for everything farther away than unit distance is "beyond infinity". Indeed, although very small triangles are hardly deformed, larger triangles become inflated and a triangle for which the vertices are at unit distance from the origin becomes a circle (Figure 14.8)! Larger configurations cannot be dealt with. Can we find an equivalent sloppy rangefinder model? We can. Consider a "visual space" that is a plane. The origin is at the center. We assume a distance function $d = r/\sqrt{1-r^2}$, that is an accelerating function (indeed, the acceleration is infinite at unit range!). In this model visual space (a complete plane) is the image of the interior of the unit circle in geometrical space. In this case there is a very simple relation between the visual space of the sloppy rangefinder model and the curved surface: we may regard the visual space as the central projection of the unit hemisphere on the tangent plane at the "north pole" ($z = 1$). It is geometrically evident that great circles on the hemisphere correspond to straight lines in visual space and *vice versa*. The predictions of these two—apparently very different—models with regard to exocentric pointing (the pregeodesics) are necessarily *identical* in all cases.

It is easy enough to construct a similar model for the hyperbolic case. However, it is *not* possible to do so for the case of non–constant curvature (Beltrami, 1865). This is perhaps less important in practice than it may seem in theory though since the observable differences will only be slight.

From the Luneburg perspective (Riemannian spaces) the metric of this model of curved space would be a spherical geometry. The sheer existence of a metric would allow additional predictions (indeed, as done by Luneburg). For instance, in the present models we are able to predict the results for "parallel alley" experiments (Figure 14.9). Indeed, the parallel alley would be curved and convergent. However, we are in no position to predict the results for "distance alley" experiments: for this a *metric* is required. In the classical literature the two cases are hardly distinguished

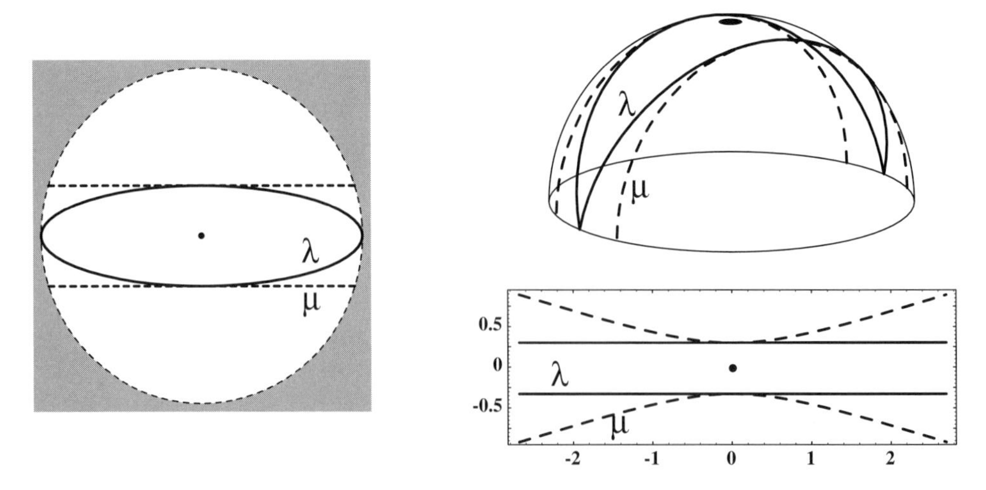

Fig. 14.9. Parallel and distance alleys in the simple elliptic model. On the left the parallel (λ) and distance (μ) alleys in geometrical space. On the top right the alleys in the curved space: here the parallel alleys are great circles (geodesics) that meet on the horizon and the distance alleys are parallel (thus equidistant) small circles. On the bottom right the alleys in the (sloppy rangefinder) visual space: here the parallel alleys are simply a pair of parallel straight lines. The distance alleys become hyperbolæ.

(*e.g.*, Hardy *et al.*, 1951; Indow & Watanabe, 1984). For instance, Luneburg notices the different predictions for these cases but does not distinguish them with regard to geometrical prerequisites. When we accept the metric of spherical geometry (Euclidean arc length along great circles), the distance alley is seen to diverge from the parallel alley, (on the hemisphere a pair of parallel small circles). In the case of the sloppy rangefinder model the issue is less clear: when we assume a metric that yields the same distance alley prediction as the spherical model, then the distance alley *in visual space* would be a pair of hyperbolæ, which is hardly intuitive. When we take the Euclidean metric in visual space (perhaps best reflecting the spirit of the sloppy rangefinder models), the parallel and distance alleys coincide. Indeed, an arbitrary sloppy rangefinder visual space with an Euclidean metric will be unable to generate different predictions for the parallel and distance alleys. When you are not prepared to accept the additional introduction of some non–Euclidean metric (in addition to the distance function that is), this class of models is falsified by the classical Hillebrand and Blumenfeld experiments (Luneburg, 1947).

It is always possible to construct a curved surface to account for the exocentric pointing data as long as the latter are consistent, *i.e.*, when the nexus of pregeodesics is isotropic. Indeed, with the observer free to turn on the spot one expects the data to be isotropic. The pregeodesics on the curved surface can be constructed by noticing that they meet the meridians in angles

α such that $\zeta \sin \alpha$ is constant, where ζ denotes the Euclidean distance (through space, not over the surface) to the axis of revolution (Clairaut's theorem: Clairaut, 1733). Conversely, we may construct the surface from the pregeodesics. Whether such a surface will also account for metrical properties (*e.g.*, distance alley data) is an empirical issue.

14.5 The egocenter

The fact that we cannot construct a geodesic surface that explains our data without a *conical point* at the origin is intriguing. From our data we know that the semi top angle of the cone is 81°–84°. The origin is thus a singularity, a concentration of elliptic curvature. Geometrically this means that we can even have a non–degenerated *biangle**, as long as it contains the origin. Operationally this means that when the observer points to a target that appears at 180° from the pointer, there exist *two distinct* pregeodesics. Either choice corresponds to an exocentric pointing, thus the observer may be expected to pick different alternatives on different trials. We find that this is indeed the case. We can influence the observer's choice by performing a series of pointings for which the apparent distance of the target from the pointer is gradually increased to 180°. The *direction* of approach determines the choice (Figure 14.10).

What happens is the following: typically observers place their feet in the direction of the bisectrix of the target and pointer directions and look back and forth between them using a combination of a twist of the body, head and eye movements. Thus the observer will face *opposite directions* when approaching the 180° condition from either side. We find that our observers mispoint by about 5° (figure 14.10). This is almost exactly what is expected from the geodesic model. This method works well, and is indeed to be preferred over the "direct" method of having observers perform the exocentric pointing task in the 180° condition: in this case, observers have no preferred stance and results are very variable. In most cases observers simply point to themselves.

Though it is interesting that the geodesic surface predicts this effect, it can hardly be said that it throws any light on what is happening. The observation that the observer has a choice of two different stances is more illuminating in this respect. A rather direct mechanistic explanation involves the assumption that the observer reckons directions from an *egocentre* (one should not speak of "apparent egocentre" since an objective correlate does not exist) that is eccentric with respect to the centre of the head (from which we reckon directions in the analysis). Indeed the egocentre may assume different positions such as the anterior nodal point of the eye (for strict

* In Euclidean geometry a biangle is always degenerated to a doubly-covered line segment, the simplest non–degenerated polygons are triangles.

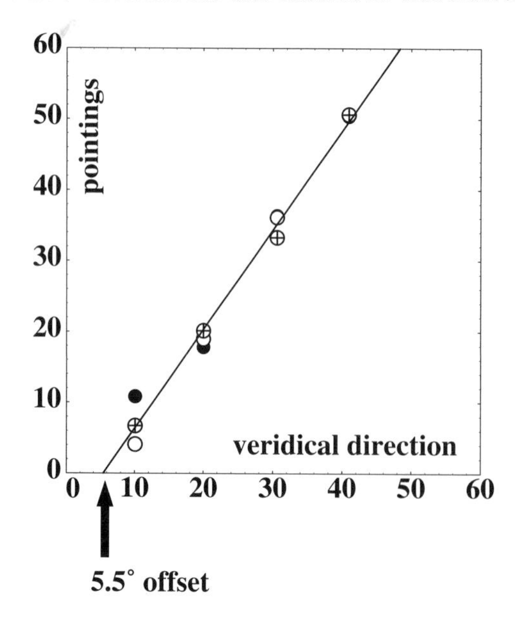

Fig. 14.10. Offsets for three observers. The 180° condition was approached from a single direction. The limiting 180° case has veridical pointing 0°. All observers end up at 5° though.

monocular fixation), the centre of the eyeball (for free monocular viewing), the centre of the binocular segment (for binocular viewing of observers without strong eye dominance), the atlas vertebra (for free head movements), and so forth. We may calculate the position of the egocentre required to account for our data. We find values of 13–26cm behind the interocular segment, that is roughly the neck area.

Can the sloppy rangefinder models may also account for this effect? The answer is a subtle one. When we assume a smooth depth function that is monotonic and such that zero depth corresponds to zero range, then (in the 180° case) the model predicts only a single pregeodesic that passes through the origin. However when we consider pregeodesics for neighbouring configurations, we find that the approach of the singular case is not a linear one. When we consider pointings with a pointer of finite size we may obtain results that are similar to the data. It is necessary to assume an accelerating depth function in order to obtain qualitative agreement. In the description of the actual data, the prediction of the distance function model need not be significantly different from that of the geodesic surface model. The models seem to be empirically indistinguishable (Figure 14.11).

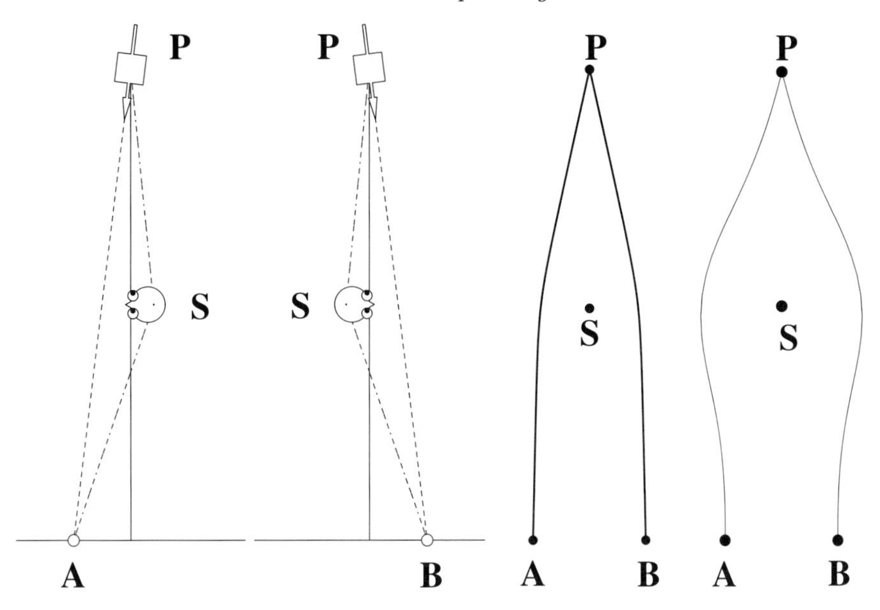

Fig. 14.11. Left: The influence of an eccentric egocentre. Two possible options for the stance of observers occur for the case that the points A and B are collinear with the pointer and the interocular midpoint. Middle: Prediction of a geodesic model. Here the model is simply a cone with apex at the origin. There exist two distinct pregeodesics in the limiting case. Right: Prediction of a sloppy rangefinder model with accelerating depth function (square). In the limiting case there exists only a single geodesic.

14.6 Conclusions

We have presented a novel paradigm—exocentric pointing—and some preliminary data. By themselves the data already suffice to falsify two large classes of models, namely the strict sloppy rangefinder models (no "visual space") and the homogeneous spaces (Riemannian spaces of constant intrinsic curvature). The data do not allow us to falsify sloppy rangefinder models of the visual space type nor models based on Riemannian metrics.

What we believe is more important here is that we have identified various strata in the structure of visual space (in the appropriate abstract sense). The present data only relate to the projective structure, quite apart from any metric. In general we argue that the constructive way of approach is to address the topological, projective and metrical aspects in their order of dependency. In the past following the pionering work of Luneburg one has tried to build the house starting from the roof so to speak. Geometricians have identified and studied many geometries with weaker structure than the full Riemannian one. Moreover, Luneburg's geometry *assumes* a strong foundation of geometrical structure for which we have no empirical evidence at all. Thus we need to stratify our experimental questions in much the same way as geometricians have been able to stratify the structure of geometries.

Unfortunately we have to go a long way here before we reach the Luneburg Olympus and—much more likely—we may never even get there.

References

Beltrami, E. (1950). (originally 1865) In *Lectures on Classical Differential Geometry*, D. J. Struik (Ed.), 178, Dover, New York, NY.

Blank, A. A. (1953). "The Luneburg theory of binocular visual space", *J. Opt. Soc. Am.*, 43: 717–727.

Blank, A. A. (1957). "A geometry of vision" *British Journal of Physiological Optics*, 14: 154–222.

Blank, A. A. (1958). "Analysis of experiments in binocular space perception", *J. Opt. Soc. Am.*, 48: 911–925.

Blank, A. A. (1959). "The Luneburg theory of binocular perception", in *Psychology: A study of a science, Vol. I, Sensory, Perceptual, and Physiological Formulations*, S. Koch (Ed.), McGraw–Hill, NewYork, NY, 395–426.

Blank, A. A. (1961) "Curvature of binocular visual space. An experiment", *J. Opt. Soc. Am.*, 51: 335–339.

Bour, E. (1950). (originally 1862) In *Lectures on Classical Differential Geometry* D. J. Struik (Ed.), 177, Dover, New York, NY.

Clairaut, A. C. (1950). (originally 1733) In *Lectures on Classical Differential Geometry* D. J. Struik (Ed.), 134-135, Dover, New York, NY.

Foley, J. M. (1964). "Desarguesian property of visual space", *J. Opt. Soc. Am.*, 54: 684–692.

Gauss, K. F. (1880). "Disquisitiones generales circa superficies curvas", in *Werke* (Göttingen) Vol 4, 217–258.

Gilinsky, A. S. (1951). "Perceived size and distance in visual space", *Psych. Rev.*, 58: 460–482.

Gilinsky, A. S. (1955). "The effect of attitude upon the perception of size", *J. Psych.*, 68: 173–192.

Hardy, L. H. (1949). "Investigation of visual space", *Arch. Ophthalmol.*, 42: 551–561.

Hardy, L. H., Rand, G. & Rittler, M. C. (1951). "Investigation of visual space: The Blumenfeld alleys", *Arch. Ophthalmol.*, 45: 53–63.

Hardy, L. H., Rand, G., Rittler, M. C., Blank, A. A., & Boeder, P. (1953). *The Geometry of Binocular Space Perception*, J Schiller, Elizabeth, NJ.

Indow, T. (1968). "Multidimensional mapping of visual space with real and simulated stars", *Percep. and Psych.*, 3: 45–53.

Indow, T., Inoue, E. & Matsushima, K. (1962a). "An experimental study of the Luneburg theory of binocular space perception (1). The 3– and 4–point experiments", *Japanese Psych. Res.*, 4: 6–16.

Indow, T., Inoue, E. & Matsushima, K. (1962b). "An experimental study of the Luneburg theory of binocular space perception (2). The alley experiments", *Japanese Psych. Res.*, 4: 17–24.

Indow, T., Inoue, E. & Matsushima, K. (1963). "An experimental study of the Luneburg theory of binocular space perception (3). The experiments in a spacious field", *Japanese Psych. Res.*, 5: 10–27.

Indow, T. & Watanabe, T. (1984). "Parallel–alleys and distance–alleys on horopter-plane in the dark", *Perception*, 13: 165–182.

Luneburg, R. K. (1947). *Mathematical analysis of binocular vision*, Princeton University Press, Princeton, NJ.

Spivak, M. (1975). *A comprehensive introduction to differential geometry Vol. III*, Publish or Perish Inc., Houston, TX.

Struik, D. J. (1950). *Lectures on Classical Differential Geometry*, Dover, New York, NY.
Zajaczkowska, A. (1956a). "Experimental test of Luneburg's theory. Horopter and alley experiments", *J. Opt. Soc. Am.*, 46: 514–527.
Zajaczkowska, A. (1956b). "Experimental determination of Luneburg's constants σ and K", *Quarterly Journal of Experimental Psychology*, 8: 66–78.

15

Vision and the level of synergies

S. Mitra, M. A. Riley, R. C. Schmidt and M. T. Turvey

15.1 Introduction: synergies in coordinated action

15.1.1 Pivotal role of the 'muscle sense': Lessons from peripheral neuropathy

Imagining or simulating the effects of losing sight or losing hearing is simple. In contrast, imagining or simulating the effects of losing touch, especially when touch is fully construed as the varied achievements of the haptic perceptual system, is extremely difficult. The haptic perceptual system is that system by which one knows the body, and objects adjacent to or attached to the body, by means of the body (Gibson, 1966). Mechanoreceptors embedded in muscle, tendon, and other tissues constitute the sensory basis for haptic perception. They are stimulated when the body's tissues are distorted, either by movement itself or by environmentally imposed contact forces. The consequences of losing haptic perception have been recently documented (Cole, 1995; Cole & Paillard, 1995). Peripheral neuropathy is the absence of afferentation from the large myelinated fibers (originating from skin, muscle, and tendon receptors) that compose the dorsal thalamic tract, but with other neural functions essentially intact (e.g., motor nerves are normal). One major documented case is a person who has no sense of the body or attachments to the body below the collar bone. His purposeful movements (e.g., of even just one finger, not all of them) rely on vision. Such movements, however, are very restricted and can only be conducted with considerable concentration and intellectual effort. A momentary lapse of concentration or an ancillary intellectual effort results in loss of coordination (e.g., sneezing when standing, and note taking while sitting, produce severe imbalance). Standing, walking, reaching, and manipulating are challenging if not impossible tasks for this sufferer of peripheral neuropathy, and the ability to perceive by eye proves to be a poor substitute for his inability to perceive by muscle.

15.1.2 Bernstein's levels of synergies and space

For Bernstein (1996), the intimate connection between haptic proprioception and muscular activation defines the relatively autonomous *level of muscular-articular links or synergies*. This functional level is responsible for assembling spatio-temporal relations among multiple body segments, relations that can be reproduced reliably in varied circumstances and sustained in the face of perturbations. It is this large-scale pattern production facility that is no longer available in the absence of haptic perception.

The level of synergies' ability to guarantee internal coherence of large-scale movements contrasts, however, with its inability to adapt coordinations to external, environmental demands. Such adjustments must be introduced by higher functional levels. In Bernstein's (1996) hierarchy, these higher functional levels are the *level of space* and the *level of actions*. Adjusting synergies to the visible spatio-temporal requirements of a task is the role of the level of space; adjusting, switching, and creating synergies to satisfy an intention (goal or plan) is the role of the level of actions. By analogy with an argument of Marr's (1977), because synergy formation and adaptation in the face of varied complexities are so fluent, reliable, and widely manifest in biological movement systems, they must express principles of the most basic and general kind. In the research reported in the present chapter, such principles at work in the interplay of the levels of synergies and space are sought within the framework of contemporary dynamical systems theory.

15.2 Coordinative structures: a dynamical systems framework

15.2.1 The concept of coordinative structures

Movement synergies are compelling instances of pattern formation. They are expressions of high-dimensional dynamical systems that come to be variously self-constraining, exhibiting rich, coherent, but essentially low-dimensional behavior. Current thinking on movement synergies draws heavily on Bernstein's (1967, 1996) "degrees of freedom problem," interpreted as the problem of how a movement system's multiple components can be encapsulated within controllable subsystems of few dimensions (Turvey, 1990). Early thinking on this problem led to the notion of the coordinative structure, defined as a group of relatively independent muscles, often spanning several joints, that is (informationally) constrained to act as a single functional unit (Turvey, Shaw, & Mace, 1978). Subsequent refinements have led to the dynamical systems framework of Figure 15.1. At the level of observable behavior, a coordination may often be characterized by a single relational, collective variable ξ (or, at most, by a few such variables) that is governed by a first-order, autonomous differential equation of motion. The variable

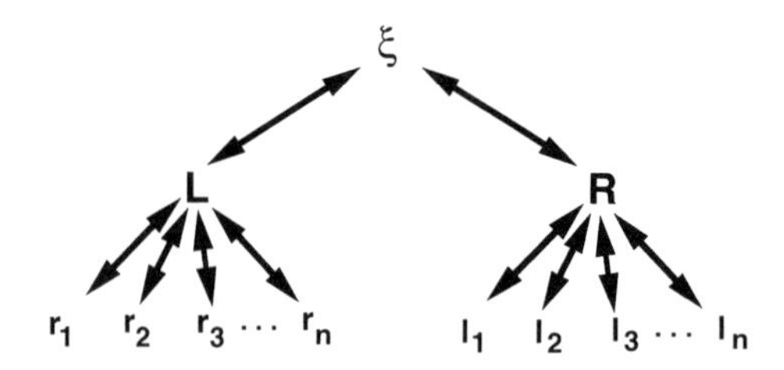

Fig. 15.1. Schematic of a coordinative structure. The symbol ξ is the relational/collective variable identifying the coordination pattern defined over the subsystems L and R (e.g., left and right upper limbs) with their component subsystems (l_i and r_i, respectively). See text for details. (Adapted from Haken, 1996.)

ξ is relational because it captures a spatio-temporal relation between movement subsystems (L and R, in Figure 15.1) and it is collective because its dynamics are realized through the cooperative or coupled activity of the subsystems that themselves express dynamics describable by first-order, autonomous differential equations. The variables characterizing the subsystems are also collective, but only with respect to the mechanical, muscular, and nervous components of the one or few *biomechanical degrees of freedom* they subsume, not with respect to the overall coordination pattern. For the subsystems L and R, each of the variables l_i and r_i, respectively, are its *active, dynamical degrees of freedom* (ADF). For the coordination as a whole, ξ is an ADF. Accordingly, the coordinative structure, the functional organization of subsystems that enables the coordination, consists of all the participating ADFs and their equations of motion.

A concrete example is presented by the extensively-studied task of one-to-one frequency-locked rhythmic coordination of two limbs or limb segments. The operative relational variable here is the difference between the left (L) and right (R) phase angles θ ($\phi = \theta_L - \theta_R$) of the oscillating segments, and the two intrinsically stable coordination modes, in-phase (segments move symmetrically) and anti-phase (segments move anti-symmetrically), correspond to $\phi = 0$ rad and $\phi = \pi$ rad, respectively. These stabilities can be obtained from a motion equation (i.e., a coordination dynamic) in ϕ (Haken, Kelso, & Bunz, 1985; Schöner, Haken, & Kelso, 1986):

$$\dot{\phi} = -a \sin(\phi) - 2b \sin(2\phi) + \sqrt{Q}\zeta_t \qquad (15.1)$$

where $\dot{\phi}$ is the first time derivative of ϕ, and the ratio b/a determines the relative strengths of the attractors at $\phi = 0$ and $\phi = \pi$. Experiments support the intepretation of b/a as inversely related to the frequency at which the coupled limbs are required to oscillate (e.g., Schmidt, Shaw, & Turvey, 1993; Sternad, Schmidt, & Turvey, 1992). The last right-hand term $\sqrt{Q}\zeta_t$ is a noise term representing fluctuations in the coordination due to underlying microstructural activity not fully subsumed by the rhythmic synergy.

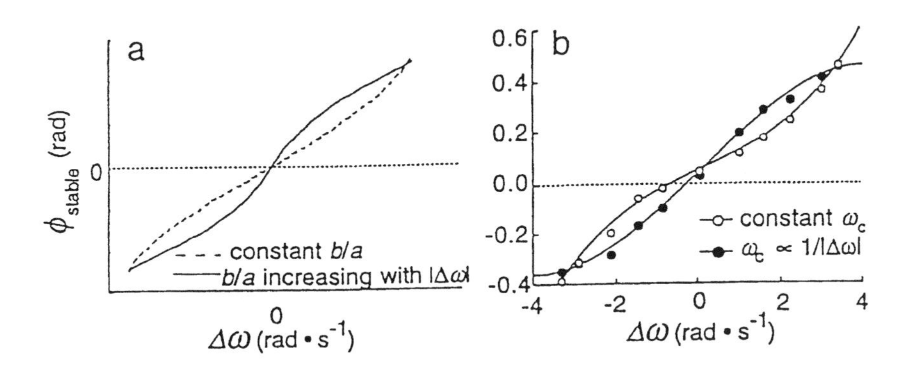

Fig. 15.2. Predictions from Equation (15.3) as a function of δ (here labeled $\Delta\omega$) and b/a (a) and supporting experimental data (b) from Amazeen, Sternad and Turvey (1996).

The fixed points (equilibria) of Equation 15.1, for a given b/a, are determined by solving for $\dot{\phi} = 0$. A zero-crossing where $d\dot{\phi}/d\phi < 0$ defines a stable fixed point (attractor), and where $d\dot{\phi}/d\phi > 0$ defines an unstable fixed point (repellor). The expected variability (standard deviation: SD) around a fixed point is:

$$SD\phi = (Q/2|\lambda|)^{1/2} \tag{15.2}$$

(Schöner *et al.*, 1986), where Q is the strength of the noise term and $\lambda = d\dot{\phi}/d\phi$ is the Lyapunov exponent of the stable state (e.g., Hillborn, 1994). Thus, a steeper negative slope at a zero crossing of Equation 15.1 means a larger absolute λ, a smaller variance in ϕ, and a fixed point that is more readily retained against perturbations.

When the two limbs or segments have different intrinsic or preferred frequencies, the symmetry of Equation 15.1 is broken by introducing an imperfection parameter δ (Kelso, Delcolle, & Schöner, 1990):

$$\dot{\phi} = \delta - a\sin(\phi) - 2b\sin(2\phi) + \sqrt{Q}\zeta_t \tag{15.3}$$

For $\delta \neq 0$, fixed points of Equation 15.3 are shifted from 0 rad and π rad in direction and magnitude dependent on the sign and size of δ. Rand, Cohen, & Holmes (1988) and Koppell (1988) have modeled δ as the uncoupled frequency (ω) difference between the individual oscillators ($\delta = \omega_{left} - \omega_{right}$). Experimental investigations usually adopt this interpretation as well (e.g., Amazeen, Sternad, & Turvey, 1996; Jeka & Kelso, 1995). Predictions of Equation 15.3 regarding fixed point locations and their corresponding variability under manipulations of b/a and δ (shown in Figure 15.2) have been

Fig. 15.3. Experimental set-up for between-persons coordination experiments. Each participant swings a hand-held pendulum about the wrist joint within a motion capture chamber. Sonic emitters are placed at the distal tips of the pendulums, and microphones within the chamber detect their position. Participants are instructed to watch the movements of the other participant's pendulum.

confirmed experimentally. Detailed reviews can be found in Kelso (1994) and Schmidt and Turvey (1995).

In the scheme of Figure 15.1, the above analysis of the basic bimanual rhythmic synergy corresponds to the coordination dynamic at the level of ξ. Clearly, the participating subsystems L and R, the left and right arms, support the assembly and retention of this dynamic by means of coordinated activity over their own muscular-articular links. Such subsystem-level coordination has its own set of ADFs, and both the number and the dynamics of these are currently under investigation (e.g., Kay, 1988; Beek & Beek, 1988; Mitra, Riley, & Turvey, 1997; Mitra, Amazeen, & Turvey, 1998). In terms of the Bernstein hierarchy, however, both these levels of dynamical analysis address organization within the level of synergies. The present chapter is primarily concerned with the roles that visual perception, a source of level of space constraints on the level of synergies, might play in shaping coordinative structures. We move, then, to the various qualitatively different ways in which vision might influence the dynamics of these systems.

15.2.2 Levels of dynamics in complex systems

Insofar as coordinative structures are complex systems, attempts to understand changes in their morphology can benefit greatly from the observation that such systems contain at least three types of dynamics: state dynamics, parameter dynamics, and graph dynamics (Farmer, 1990; Saltzman & Munhall, 1992). State dynamics refer to the evolution rules that directly govern motions of the system's state variables, such as position or velocity of a pendulum, or activation levels of individual neurons in a neural network.

By the same token, changes in parameters such as damping or stiffness in a mechanical mass-spring system, or inter-neuron connection weights in a neural network, are governed by the respective systems' parameter dynamics. Finally, graph dynamics refers to processes that govern changes in a system's dynamical 'architecture,' which may involve changes in both the type and number of its state variables and parameters. A mechanical system of coupled oscillators that systematically loses or gains oscillators, or a neural network with a rule that 'prunes' neurons, both exhibit graph dynamics. In systems that are as adaptive and malleable as biological movement synergies, all three types of dynamics are ubiquitous. In the present context, the latter implies that vision may affect the level of synergies in any or all of these three ways.

15.3 Possible types of visual influence on the level of synergies

Equation 15.3 describes the state dynamics at the highest relational level of the bimanual rhythmic synergy. Since the state variable ϕ is an informational quantity, success of the coordination is critically dependent upon its accurate detection and continued availability, normally achieved by means of haptic proprioception. It is also possible, however, to determine relative phase visually, and, as such, visual proprioception may affect or even enable coordination at the level of relational state dynamics. If, additionally, visual determination of relative phase produces such effects as changes in the strength of L-R coupling or of functional L-R asymmetry, then these would constitute visual influences at the level of relational parameter dynamics. Experimental observations of these two types of visual influence are described in §15.4.

Coordination of rhythmic oscillations of a single arm or arm segment also presents possibilities for visual influences on all three types of dynamics. Recent work on the dynamical structure of single segment rhythmic synergies in both isolated and bimanually coupled movement settings indicates that their dynamics involve between three and five ADFs (Mitra, Riley, & Turvey, 1997; Mitra, Amazeen, & Turvey, 1998). Since the precise form of state dynamics that can accommodate these data are not currently known, a rigorous demonstration of state or parameter dynamical effects on such synergies due to vision is not yet possible. However, empirical methods that can track changes in the number of ADFs in such systems are available. It is possible, therefore, to determine whether deployment of such synergies in visually guided tasks can induce changes in their graphs. §15.4 presents some results to this effect from a recently conducted study on a visually-guided single-segment precision task.

15.4 Studies of visual influences on the level of synergies

15.4.1 Visual influences on state dynamics

Experiments on between-persons rhythmic coordination provides evidence for the visual assembly of state dynamics at the level of relational variables. Two seated subjects can synchronize the swinging of their forelegs (the right foreleg for one subject and the left foreleg for the other subject) in either in-phase or anti-phase coordination as the required frequency of the coupled oscillations is increased (Schmidt, Carello, & Turvey, 1990). At higher coupled frequencies, coordinations begun in anti-phase break down, and transitions to the in-phase pattern, preceded by a magnification of SDϕ, are observed. These observations replicate those for within-person inter-limb coordination (Kelso, 1984; Scholz, Kelso, & Schöner, 1986). Decreasing the ratio b/a in Equation (15.1) to critical values (e.g., $b/a = 0.25$, when the noise term is 0) causes the stable state at π rad to disappear (coordination breakdowns are predicted). If, as noted in §15.2, b/a is inversely related to the frequency of the coupled oscillations, then predictions of Equation (15.1) apply in qualitatively identical fashion to the phase transitions and the augmented $SD\phi$ induced by the scaling of coupled frequency in both haptically- and visually-assembled coordinations.

Further indication that the state dynamics of Equation (15.1) can be just as well assembled at the level of space, comes from between-persons visual coordination experiments in which δ was manipulated (Figure 15.3; Schmidt & Turvey, 1994; Amazeen, Schmidt, & Turvey, 1995; Schmidt, Bienvenu, Fitzpatrick, & Amazeen, in press). The steady state ϕ and $SD\phi$ attained through visual coordination were found to be dependent upon δ; deviations from perfect in-phase and anti-phase increased, as did $SD\phi$, with increases in $|\delta|$. Results obtained by manipulating the frequency of the coupled oscillations were similarly consonant with prediction (Amazeen *et al.*, 1995; Schmidt *et al.*, in press); exaggerated deviations from intended relative phase and exaggerated fluctuations in relative phase were observed at higher coupled frequencies for which Equation (15.3) predicts weaker attractors for both in-phase and anti-phase modes.

In sum, the preceding results reveal that vision can provide the informational support needed to establish the essential relational state dynamic for 1:1 frequency-locked coordination, and that it can do so in a manner that is identical in many details to the manner in which haptic information performs the same function. It is worth noting, however, that visually assembled relational dynamics of this kind are weaker in overall strength than their haptically assembled counterparts. Schmidt *et al.* (in press) used identical conditions in within-person and between-persons tasks and found greater instability of relative phase (i.e., greater number of anti- to in-phase transitions and higher $SD\phi$) in the between-persons task. Despite profound

Fig. 15.4. Experimental set-up for the bimanual visual attention task of Amazeen *et al.* (1997) and Riley *et al.* (1997). Participants swing both pendulums in phase while controlling the contact of the upper portion of one pendulum to the paper targets suspended in front of and behind one hand.

similarities between the visually based and haptically based state dynamics of Equation (15.3), the indications are that visual coupling is weaker than haptic coupling. The latter observation provides a point of entry to influences of vision that operate on parameter dynamics.

15.4.2 Visual influences on parameter dynamics

To address visual influences on the parameterization of relational dynamics at the level of synergies we can add to a haptically based coordination, aspects that require visual guidance for successful execution. Imagine a task, for example, in which visual attention is needed to direct the behavior of one hand during a 1:1 bimanual rhythmic coordination (Amazeen *et al.*, 1997). In these experiments, visual attention was operationalized by having left- and right-handed (LH and RH) subjects control the top of a hand-held pendulum's contact with paper targets suspended in the front and back of either the right or left hand. Subjects had to perform this visual targeting task with one hand while coordinating the two pendulums in an in-phase pattern (Figure 15.4). Amazeen *et al.* (1997) showed that the normal handedness bias present in coordinated rhythmic movement* can be increased by visually attending to the preferred hand or diminished by attending to the non-preferred hand. An extension of Equation (15.3) can incorporate the body's functional asymmetry by introducing asymmetric coupling terms which slightly deflect the fixed points away from $\phi = 0$ and $\phi = \pi$ rad (Treffner & Turvey, 1995):

$$\dot{\phi} = \delta - a\sin(\phi) - 2b\sin(2\phi) - c\cos(\phi) - 2d\cos(2\phi) + \sqrt{Q}\zeta_t \qquad (15.4)$$

* The preferred hand completes its cycle slightly ahead of the non-preferred–for LH, $\phi > 0$ (left hand lead), and for RH, $\phi < 0$ (right hand lead); see Treffner & Turvey (1995, 1996).

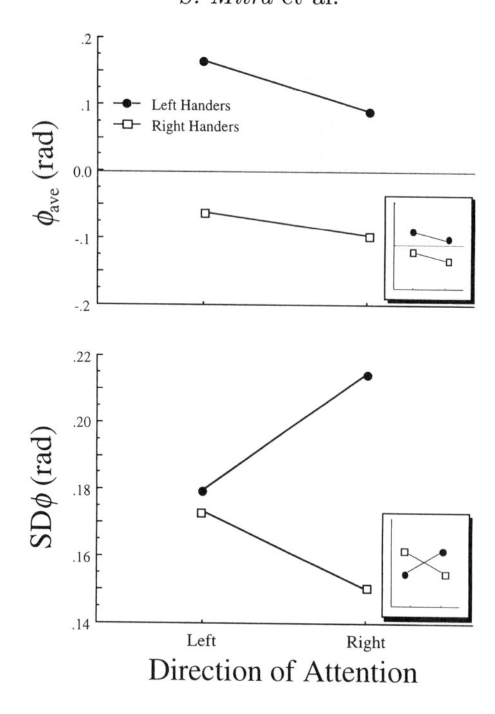

Fig. 15.5. Predictions (insets) and results of Amazeen *et al.* (1997). The top panel shows mean ϕ, and the bottom panel shows $SD\phi$.

Here, d is the asymmetric coupling coefficient of interest. Attending to the left hand corresponds to a slight increase in d, and attending to the right hand corresponds to a slight decrease in d. Figure 15.5 (top) shows both the predictions (inset) of Equation (15.4) for reasonable values of the coefficients and the data of Amazeen *et al.* (1997). The data supported, additionally and importantly, the counter-intuitive prediction that increased deviation from in-phase associated with attending to the preferred hand should be accompanied by an increase in stability (a decrease in $SD\phi$) (Figure 15.5, bottom).

Riley *et al.* (1997) had subjects perform the same targeting task at one of three coupled movement frequencies (low-, preferred-, or high frequency, paced by auditory metronome). Based on Treffner and Turvey (1996), Riley *et al.* (1997) assumed that as movement frequency increased (i.e., b/a decreased), d would remain invariant, thereby increasing the influence of d relative to the influence of b/a. Thus, if movement frequency were increased, deviation from in-phase due to attending to one hand or the other would increase. The data (Figure 15.6) generally supported this prediction (inset): when attending to the preferred hand, deviation from in-phase increased with increasing movement frequency. However, when attending to the non-preferred hand, the opposite occurred; deviations from in-phase *decreased* with increasing frequency, and in high frequency conditions, the sign

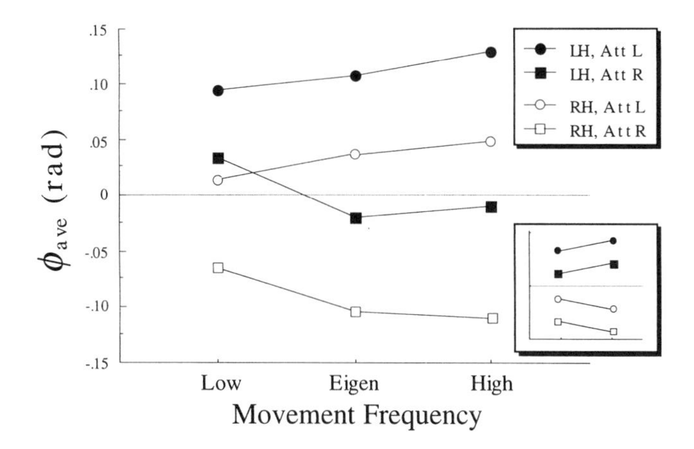

Fig. 15.6. Predictions (inset) and results of Riley *et al.* (1997). The graph shows mean ϕ as a function of movement frequency and direction of overt visual attention for LH and RH participants. Eigenfrequency refers to the natural frequency of the system of coupled pendulums depicted in Figure 15.4.

of phase deviation magnitudes was reversed so that the hand that was attended to, rather than the preferred hand, led the coordination. That is, the effects of visual attention were not invariant over changes in movement frequency. Taken together, these results indicate that visually attending to one hand renders interlimb coupling more asymmetric, with the direction of the asymmetry depending, in part, upon the hand (preferred or non-preferred) to which visual attention is directed. Further, the magnitude of asymmetry depended upon coupled movement frequency. All are clear examples of effects of parameter-dynamics.

While the above visual influences on state and parameter dynamics leave the basic form of the elementary rhythmic synergy [Equations (15.1), (15.3), (15.4)] essentially intact, not all effects of vision on bimanual rhythmic coordinations are so benign. In combination with another factor ascribable to the level of space, such as the orientation of a coordination with respect to the major body axes, visual guidance can not only affect the parameter-dynamics but also introduce properties into the relational dynamics that are not easily reconciled with their given form. Mitra, Amazeen, and Turvey (1997) asked subjects to produce a bimanual rhythmic coordination task similar in all respects to the ones above except that it was oriented frontally parallel to the body's coronal plane (Figure 15.7, top). In half the trials, subjects guided in-phase and anti-phase coordinations by watching the swinging pendulums, and in the other half, the entire coordination was occluded from sight. By Equation (15.3), when $\delta \neq 0$, haptic coordination deviates from perfect in-phase and anti-phase. For vision, this fixed point shift is an apparent failure to maintain intended phase and invites 'correc-

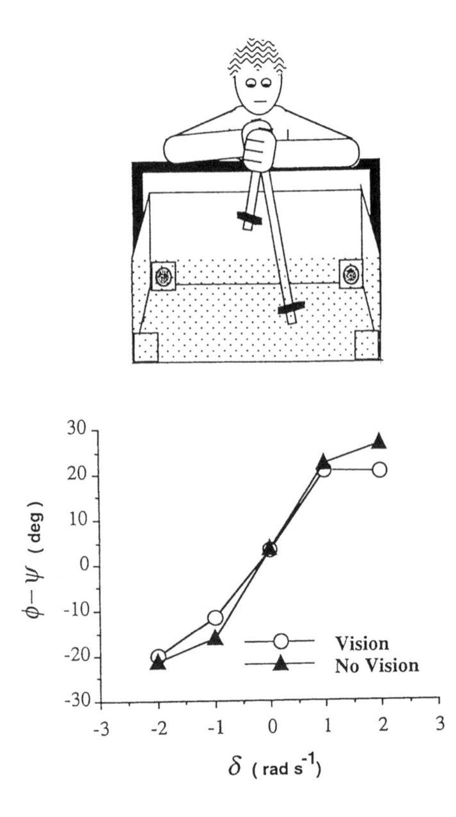

Fig. 15.7. Experimental set-up (top) and results (bottom) of Mitra *et al.* (1997). The data are of the difference between mean phase (ϕ) and intended phase (Ψ).

tion.' For this task, the corrective tendency can be modeled as a 'fictive force' of magnitude proportional to the visually apparent deviation from required phasing (e.g., Schöner & Kelso, 1988), an additional coupling term reflecting the superimposition of a required (visual) pattern on an intrinsic (haptic) pattern:

$$\dot{\phi} = \delta - a\sin(\phi) - 2b\sin(2\phi) - c\sin(\phi - \Gamma) + \sqrt{Q}\zeta_t \tag{15.5}$$

Here Γ stands for relative phase in visual-spatial coordinates, so that when $\phi = 0$ and $\phi = \pi$ rad in muscular coordinates, Γ is made equal to π and 0, respectively. Equation (15.5) can be solved numerically for an intended coordination of muscular in-phase (in which case $\Gamma = \pi$) and for an intended coordination of muscular anti-phase (in which case $\Gamma = 0$). It predicts that, as δ moves away from 0 (i.e., the two movement oscillators are increasingly asymmetric), deviation of the stable states of the coordination from intended phase should become lower with visual guidance than without, and that with vision $SD\phi$ should be lower for both stable states.

As shown in Figure 15.7 (bottom), the predicted attenuation of fixed point shift in the presence of vision was observed. This result indicates that visual information entered into the relational dynamics through a change in coupling: a parameter-dynamics effect. Interestingly, however, the prediction regarding the strength of the stable states that should have necessarily accompanied the observed fixed-point shift effect was not upheld by the data. In no case did visual monitoring increase the stability of either phase relation, a result contrary to predictions of Equation (15.5). This result, in conjunction with certain others (not related to vision), raises the possibility that vision may combine with spatial aspects of movement (e.g., coordination orientation) to induce more fundamental changes in coordinative structures. Such changes might ultimately involve the 'architectures' of coordinative structures–expressions of visual influences on graph dynamics at the level of synergies. It is currently an open question whether such influences operate on interlimb synergies. An effect of vision on the graph dynamics of an intra-limb synergy has, however, been observed.

15.4.3 Visual influences on graph dynamics

The preceding section demonstrated that adding a task component that requires visual guidance can produce parameter-level effects on the relational dynamics of a rhythmic synergy. Dynamics of interlimb synergies at the level of ξ (see Figure 15.1) emerge, however, from coupling the state dynamics of the L and R subsystems in specific ways. Either of these systems, by itself, is a dynamical system written over variables (i.e., ADFs) that are collective with respect to the skeletal and neuro-muscular degrees of freedom they subsume (see Figure 15.1). As such, a rhythmic synergy assembled in a single arm or segment has state, parameter, and graph dynamics of its own. Until recently, single-joint rhythmic movements had been dynamically modeled as limit cycle oscillators (with added noise) (e.g., Haken *et al.*, 1985; Rand *et al.*, 1988), which yield, by our adopted definition (see §15.2.1), state dynamics with two ADFs-position and velocity. Recent work, however, raises the possibility that these movements may not be limit cycles, but three- to five-variable systems parameterized to produce weakly chaotic behavior (Mitra, Riley, & Turvey, 1997; Mitra *et al.*, 1998). If this turns out to be the case, then the precise form of such state dynamics is not currently known. It is reasonable to think however that the additional ADFs may have their origin in mass-spring parameters, such as damping or stiffness that become, in real movements, active variables that contribute dimensions to the state space.

Even though details of the state dynamics are unknown, the number of ADFs in the state dynamics (i.e., the size of the graph) can now be determined empirically. Of interest here is the possibility that single-joint rhythmic synergies deployed in visually guided precision tasks might undergo changes in graph size when conditions of visual monitoring are manipulated. In an experiment exploring this possibility, Mitra, Riley, Santana, and Turvey (in preparation) asked seated subjects to wield horizontally a thin wooden rod with a motion sensor attached to the tip, and to oscillate the rod from the middle of one vertically placed visual target to the middle of another at the highest possible frequency that would not compromise accuracy (i.e., enable the subject to keep peak amplitude points within the respective targets). Both the gap between the targets and target widths were manipulated to produce four indices of task difficulty (ID). In half the trials, subjects could see both the targets as well as the entire hand-rod system moving between them. In the remaining trials, the hand-rod system was hidden from view, leaving only the targets visible. Data analysis was directed, among other things, to the question of how many ADFs were required to describe the dynamics of the hand-rod oscillator as it moved through its state space. In the following, we provide a very brief and admittedly cursory description of the analysis technique. Comprehensive and accessible discussions can be found in Abarbanel (1996) and Mitra *et al.* (1998).

Any measurement of the time series of a (scalar) variable from a given dynamical system gives a one-dimensional projection $x(t)$ of the system's full dynamics in its true state space. If an attractor exists in the system's true state space, it is often possible to recover some of the invariants of the system's motion on the attractor by reconstructing it from the measured (scalar) time series. In a common procedure called time-delay embedding, the series of scalar measurements is converted to a vector series $\vec{y}(t)$ by adding time-delayed copies of the measured series itself as additional coordinates. If done with a proper time-delay, this process keeps removing 'false neighbors' in the reconstructed space; datapoints that are close together in, say, m-dimensional reconstructed space move far apart in $m+1$-dimensional reconstructed space if they were close not by virtue of dynamic relatedness but because of projection from the higher-dimensional, true state space. Once the number of dimensions in reconstructed space is large enough to contain the (reconstructed) attractor within it, the embedding dimension d_E has been obtained, and the geometric properties of the reconstructed attractor are such that certain invariants of the system's dynamics in true state space can be extracted from it. Among them is the number of dimensions d_L (the number of ADFs) required to describe adequately the system as it moves locally along the attractor. Determining this number in reconstructed space involves starting in a space of dimension $d_W \geq d_E$ and finding sub-spaces of dimension $d_L \leq d_E$ in which accurate local neighborhood-to-neighborhood maps of the data can be constructed. Neighborhoods of several sizes are obtained by taking N neighbors of a given point $\vec{y}(t)$, and a local rule is abstracted for how these points evolve to the same N points in the neighborhood of $\vec{y}(t+1)$. The success rate of the rule is measured as percent bad

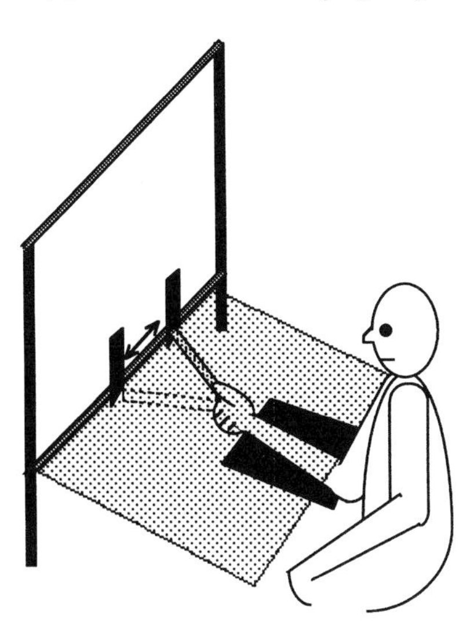

Fig. 15.8. Experimental set-up for a rhythmic targeting task. Participants moved a rod back and forth between two vertically-oriented targets. On half of the trials, vision of the arm and rod was occluded by a screen, leaving only the targets visible; on remaining trials, full vision of the arm, rod, and targets was allowed.

predictions (%bad), and the minimum number of dimensions in which %bad becomes independent of both N and dimensions is the number of ADFs of the system. For our present purpose, as mentioned above, this number is a measure of the system's graph size.

For all IDs tested in the experiment, the number of ADFs obtained was uniformly four when subjects could not visually monitor the hand-rod system's motion. With vision, however, the number of ADFs dropped to three with an increase in ID; the size of the system's graph actually became smaller as task difficulty increased (Figures 15.9 and 15.10). Although ascertaining the full implications of this result will require further study, it seems clear that guiding the hand visually in this type of task leads to the assembly of a qualitatively different synergy than guiding the hand non-visually. This task, essentially the same as Fitts' (1959) aiming task with vision, is a basic example of an organization at the level of synergies being deployed with strong constraints from the level of space. Also, this task is simply analogous to the (visual) attentional component of the bimanual in-phase coordination task of Amazeen *et al.* (1997) and Riley *et al.* (1993) (§15.4.2) which was found to introduce parameter-level changes into the interlimb relational dynamics. We may ask, therefore, whether changes in graph dynamics at the L and R subsystem level of Figure 15.1 generally underlie (or lead to) changes in the parameter dynamics of ξ. More generally, the question is whether there is a pattern of dependence among the changes in state, parameter

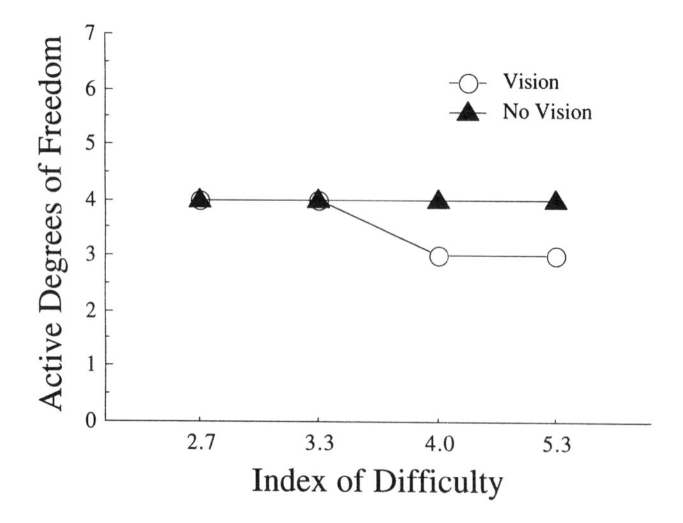

Fig. 15.9. Number of active degrees of freedom (ADFs) as a function of index of difficulty (ID) and vision condition. Number of ADFs remained at 4 for conditions in which visual proprioception of the arm was not allowed, but in full vision conditions dropped to 3 at high levels of ID.

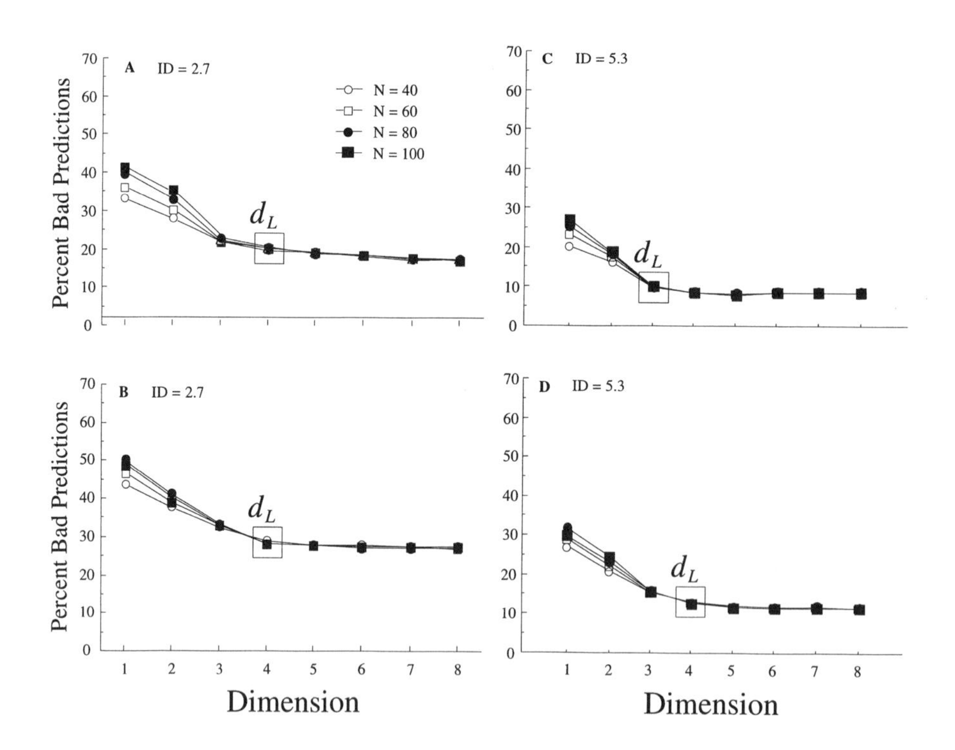

Fig. 15.10. Neighborhood size (N) by dimension interactions for %bad for four experimental conditions. ID = 2.7, full vision (a); arm occluded (b); ID = 5.3, full vision (c); arm occluded (d). The number of ADFs is the dimension d_L at which %bad becomes independent of both N and dimensions.

and graph dynamics occurring at different levels of a coordinative structure. While a precise answer is not currently available, simultaneous changes in the state, parameter and graph dynamics of a synergy may well be the rule rather than the exception when strong constraints from the level of space (and the level of actions) are prevalent.

15.5 Conclusions

In this chapter, the question of how vision modulates the functional organization of movement synergies was posed using a dynamical systems framework. It was shown that phrasing the question in terms of vision's role in assembling and modifying the state, parameter, and graph dynamics of synergies has enabled rigorous experimental investigation of the problem using elementary inter-limb coordination and simple aiming tasks. The general indication is that casting the problem in these terms (a) introduces a technical language that is abstract enough to accommodate questions pertaining to a large variety of visually influenced coordinations, (b) makes it easier to intuit connections between apparently disparate phenomena, and (c) creates opportunities for profitably applying methods from nonlinear dynamics to a range of problems in perception and action.

Authors' Notes

Preparation of this chapter was supported by NSF Grant SBR-9728970. R. C. Schmidt is also at the College of Holy Cross and M. T. Turvey is also at Haskins Laboratories. We thank Claudia Carello for her conceptual and artistic contributions.

References

Abarbanel, H. D. I. (1996). *Analysis of observed chaotic data*, Springer, New York, NY.

Amazeen, E. L., Amazeen, P. G., Treffner, P. J., & Turvey, M. T. (1997). "Attention and handedness in bimanual coordination dynamics", *J Exp. Psych.: Human Percept. and Perf.*, 23: 1552-1560

Amazeen, E. L., Sternad, D., & Turvey, M. T. (1996). "Predicting the nonlinear shift of stable equilibria in interlimb rhythmic coordination", *Human Movement Science*, 15: 521-542.

Amazeen, P. G., Schmidt, R. C., & Turvey, M. T. (1995). "Frequency detuning of the phase entrainment dynamics of visually coupled rhythmic movements", *Biol. Cybern.*, 72: 511-518.

Beek, P. J., & Beek, W. J. (1988). "Tools for constructing dynamical models of rhythmic movement", *Human Movement Science*, 7: 301-342.

Bernstein, N. A. (1967). *Coordination and regulation of movements*, Pergamon Press, New York, NY.

Bernstein, N. A. (1996). "On dexterity and its development", In *Dexterity and its development*, M. Latash & M.T. Turvey (Eds.), 2-244, Lawrence Erlbaum Associates, Hillsdale, NJ.

Cole, J. (1995). *Pride and a daily marathon*, MIT Press, Cambridge, MA.

Cole, J., & Paillard, J. (1995). "Living without touch and peripheral information about body position and movement: Studies with deafferented subjects", in *The body and the self*, J.L. Bermúdez, A. Marcel, & N. Eilan (Eds.), 245-266, MIT Press,. Cambridge, MA.

Farmer, J. D. (1990). "A Rosetta Stone for connectionism", *Physica D*, 42: 153-187.

Fitts, P. M. (1954). "The information capacity of the human motor system in controlling the amplitude of movement", *J. Exp. Psych.*, 47: 381-391.

Gibson, J. J. (1966). *The senses considered as perceptual systems*, Houghton Mifflin, Boston, MA.

Gibson, J. J. (1986). *The ecological approach to visual perception*, Lawrence Erlbaum Associates, Hillsdale, NJ.

Haken, H. (1996). *Principles of brain functioning*, Springer, Berlin.

Haken, H., Kelso, J. A. S., & Bunz, H. (1985). "A theorectical model of phase transitions in human hand movements", *Biol. Cybern.*, 51: 347-356.

Hillborn, R. C. (1994). *Chaos and nonlinear dynamics*, Oxford University Press, New York, NY.

Jeka, J. J., & Kelso, J. A. S. (1995). "Manipulating symmetry in the coordination dynamics of human movement", *J. Exp. Psych.: Human Percept. and Perf.*, 21: 360-374.

Kay, B. A. (1988). "The dimensionality of movement trajectories and the degrees of freedom problem: A tutorial", *Human Movement Science*, 7: 343-367.

Kelso, J. A. S. (1994). "Elementary coordination dynamics", in *Interlimb coordination: Neural, dynamical, and cognitive constraints*, S. P. Swinnen, J. Massion, H. Heuer, & P. Casaer (Eds.), 301-318, Academic Press, San Diego, CA.

Kelso, J. A. S., Delcolle, J. D., & Schöner, G. (1990). "Action-perception as a pattern formation process" In *Attention and performance XIII*, M. Jeannerod (Ed.), 139-169, Erlbaum, Hillsdale, NJ.

Koppell, N. (1988). "Toward a theory of modelling pattern generators. In *Neural control of rhythmic movement in vertebrates*, A. H. Cohen, S. Rossignol, & S. Grillner (Eds.), 369-413, John Wiley & Sons, New York, NY.

Marr, D. (1977). "Artificial intelligence – a personal view", *Art. Intel.*, 9: 37-48.

Mitra, S., Amazeen, P. G., & Turvey, M. T. (1997). "Dynamics of bimanual rhythmic coordination in the coronal plane", *Motor Control* 1: 44-71.

Mitra, S., Amazeen, P. G., & Turvey, M. T. (1998). "Intermediate motor learning as decreasing active (dynamical) degrees of freedom", *Human Movement Science*, 17: 17-65.

Mitra, S., Riley, M. A., Santana, M-V., & Turvey, M. T. (in preparation). "Dynamical analysis of a Fitts' task: The role of visual proprioception".

Mitra, S., Riley, M. A., & Turvey, M. T. (1997). "Chaos in human rhythmic movement", *J. Motor Behavior*, 29: 195-198.

Rand, R. H., Cohen, A. H., & Holmes, P. J. (1988). "Systems of coupled oscillators as models of central pattern generators", In *Neural control of rhythmic movement in vertebrates*, A.H. Cohen, S. Rossignol, & S. Grillner (Eds.), 333-367, John Wiley & Sons, New York, NY.

Riley, M. A., Amazeen, E. L., Amazeen, P. G., Treffner, P. J., & Turvey, M. T. (1997). "Effects of temporal scaling and attention on the asymmetrical dynamics of bimanual coordination", *Motor Control*, 1: 263-283.

Saltzman, E. L., & Munhall, K. G. (1992). "Skill acquisition and development: The roles of state-, parameter-, and graph-dynamics", *J. Motor Behav.*, 24: 49-57.

Schmidt, R. C., Bienvenu, M., Fitzpatrick, P., & Amazeen, P. G. (in press). "A comparison of within- and between-person coordination: Coordination breakdown and coupling strength", *J. Exp. Psych.: Human Percept. and Perf.*.

Schmidt, R. C., Carello, C., & Turvey, M. T. (1990). "Phase transitions and critical fluctuations in the visual coordination of rhythmic movements between people", *J. Exp. Psych.: Human Percept. and Perf.*, 16: 227-247.

Schmidt, R. C., Shaw, B. K., & Turvey, M. T. (1993). "Coupling dynamics in interlimb coordination", Journal of Experimental Psychology: Human Perception and Performance, 19, 397-415

Schmidt, R. C., & Turvey, M. T. (1994). "Phase entrainment dynamics of visually coupled rhythmic movements", *Biol. Cybern.*, 70: 369-376.

Schmidt, R. C., & Turvey, M. T. (1995). "Models of interlimb coordination: Equilibria, local analyses, and spectral patterning", *J. Exp. Psych.: Human Percept. and Perf.*, 21: 432-443.

Scholz, J., Kelso, J. A. S., and Schöner, G. (1986). "Nonequilibrium phase transitions in coordinated biological motion: Critical slowing down and switching time", *Physics Letters A*, 123: 390-394.

Schöner, G., & Kelso, J. A. S. (1988). "A synergetic theory of environmentally-specified and learned patterns of movement coordination: I. Relative phase dynamics", *Biol. Cybern.*, 58: 71-80.

Schöner, G., Haken, H., & Kelso, J. A. S. (1986). "A stochastic theory of phase transitions in human hand movement", *Biol. Cybern.*, 53: 442-452.

Sternad, D., Turvey, M. T., & Schmidt, R. C. (1992). "Average phase difference theory and 1:1 phase entrainment in interlimb coordination", *Biol. Cybern.*, 67: 223-231.

Treffner, P., & Turvey, M. T. (1995). "Handedness and the asymmetric dynamics of bimanual rhythmic coordination", *J. Exp. Psych.: Human Percept. and Perf.*, 21: 318-333.

Treffner, P., & Turvey, M. T. (1996). "Symmetry, broken symmetry, and handedness in bimanual coordination dynamics", *Exp. Brain Res.*, 107: 463-478.

Turvey, M.T., Shaw, R.E., & Mace, W. (1978). "Issues in the theory of action: Degrees of freedom, coordinative structures, and coalitions", In *Attention and performance VII*, J. Requin (Ed.), 557-595, Lawrence Erlbaum Associates, Hillsdale, NJ.

16

Movement planning: kinematics, dynamics, both or neither?

John F. Soechting and Martha Flanders

16.1 Introduction

The path that the hand takes during a reaching movement is highly consistent from trial to trial. This is true for any one subject; it is also true across subjects (Georgopoulos, Kalaska & Massey, 1981; Soechting & Lacquaniti, 1981) and it is independent of the time it takes the hand to reach the target. Generally, the path of the hand is close to a straight line (Morasso, 1981), although there can be instances in which the hand path exhibits a considerable amount of curvature (Atkeson & Hollerbach 1985; Lacquaniti, Soechting & Terzuolo, 1986). Furthermore, the curvature of the hand path, even when small is not arbitrary. For example, if one varies the direction of the movement from a fixed starting point (i.e. the now classic centre-out task introduced by Georgopoulos), one finds that the curvature varies in an orderly fashion with the direction of the movement (Flanders, Pellegrini & Geisler, 1996; Pellegrini & Flanders, 1996).

Theoretically, the hand could follow any number of paths to achieve a particular target. In fact, if an obstacle is placed in the way, subjects do alter the hand's trajectory to circumvent the obstruction (Abend, Bizzi & Morasso, 1982). Furthermore, if visual feedback is altered, a subject can be induced to alter the hand's trajectory. For example, if a distorted replica of the hand path is displayed on a computer monitor, a subject will alter the arm's trajectory over repeated trials so that the displayed path appears to be straight (Wolpert, Ghahramani & Jordan, 1994; 1995). Subjects can even be induced to produce hand paths that are straight when the paths are displayed in the space of joint angles, i.e. as a plot of elbow angle versus shoulder angle (Flanagan & Rao, 1995).

16.2 Movement planning

So the question arises: why do arm movements normally follow one particular trajectory and not any of the others that are possible? Since subjects can in fact produce alternate trajectories, it is clear that the observed behavior results from constraints that reflect natural modes of neural processing. The question has generally been posed in the following manner: are movements planned in extrinsic coordinates (the path that the hand takes in extrapersonal space) or in intrinsic coordinates (the space of joint angles)? Underlying this question is a tacit assumption that has been borrowed from the field of robotics: movement planning and execution is a sequential process in which movement kinematics and dynamics* (or more precisely, kinetics) are expressed separately at different stages in the process (cf. Brady, 1982; Taylor, 1982). This assumption can be recast along the following lines: movements are planned at a kinematic level (in either joint or extrinsic coordinates) and, once such a plan of movement exists, forces adequate to produce the desired movement are first computed and then generated by the neuromuscular system.

16.3 The distinction between kinematics and dynamics

Why is it necessary even to make a distinction between kinematics and dynamics? In a one degree of freedom system, the force 'F' and the acceleration 'A' will be proportional to each other by Newton's second law of motion and the distinction between the two is academic. This is not the case for the human arm, however, or any mechanical system with more than one degree of freedom. The force developed at the hand, if motion were to be impeded, is not parallel to the direction of the hand's motion (Soechting & Flanders, 1992), nor are the torques at the shoulder and elbow proportional to the respective angular accelerations at the two joints. Thus, for the arm, there is a real distinction between kinematics and dynamics, the two being related through coupled nonlinear differential equations (Hollerbach & Flash, 1982). Given a desired motion of the hand, one can compute the torques that will produce it. Conversely, given a pattern of torques one can compute the motion of the hand that it will generate.

Given the distinction between kinematics and dynamics, the hypothesis that movements are planned at a kinematic level appears to be a reasonable one. However, other schemes have been proposed as well. Perhaps the one that is best known is the "equilibrium point hypothesis", put forward in two different forms by Feldman (1986) and by Bizzi and his colleagues (Hogan *et al.*, 1987). According to this hypothesis, the CNS controls muscle stiffness and muscle rest length (the length at which the muscle generates zero force) and by adjusting the values of these two parameters, the arm can be made to move from one resting state to another. Note that this hypothesis avoids the need to compute torques and muscles forces from

* Kinematics refers to a description of the displacements, velocities and accelerations of the motion whereas dynamics refers to the forces and torques that produce the motion.

kinematics and in fact, movement is no more than a progression of static postures. The merits of the hypothesis have been debated at length in a variety of forums (cf. commentary on Bizzi *et al.*, 1992), and we will not repeat the various arguments here. Muscle stiffness must be appreciable for the hypothesis to be viable, especially for fast movements, and the best measurements to date (Bennett *et al.*, 1992; Gomi & Kawato 1996) indicate that the actual stiffness is far below the values required by the "equilibrium point" hypothesis.

Neural network solutions for the control of reaching movements also avoid the need to separate kinematic and dynamic aspects of movement control. In this scheme, there is an input layer (typically encoding the location of the target and the starting posture of the hand, i.e. kinematics) and an output layer (typically encoding muscle forces). Given the properties of neural networks, however, the information encoded by "neurons" need not be related simply to any one particular parameter (Alexander *at al.*, 1992; Robinson, 1992; but see Lukashin & Georgopoulos, 1993). Thus, one can envisage the possibility that kinematics are encoded by sensory receptors (the input level) and dynamics by the effectors (the output level), but that either kinematics or dynamics, or both, or neither are encoded anywhere in the brain circuits responsible for generating a goal-directed movement.

16.4 Optimality criteria

One way to make the question that we have posed at the outset of this chapter more tractable is to restate it in the following way: in what sense are arm movements optimized? The general procedure is to establish a "cost function" and then to compute the trajectory that minimizes that particular cost function (Nelson, 1983). At first glance, this approach seems far removed from the question: are movements planned at a kinematic or dynamic level. However, as we will show, the two approaches are equivalent to each other.

Consider for example neural network models. In this approach, the "synaptic weights" in the connections between the various levels are changed in an iterative manner to arrive at the ultimate solution. For example, the weights may be changed to minimize the errors at the end of the movement to targets distributed within a certain region of space. Note that in so doing, we have established an optimality criterion, namely to optimize movement accuracy. We could add other criteria as well: for example, to minimize the movement time or the distance that the arm moves, or the energy that is expended in moving the arm. Now some of these criteria, such as maximum accuracy or minimum time, do not lend themselves to a distinction between kinematic or dynamic planning. Others, such as minimum distance (kinematic) or minimum energy (kinetic), clearly do.

In fact, such an approach has provided the most compelling evidence in favor of movement planning at a kinematic level. Hogan and Flash (1987) proposed that arm motion was regulated so that the hand followed a trajectory that was maximally smooth, or to state it more precisely, that minimized jerk. According to this "minimum jerk" criterion, the hand should follow a straight path, with a bell-shaped velocity profile. In fact, the velocity profiles of hand trajectories generally do conform to this prediction. Recall however that hand paths, while often times being close to straight, nevertheless do exhibit curvature that is sometimes appreciable and is always predictable, contrary to the model. In the context of the equilibrium point hypothesis, Flash (1987) found a way around this dilemma by assuming that it was virtual trajectory that was optimally smooth, while the actual hand path was determined by muscle elasticity. Given sufficiently high values for muscle stiffness (and viscosity) this model was able to account for a wide range of experimental data.

Kawato and his colleagues (Uno, Kawato & Suzuki, 1989; Kawato *et al.*, 1990; Dornay *et al.*, 1996) also supposed that arm movements obeyed an optimality criterion, but one that was related to kinetics rather than to kinematics. In particular, they assumed that the rate of change of muscle torque was minimized ("minimum torque-change"), or more precisely, the integral over time of the sum of the squares of the rates of change of joint torques. The question arises: why the rate of change of torque and not the squares of the torques? The answer is that a "minimum torque" criterion has a solution where torque starts out with a large, positive (propulsive) value and ends with a large, negative (decelerative) value and changes smoothly in between. (Similarly, a "minimum acceleration" criterion has large values of acceleration at the start and large values of deceleration at the end.) Experimentally, torques and accelerations build up gradually as the movement begins and they decay gradually as the movement is arrested. Consequently, "minimum torque change" or a "minimum jerk" criteria are most parsimonious.

Uno *et al.*, (1989) reported that the "minimum torque change" criterion was able to fit experimental data (Atkeson & Hollerbach, 1985) better than did the "minimum jerk" criterion described above. Subsequent work (Kawato *et al.*, 1990) was aimed at demonstrating the biological feasibility of this approach by implementing the solution to this problem successfully by means of a cascaded neural network. In order to fit the experimental data, these investigators needed to add a viscous term to the equations for the torques. Since the biological motivation for this adjustment is not immediate, the result is not as clear cut as it may appear at first glance.

16.5 Is the problem resolvable?

From the brief survey we have just given, it should be clear that the question "which parameters are the subject of movement planning"; has received a considerable amount of attention. Furthermore, it should also be evident that no clear answer has emerged. On a purely qualitative level, one finds straight hand paths, but one can also finds instances of straight paths in joint space (cf. Soechting & Lacquaniti, 1981; Hollerbach & Atkeson, 1986; Shen & Poppele, 1995). Similarly, one finds adaptation to altered displays of movement kinematics in extrapersonal space or in joint space. Finally, quantitative models have given conflicting results.

Despite the slow progress on this problem, we argue that the question is a valid one. Furthermore, we argue that it is a well-posed question, as long as it is framed more precisely as: what parameter(s) is (are) optimized for arm movements? In the following sections, we will summarize some of our own recent work on this topic. We will present experimental data that indicate that kinetic considerations do indeed influence the trajectory of the arm, contradicting schemes in which movement planning is purely kinematic. We will also present the results of some modelling studies in which compare the relative importance of various "kinetic considerations".

16.6 Donders' Law and arm movements

According to Donders' Law, for any particular gaze direction, there is a unique orientation of the eyes (Nakayama & Balliet, 1977; Tweed & Vilis, 1990), independent of previous gaze directions. This is a powerful result and considerable work has been done to search for its neural substrates (cf. Crawford, 1994; Chapter 7 of this volume). From the perspective of this review, its significance is that it points to movement planning at the level of kinematics. Several investigators have tested the validity of Donders' Law for head and arm movements and have reported in the affirmative (Straumann *et al.*, 1991; Hore, Watts & Vilis, 1992; Miller, Theeuwen & Gielen, 1992). Since arm movements were tested only under the very special condition of the fully extended arm, we wanted to know whether Donders' Law would also holds when there are varying degrees of flexion at the elbow (Soechting *et al.*, 1995).

We found that it does not, but that we could account for the dependence of the final posture of the arm on its starting posture by a simple optimality criterion. In our task, subjects made pointing movements from one of 17 starting locations to five target locations arrayed throughout the three-dimensional workspace. The results from one subject are illustrated in Figure 16.1. The plots are arrayed in the same manner as the targets appeared to the subject: target 5 was in the middle, at shoulder level, target

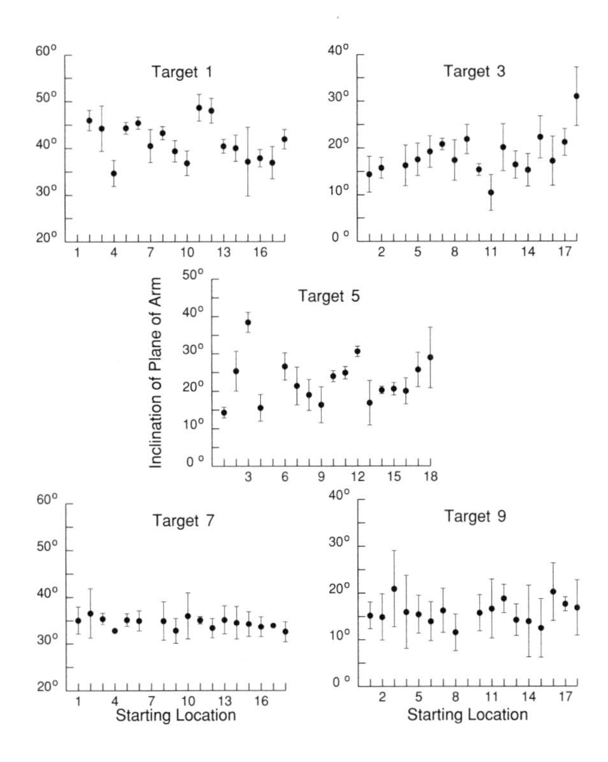

Fig. 16.1. Arm posture at the end of a reaching movement depends on the initial posture of the arm. Subjects made reaching movements to 5 targets from each of 17 starting locations. The plots depict the average ($\pm$ 1 SD) of the inclination of the plane of the arm for each of these combinations, from one subject. Note that for most of the targets, this parameter depends on the starting location.

3 was at the upper right, etc. In Figure 16.1 we show how the posture of the arm at each of the target locations varied as a function of the starting point of the movement. While there are four angles that are required to describe arm posture fully (three at the shoulder and one at the elbow), one variable suffices to characterize variations in arm posture with the hand at a particular location. We used the inclination of the plane defined by the upper arm and the forearm as this variable. Specifically, we computed the angle between the perpendicular to the arm plane and the horizontal plane and plotted the mean ($\pm$1 SD) of this angle in Figure 16.1.

If Donders' Law were obeyed, the inclination of the plane of the arm should not depend on the starting location of the movement. For Target 7, the one at the lower left, this prediction held. However, it did not hold true for the other four targets. For each of them, there was a statistically significant dependence of arm inclination on the starting location. Furthermore, the patterns were largely similar for the four subjects who were tested (see Figures 3 and 4, Soechting *et al.*, 1995). We then set out to determine whether we could account for the data by some simple criterion. Our initial hypothesis was one based on movement kinematics, namely that the final posture was the one that was closest to the starting posture, in the space of joint angles. The results did not conform to that hypothesis at all. In fact, when starting points and end points are plotted in joint space (Figure 5 of Soechting *et al.*, 1995), one finds that there is no consistent relation between the two.

Other kinematic criteria were equally unpromising. However, finally we noted a characteristic that provided a clue to the answer: for some combinations of starting locations and target locations, the target could be acquired by simply rotating the arm about the long axis of the humerus. In fact, this is what the subjects did. The inertia of the upper arm is much less for rotations about the long axis of the humerus than it is for rotations about any other axis. Since joint torques are proportional to rotational inertia, this consideration suggests an optimality criterion based on kinetic rather than on kinematic considerations, for example the minimum torque change criterion mentioned above.

Generating predictions of the "minimum torque change" criterion is a considerable computational challenge for the case of a four-degree-of-freedom arm. Furthermore, from the point of view of testing the predictions of a model, we were primarily concerned with the final posture of the arm and less so with the path the arm took to get there. Accordingly, we developed a criterion that was much simpler computationally, but one that retained a kinship to the "minimum torque change" criterion. In particular, we developed a criterion based on the amount of work expended to move the arm from starting location to the target. Since work is equal to the change in kinetic energy, the total work is zero. (It is positive during acceleration and negative during deceleration.) Thus, total work done is not an appropriate criterion, but the peak positive work done during the movement could be. The peak positive work during the movement is equal to the peak kinetic energy, and this can be derived quite simply. (Soechting *et al.*, 1995).

Figure 16.2 illustrates that this simple criterion was able to account for a large amount of the variability in the experimental data. In this illustration, we have combined the data for all four subjects, plotting the amount of humeral rotation (i.e. rotation about the long axis of the humerus) measured experimentally against the amount predicted by our "minimum peak work" criterion. Each of the symbols denotes a different target location. If the model's predictions were satisfied, all of the data points should fall on a 45° line. Since this criterion was able to account for the data, whereas others that were based on purely kinematic considerations could not, we concluded that kinetic considerations influenced the final posture, and by inference, the arm's trajectory. Thus, these results are not compatible with models in

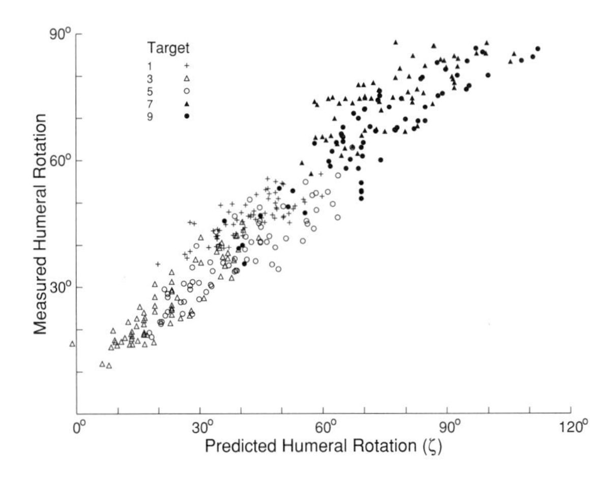

Fig. 16.2. Arm posture at the end of a reaching movement can be predicted by a "minimum peak work" criterion. The humeral rotation (rotation about the long axis of the humerus) measured experimentally is plotted against the value predicted according to the criterion that the peak positive work is minimized. The different symbols represent the five different targets and the data are from all four subjects.

which a purely kinematic plan of movement is formulated first, and then the requisite torques are computed.

16.7 Evaluating a minimum-torque-change model

The results presented in the previous section suggested that energetic considerations did have an influence on the arm's trajectory during pointing movements. The approach we used gave no predictions concerning the path however, only about the final posture of the arm. However, ultimately we hoped to gain some insight into what the advantages might be of moving along one particular path rather than another one. Simulation studies can be helpful in this regard, but the challenge of finding optimal solutions to the full four-degree of motion problem of the upper arm appeared too complex at the outset. Therefore, we began with a simpler case: planar motion of the arm, involving only two degrees of freedom - one at the shoulder and one at the elbow.

In particular, we considered the case of arm motion in a parasagittal plane, starting from a posture in which the upper arm is vertical and the forearm is horizontal. We considered motion to each of 20 targets arrayed around the perimeter of a circle centered at the hand. We chose this particular case because we had acquired a considerable amount of data concerning both the kinematic patterns and the patterns of EMG activity that characterized movements to this set of targets (Flanders, 1991; Flanders, Pellegrini & Soechting, 1994; Flanders, Pellegrini & Geisler 1996). The hand paths from one subject are shown at the top of Figure 16.3. Note that the paths are gently curved and that most of them are curved in a clockwise direction. Paths directed to a quadrant that is centred about the upward direction, show the opposite curvature. This pattern is consistent from subject to subject (cf. Figure 16.6A).

There is also a consistent pattern of EMG activation. For relatively fast movements (i.e. movement times $\approx$ 400 ms), EMG activity in individual muscles is similar to the pattern for single joint movements, namely a burst of activity either around movement onset (agonist) or around the time the movement has reached its peak speed (antagonist). However, the timing of these bursts is not consistent from muscle to muscle. For any given muscle, the timing varies in a continuous manner with the direction of the movement (Flanders, Pellegrini & Geisler, 1996).

Simulation studies by Buneo *et al.* (1995) suggest that these two phenomena (hand path curvature and staggered timing of EMG bursts) are linked. They computed shoulder and elbow torques for hypothetical straight movements and found that the timing of the maxima and minima of the torques did not vary with movement direction. Such a temporal invariance is not found experimentally, however, as can be appreciated in the lower part of Figure 16.3 which shows contour plots of shoulder and elbow torques for one subject (Soechting & Flanders, 1997). The comparison of simulations and experimental results suggests that subjects could in fact generate straight movements if only they could activate synergistic muscles in unison. Since it is commonly believed that muscles synergies lead to a simplification of the control problem (cf. Macpherson, 1991), one is left with the quandary that subjects appear to invoke a control strategy that is more complicated than it would need to be. Clearly there should be some advantage to the experimentally observed trajectories that makes up for the increased complexity. Saving energy is a likely candidate.

Therefore, we began by computing the hand paths that would satisfy the "minimum torque change criterion" Specifically we computed trajectories for which

$$J = \int [(dT_s/dt)^2 + (dT_e/dt)^2]dt \qquad (16.1)$$

was a minimum. The shoulder and elbow torques (T_s and T_e) were computed from the angular motions at the shoulder and elbow. The motions in turn were computed from the path of the hand, using trigonometry. Thus, the procedure was to vary the hand's trajectory and finding the particular trajectory for which J was a minimum, keeping the movement time fixed. Specifically, we defined x as the distance along a straight line from the start

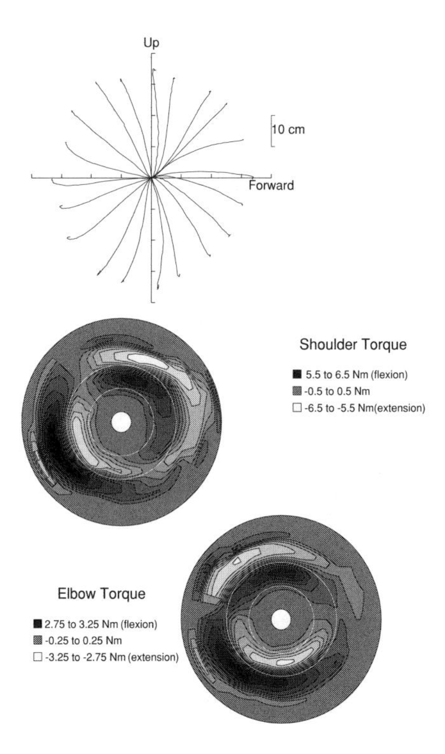

Fig. 16.3. Hand paths and joint torques during reaching movements. The top panel depicts hand paths for reaching movements to 20 targets spaced at 18° intervals on the circumference on a circle. The movements were performed with the arm in a parasagittal plane, starting with the upper arm vertical and the forearm horizontal. Note the curvature of the hand paths. The lower two panels depict contour plots of the temporal variation in shoulder and elbow torques for these movements. In these plots, time increases starting from the center. The inner white circle denotes the time of movement onset and the second circle denotes the midpoint of the movement (200 ms later).

to the target and y as perpendicular to x. We then used polynomials in time to characterize the movement of the hand:

$$\begin{aligned} x(t) &= P(t) \\ y(t) &= P(t)s(t) \end{aligned} \tag{16.2}$$

where $P(t)$ and $s(t)$ are polynomials, subject to

$$\begin{aligned} &P(0) = dP(0)/dt = dP(T)/dt = 0 \\ &P(T) = D \\ &s(0) = s(T) = 0 \end{aligned} \tag{16.3}$$

where T is the movement time and D the distance from the start to the target. We used an iterative procedure (Nelder & Mead, 1964; see also Press *et al.*, 1992) to find the coefficients of the polynomial that minimized

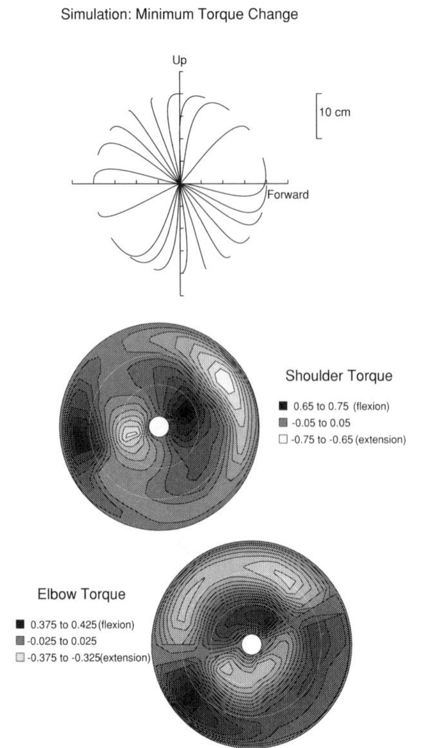

Fig. 16.4. Predictions of the "minimum torque change" criterion. The format of this figure is the same as that of Figure 16.3. In this simulation, hand paths and torques were computed based on the criterion that the rate of change of torque at the shoulder and elbow was minimized. The components of the torques that counteract the effects of gravity were not included in this simulation.

J. We used five terms for P (beginning with $a_P t^2$) and four terms for s (beginning with $a_s t$). There were six independent variables, the other three coefficients being determined by the boundary conditions (equation 16.3).

Figure 16.4 illustrates the results of this simulation. In this instance, we did not include the gravitational contributions to the shoulder and elbow torques because previous research had suggested that the static (gravitational) and the dynamic (speed-related) components of muscle activations were controlled separately (Flanders & Herrmann, 1992). The model clearly does not match the experimental data (Figure 16.3). The hand path curvatures predicted by the "minimum torque change" criterion were generally much larger than the experimental ones. Even worse, for most of the movement directions, the paths curved in the wrong direction. This can also be appreciated by comparing the results in Figure 16.6B with those in Figure 16.6A.

The fact that we ignored the static (gravitational) torque components did not account for the discrepancy between the model and the experimental data. We repeated the simulation, now including gravitational torques, for a movement time of 400 ms and found almost exactly the same hand paths as those shown in Figure 16.4 (compare heavy and light lines in Figure 16.6B).

An examination of the torque profiles (lower part of Figure 16.4) provided a suggestion as to why the model failed. Note that for movements directed slightly backward and upward and slightly forward and downward, there is little modulation in shoulder torque for the optimal hand paths. Similarly, for movements directed slightly upward and forward or slightly downward and backward there is little modulation in elbow torque. That is to say, the shoulder and elbow torques are modulated maximally in approximately orthogonal directions. For the experimental data (Figure 16.3), shoulder and elbow torques are much more congruent. The pattern predicted by the model (Figure 16.4) could be implemented economically by mono-articular muscles, but bi-articular muscles (such as biceps and the long head of triceps) could not be used effectively to generate the motions depicted in Figure 16.4. This is because, when a bi-articular muscle is recruited to generate shoulder torque, its action at the elbow would have to be negated by monoarticular antagonists, such as brachioradialis and the medial or lateral head of triceps. A similar situation would hold for movement directions in which a large amount of elbow torque is required.

This observation suggested that an appropriate optimality criterion would have to be applied at the level of muscle forces, rather than for the overall torques generated at each joint. We set out to test the viability of this hypothesis by additional simulations. A realistic model, that incorporated all the major muscles acting at the shoulder and elbow, and also the viscoelastic properties of muscles (Soechting and Flanders 1997) appeared to be too complicated to implement. However, the techniques that we used to generate the predictions in Figure 16.4 could be extended readily to a simpler model, restricted to only six muscles (a mono-articular agonist/antagonist pair at the shoulder and elbow and bi-articular muscles). This model ignored muscle viscoelasticity (Zajac, 1989), but did incorporate muscle geometry, i.e. the effective insertions and origins of each of the muscles (Wood, Meek & Jacobsen, 1989; Buneo, Soechting & Flanders, 1997).

If there are only six muscles, as in our model, one can obtain a unique solution for the distribution of force among the muscles that minimizes the sum of the squares of muscle forces, F_m. First of all, this criterion predicts there should be no muscle coactivation. Therefore, for any combination of shoulder and elbow torques, one need only take into account three muscles. For example, if there is flexor torque at the shoulder and elbow, one need only consider anterior deltoid, biceps and brachioradialis - forces in the shoulder and elbow extensors (posterior deltoid, and the long and medial head of triceps) should be zero. Then, for the three remaining muscles, there

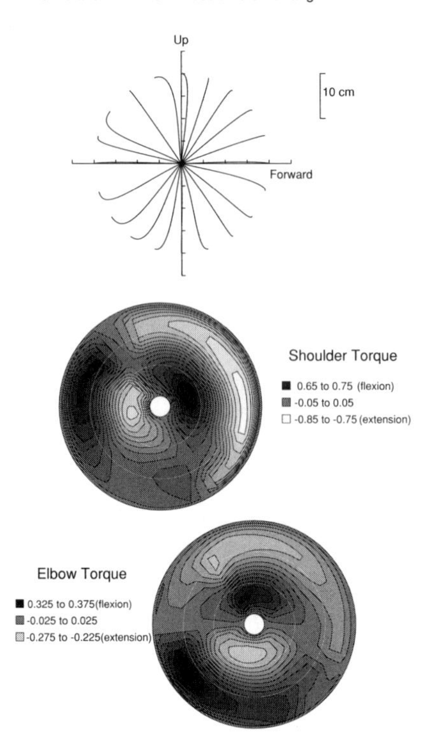

Fig. 16.5. Predictions of the "minimum muscle force change" criterion. In this simulation, hand paths and torques were computed according to the criterion that the sum of the rates of change of muscle forces (including mono-articular and bi-articular muscles) was minimized. Note that the predictions of this simulations correspond much more closely to the experimental data than do those depicted in Figure 16.4.

is a unique analytical solution for which

$$J = \sum F_m^2 \tag{16.4}$$

is a minimum. Accordingly, this model required only small changes to the procedures we had used to find the optimal hand path. We again did an iterative search, first defining a hand path, then computing shoulder and elbow torques for that path, then the corresponding muscle forces and finally the criterion to be minimized:

$$J = \int \sum (dF_m/dt)^2 dt \tag{16.5}$$

Figure 16.5 shows the results of this simulation for the case in which gravitational torques are excluded. These results were much more encouraging in that the predicted hand paths resembled much more closely the experimental data than did the joint torque model. Generally, the paths were much straighter. Furthermore, for most paths, the curvature was in the correct

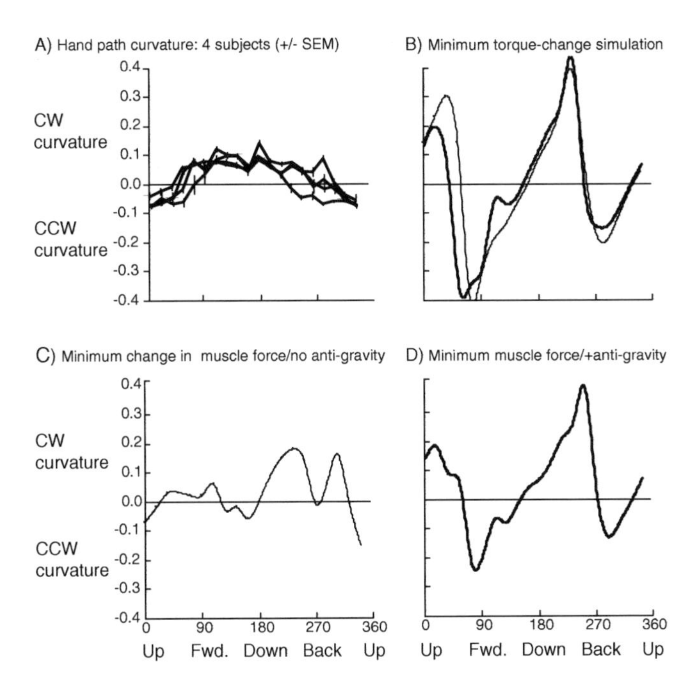

Fig. 16.6. Variations in hand path curvature. The panels depict: A) experimental data for four subjects, B) the results of simulations based on the "minimum torque change" criterion, and C) and D) the predictions based on the minimum "muscle force change" criterion. In each case the heavy line shows results in which gravitational effects were included; the lighter line shows results in which gravitational effects were not included. Curvature was computed as the ratio of the maximum deviation from a straight path to the path length (see Pellegrini & Flanders, 1996 for details).

direction (compare Figure 16.6C with Figure 16.6A). While one would be reluctant to take the results as proof of the model's validity, neither is there an inducement to reject the model outright (especially in view of the major simplifications introduced in the analysis).

We performed one additional simulation, in which we included static (gravitational) torques in the minimum muscle force change model. The results of this simulation (Figure 16.6D) were very similar to those obtained from the minimum torque change model, that is to say, they did not fit the experimental data at all. This observation is consonant with experimental data that suggest that static (gravitational, speed-insensitive) and dynamic (speed-sensitive) components of muscle torques are controlled separately and that only the dynamic components influence the selection of the arm trajectory (Flanders & Herrmann, 1992; Soechting *et al.*, 1995).

16.8 Conclusion

We have shown that the final posture of the arm in a reaching movement can be predicted according to a criterion that is related to energy expenditure (minimum peak work) and that the hand paths during planar arm movements can be at least partly predicted by another criterion (minimum muscle force change) that is also related to energy expenditure. There are differences between the two models - the major one being that the simulations related to arm posture at the end of a pointing movement (Figure 16.2) reflected global parameters (such as joint torque) whereas the second set of simulations indicated that an evaluation of energy expenditure that is muscle-based would be more appropriate. We do not know whether or not a muscle-based criterion could also account for the variations in arm posture depicted in Figure 16.1.

In both instances, we made major simplifying assumptions. We did so because more realistic models that more accurately reflected muscle properties such as force-velocity and length-tension would have made the analysis computationally intractable. We also refrained from making the models more complicated because there is a large amount of uncertainty in the values of many of the parameters needed to characterize the properties of each of the muscles. Nevertheless, with regard to the question that we posed in the title of this article, we believe that we have demonstrated that movement planning involves more than just kinematics. That is to say, the data are not compatible with a scheme whereby movement kinematics are worked out first, and then the required torques and muscle forces are computed. Rather, movement planning must involve dynamics.

Does movement planning involve kinematics, as well? Of course! Consider for example the case where there are obstacles in the path. Clearly, the hand path must be planned to avoid the obstacle. But what if the obstacle impedes the motion of the elbow, but not the hand? We have no difficulty altering the configuration of our arm to avoid such obstacles. Thus kinematic planning can't be restricted purely to the trajectory of the hand (i.e. its motion in extrapersonal space), but must be concerned with the angular motion at each of the joints as well. What if there are no obstacles? In all of the movements that have been studied, the hand moves in a fairly direct path towards the target. It would be incorrect to characterize the motion as rectilinear but it would also be incorrect to characterize it as being highly curved. It is not unreasonable to suggest minimizing path length is part of the plan for a movement.

This discourse suggests that the answer to the title question is "both". Could it be "neither"? We cannot exclude that possibility, but we have trouble finding some criterion that is not related to kinematics or to kinetics. Certainly, the fact that individual subjects make movements that are highly consistent from one trial to another and the fact that many of these characteristics are preserved from one subject to another suggest that there are strong constraints that shape the manner in which movements are normally performed. We suggest that these constraints can be defined by optimality criteria, but there is no one single such criterion that determines behavior.

Acknowledgments

This work was supported by USPHS Grants NS 15018 (JFS) and NS 27484 (MF).

References

Abend, W. K., Bizzi, E. & Morasso, P. (1982). "Human arm trajectory formation", *Brain*, 105: 331-348.

Alexander, G. E., DeLong, M. R. & Crutcher, M. D. (1992). "Do cortical and basal ganglionic motor areas use 'motor programs' to control movement?", *Behav. Brain Sci.,* 15: 656–665.

Atkeson, C. G. & Hollerbach, J. M. (1985). "Kinematic features of unrestrained vertical arm movements", *J. Neurosci.*, 5: 2318-2330.

Bennett, D. J., Hollerbach, J. M., Xu, Y. & Hunter, I. W. (1992). "Time-varying stiffness of human elbow joint during cyclic voluntary movement", *Exp. Brain Res.*, 88: 433-442.

Bizzi, E., Hogan, N., Mussa-Ivaldi, F. A. & Giszter, S. (1992). "Does the nervous system use equilibrium-point control to guide single and multiple joint movements?", *Behav. Brain Sci.*, 15: 603-613.

Brady, M. (1982). "Trajectory planning", In *Robot Motion: Planning and Control*, M. Brady, J. M. Hollerbach, T. L. Johnson, T. Lozano-Perez & M. T. Mason (Eds.), MIT Press, Cambridge, MA, 221-244.

Buneo, C. A., Boline, J., Soechting, J. F. & Poppele, R. E. (1995). "On the form of the internal model for reaching", *Exp. Brain Res.,* 104: 467-479.

Buneo, C. A., Soechting, J. F. & Flanders, M. (1997). "Postural dependence of muscle actions: Implications for neural control", *J. Neurosci.* 17: 2128-2142.

Crawford, J. D. (1994). "The oculomotor neural integrator uses a behavior-related coordinate system", *J. Neurosci.,* 14: 6911-6923.

Dornay, M., Uno, Y., Kawato, M. & Suzuki, R. (1996). "Minimum muscle tension change trajectories predicted by using a 17-muscle model of the monkey's arm", *J. Mot. Behav.*, 28: 88-100.

Feldman, A. G. (1986). "Once more on the equilibrium-point hypothesis (l model) for motor control", *J. Motor Behav.*, 18: 17-54.

Flanagan, J. R. & Rao, A. K. (1995). "Trajectory adaptation to a nonlinear visuomotor transformation: evidence of motion planning in visually perceived space", *J. Neurophysiol.*, 74: 2174-2178.

Flanders, M. (1991). "Temporal patterns of muscle activation for arm movements in three-dimensional space", *J. Neurosci.*, 11: 2680-2693.

Flanders, M. & Herrmann, U. (1992). "Two components of muscle activation: scaling with the speed of arm movement", *J. Neurophysiol.*, 67: 931-943.

Flanders, M., Pellegrini, J. J. & Geisler, S. D. (1996). "Basic features of phasic activation for reaching in vertical planes", *Exp. Brain Res.*, 110: 67-79.

Flanders, M., Pellegrini, J. J. & Soechting, J. F. (1994). "Spatial/temporal characteristics of a motor pattern for reaching", *J. Neurophysiol.*, 71: 811-813.

Flash, T. (1987). "The control of hand equilibrium trajectories in multi-joint arm movements", *Biol. Cybern.*, 57: 257-274.

Georgopoulos, A. P., Kalaska, J. F. & Massey, J. T. (1981). "Spatial trajectories and reaction times of aimed movements: effects of practice, uncertainty and change in target location", *J. Neurophysiol.*, 46: 725-743.

Gomi, H. & Kawato, M. (1996). "Equilibrium-point control hypothesis examined by measured arm stiffness during multijoint movement", *Science*, 272: 117-120.

Hogan, N., Bizzi, E., Mussa-Ivaldi, F. A. & Flash, T. (1987). "Controlling multijoint motor behavior", *Exerc. Sports Sci. Rev.*, 15: 153-190.

Hogan, N. & Flash, T. (1987). "Moving gracefully: quantitative theories of motor coordination", *Trends Neurosci.*, 10: 170-174.

Hollerbach, J. M. & Atkeson, C. G. (1986). "Characterization of joint interpolated arm movements", *Exp. Brain Res. Ser.*, 15: 41-54.

Hollerbach, J. M. & Flash, T. (1982). "Dynamic interactions between limb segments during planar arm movement", *Biol. Cybern.*: 44, 67-77.

Hore, J., Watts, S. & Vilis, T. (1992). "Constraints on arm position when pointing in three dimensions: Donders' law and the Fick gimbal strategy", *J. Neurophysiol.*, 68: 374-383.

Kawato, M., Maeda, Y., Uno, Y. & Suzuki, R. (1990). "Trajectory formation of arm movement by cascade neural network model based on minimum torque-change criterion", *Biol. Cybern.*, 62: 275-288.

Lacquaniti, F., Soechting, J. F. & Terzuolo, C. A. (1986). "Path constraints on point-to-point arm movements in three-dimensional space", *Neurosci.*, 17: 313-324.

Lukashin, A. V. & Georgopoulos, A. P. (1993). "A dynamical neural network model for motor cortical activity during movement: population coding of movement trajectories", *Biol. Cybern.*, 69: 517-524.

Macpherson, J. M. (1991). "How flexible are muscle synergies?", In *Motor Control: Concepts and Issues*, D. R. Humphrey & H.-J. Freund (Eds.), 33-48, John Wiley and Sons, Chichester.

Miller, L. E., Theeuwen, M. & Gielen, C. C. A. M. (1992). "The control of arm pointing movements in three dimensions", *Exp. Brain Res*, 90: 415-426.

Morasso, P. (1981). "Spatial control of arm movements", *Exp. Brain Res.*, 42: 223-237.

Nakayama, K. & Balliet, R. (1977). "Listing's law, eye position sense and perception of the vertical", *Vision Res.*, 17: 453-457.

Nelder, J. A. & Mead, R. (1964). "A simplex method for function minimization", *Comp. J.*, 7: 308-313.

Nelson, W. L. (1983). "Physical principles for economies of skilled movements", *Biol. Cybern.*, 46: 135-147.

Pellegrini, J. J. & Flanders, M. (1996). "Force path curvature and conserved features of muscle activation", *Exp. Brain Res*, 100: 80-90.

Robinson, D. A. (1992). "Implications of neural neworks for how we think about brain function", *Behav. Brain Sci.*, 15: 644-655.

Shen, L. & Poppele, R. E. (1995). "Kinematic analysis of cat hindlimb stepping", *J. Neurophysiol.*, 74: 2266-2280.

Soechting, J. F., Buneo, C. A., Herrmann, U. & Flanders, M. (1995). "Moving effortlessly in three dimensions: Does Donders' law apply to arm movement?", *J. Neurosci.*, 15: 6271–6280.

Soechting, J. F. and Flanders, M. (1992). "Moving in three-dimensional space: frames of reference, vectors and coordinate systems", *Ann. Rev. Neurosc*,i. 15: 167-191.

Soechting, J. F. & Flanders, M. (1997). "Evaluating an integrated musculoskeletal model of the human arm", *J. Biomechan. Engrg.*, 119: 93-102.

Soechting, J. F. & Lacquaniti, F. (1981). "Invariant characteristics of a pointing movement in man", *J. Neurosci.*, 1: 710-720.

Straumann, D., Haslwanter, T., Hepp-Reymond, M.-C. & Hepp, K. (1991). "Listing's law for eye, head and arm movements and their synergistic control", *Exp. Brain Res.*, 86: 209–215.

Taylor, R. H. (1982). "Planning and execution of straight-line manipulator trajectories", In *Robot Motion: Planning and Control*, M. Brady, J. M. Hollerbach, T. L. Johnson, T. Lozano-Perez & M. T. Mason (Eds.), 265–286, MIT Press, Cambridge, MA.

Tweed, D. & Vilis, T. (1990). "Geometric relations of eye position and velocity vectors during saccades", *Vision Res.*, 30: 111-127.

Uno, Y., Kawato, M. and Suzuki, R. (1989). "Formation and control of optimal trajectory in human multijoint arm movement. Minimum torque change model", *Biol. Cybern.*, 61: 89-102.

Wolpert, D. M., Ghahramani, Z. & Jordan, M. I. (1994). "Perceptual distortion contributes to the curvature of human reaching movements", *Exp. Brain Res.*, 98, 153-156.

Wolpert, D. M., Ghahramani, Z. & Jordan, M. I. (1995). "Are arm trajectories planned in kinematic or dynamic coordinates? An adaptation study", *Exp. Brain Res.*, 103, 460-470.

Wood, J. E., Meek, S. G. & Jacobsen, S. C. (1989). "Quantitation of human shoulder anatomy for prosthetic arm control. I. Surface modeling", *J. Biomech.*, 22, 273-292.

Zajac, F. E. (1989). "Muscle and tendon: properties, models, scaling, and application to biomechanics and motor control", *Crit. Rev. Biomed. Engrg.*, 17, 359-411.

Author index

Subject index